Device-to-Device based Proximity Service

T0199588

Device-to-Device based Proximity Service

Architecture, Issues, and Applications

Yufeng Wang
Athanasios V. Vasilakos
Qun Jin
Hongbo Zhu

CRC Press
Taylor & Francis Group
Boca Raton London New York

CRC Press is an imprint of the
Taylor & Francis Group, an **informa** business

CRC Press
Taylor & Francis Group
6000 Broken Sound Parkway NW, Suite 300
Boca Raton, FL 33487-2742

First issued in paperback 2020

ISBN 13: 978-0-367-57334-8 (pbk)
ISBN 13: 978-1-4987-2417-3 (hbk)

Library of Congress Cataloging-in-Publication Data

Names: Wang, Yufeng (Computer scientist), author. | Vasilakos, Athanasios, author. | Jin, Q. (Qun), 1962- author. | Zhu, Hongbo (Computer scientist), author.
Title: Device-to-device-based proximity service : architecture, issues, and applications / Yufeng Wang, Athanasios V. Vasilakos, Qun Jin, Hongbo Zhu.
Description: Boca Raton : Taylor & Francis, CRC Press, 2017. | Includes bibliographical references.
Identifiers: LCCN 2017003187 | ISBN 9781498724173 (hardback : alk. paper)
Subjects: LCSH: Machine-to-machine communications.
Classification: LCC TK5105.67 .W36 2017 | DDC 004.6--dc23
LC record available at https://lccn.loc.gov/2017003187

Visit the Taylor & Francis Web site at
http://www.taylorandfrancis.com

and the CRC Press Web site at
http://www.crcpress.com

Contents

PART II   FUNDAMENTAL ISSUES

CHAPTER 6   SMARTPHONE-BASED HUMAN ACTIVITY RECOGNITION

Preface

The pervasive presence of mobile personal devices with sensing, communication, and computation capabilities together with the massive use of online social network systems is increasingly creating a cyber–physical–social space where users can interact, exploite, and generate information. Moreover, enriched with several sensing capabilities and networking interfaces, today's portable devices are enabling new types of interactions, especially proximity services (ProSe). Proximity awareness means the ability to actively/passively and continuously seek relevant value in one's physical proximity.

By the end of 2025, ProSe Market for long term evolution (LTE) & 5G Networks 2017–2030 research estimates that mobile operators can pocket as much $17 billion in ProSe based annual service revenue. This trend not only provides new opportunities for application vendors but also has the potential to disrupt the current social networking market and the architecture of the web.

Existing technologies used to serve the proximity awareness can be broadly divided into over-the-top (OTT) and device-to-device (D2D) peer-to-peer (P2P) modes.

In the OTT mode, the centralized infrastructure (usually servers located in the cloud) receives periodic location updates from user mobile devices. The server then determines the proximity based on location updates and interests. The constant location updates not only result in significant battery impact because of GPS power consumption and the periodic establishment of cellular connections, but also cause a serious privacy problem.

Different from the OTT mode, in local area, D2D-based ProSe schemes forgo centralized processing in identifying relevancy matches, instead autonomously determining the relevance at the device level by transmitting and monitoring for relevant attributes. This approach offers crucial privacy benefits. In addition, by keeping discovery on the device rather than in the cloud, it allows for user-level controls over what are shared.

D2D-based schemes cannot be seen as the replacements to OTT-based schemes. Intuitively, using OTT, ProSe users can link to online social networks to acquire others' profiles over Internet and enable common profile exchange. With this feature, proximity services are capable of performing common interest matchmaking and content recommendation for users. Naturally, OTT and D2D paradigms are complementary with each other, enabling a proximity service that is more efficient and robust.

Nowadays, cloud computing provides a large range of services and virtually unlimited available resources for users. New applications, such as virtual reality and smart building control, have emerged due to the abundance of resources and services brought by cloud computing. However, the delay-sensitive applications face the problem of long latency, especially when many smart devices and objects are getting involved in people's lives such as the case of smart cities or the Internet of Things. Therefore, cloud computing is unable to meet the requirements of low latency, location awareness, and mobility support. Our book is closely related to the emerging field of mobile edge computing (fog computing), which facilitates the leveraging of available services and resources in the edge networks. The purpose of fog and mobile edge computing is to run the heavy real-time applications at the network edge directly using the billions of connected mobile devices.

In a sense, our book could be regarded a discussion on D2D-enabled fog computing from service and interdisciplinary viewpoints, so-called ProSe.

Traditionally, D2D communication is a technology component for long term evolution-advanced (LTE-A), working in licensed spectrum, which serves as an underlay to the cellular network as a means to increase spectral efficiency. Here, our book explicitly regards various methods as D2D communications technologies, which support infrastructure-free and self-organized proximity area meaning that there is no communication station and a pair of devices can send and receive messages only when they move into each other's communication range.

Unlike existing books on D2D communications, which mainly focus on some physical layer issues, including power and interference management, resource allocation, and mode selection, especially in LTE Direct underlaying cellular networks, besides fundamental D2D communications technologies (Wi-Fi Direct and LTE Direct), our book thoroughly presents architecture, issues, and applications in D2D networking environment.

First, the unified architecture from bottom to top, is composed of the following layers/components: D2D communications layer introduces the basic communications technologies and their issues. Networking layer focuses on peer and service discovery. Topological layer discusses the store-carry-forward based message forwarding schemes (utilizing users' mobility to disseminate content through opportunistic contacts). Profile matching component includes two parts: privacy-preserving profile-matching schemes and OTT-based recommendation systems. Finally, ProSe development framework investigates the requirements, challenges, and typical frameworks of ProSe development and deployment.

Second, the fundamental issues are organized from realistic application and interdisciplinary viewpoints. Specifically, the book will pay special attention to the following issues: smartphone-based human activity recognition, indoor localization and tracking systems, incentive mechanisms (i.e., mobile crowdsourcing systems), and energy-efficient ProSe technologies.

Finally, typical applications of D2D-based ProSe are given, including vehicular social networks (VSNs), and D2D-based cellular traffic offloading.

In brief, D2D-based proximity service is a very hot topic and has great commercial potential from an application viewpoint. Unlike existing books that focus on D2D communications technologies, our book fills a gap: by summarizing and analyzing the latest application and research results in academic, industrial fields and standardization, we comprehensively present the architecture, fundamental issues, and applications in D2D networking environment from application and interdisciplinary viewpoints.

The features of this book are given as follows:

- Proximity awareness, the ability to actively/passively and continuously search for relevant value in one's physical proximity.
- The unified architecture including D2D communications technologies, networking layer, topological layer, profile matching in D2D- and OTT-based ProSe, and the proposed ProSe development framework.
- Fundamental problems in D2D-based proximity service from application and interdisciplinary viewpoints.
- Typical and hot applications of D2D-based ProSe.

Authors

Yufeng Wang earned his PhD in state key laboratory of networking and switching technology, Beijing University of Posts and Telecommunications, Beijing, People's Republic of China. He is a full professor in Nanjing University of Posts and Telecommunications, Nanjing, People's Republic of China. From March 2008, he has been an expert researcher in the National Institute of Information and Communications Technology, Tokyo, Japan. He is a guest researcher at Media Lab, Waseda University, Tokyo, Japan. He has published nearly 100 English journal/conference papers. His research interests focus on multidisciplinary inspired networks and systems. E-mail: wfwang@njupt.edu.cn or wfwang1974@gmail.com

Athanasios V. Vasilakos is currently a distinguished professor of computer science at Luleå University of Technology, Luleå, Sweden. He has authored or coauthored more than 250 technical papers in major international journals and conferences. He is author/coauthor of 5 books and 20 book chapters. His works have received more than 16,000 citations. He has an h-index of 75 and is an ISI Highly Cited researcher. He served or is serving as an editor for many technical journals, such as *IEEE Transactions on Network and Service Management, IEEE Transactions on Cloud Computing, IEEE Transactions on Information Forensics and Security, IEEE*

Transactions on Cybernetics, IEEE Transactions on Nanobioscience, IEEE Transactions on Information Technology in Biomedicine, ACM Transactions on Autonomous and Adaptive Systems, and *IEEE Journal on Selected Areas in Communications.*

Qun Jin is currently a tenured full professor of human informatics and cognitive sciences, Faculty of Human Sciences, Waseda University, Tokyo, Japan. He has been engaged extensively in research works in the fields of computer science, information systems, and social and human informatics. He seeks to exploit the rich interdependence between theory and practice in his work with interdisciplinary and integrated approaches. Dr. Jin has published more than 200 refereed papers in world-renowned academic journals, such as *ACM Transactions on Intelligent Systems and Technology, IEEE Transactions on Learning Technologies, IEEE Systems Journal,* and *Information Sciences* (Elsevier), and international conference proceedings in the related research areas. He has served as a general chair, program chair and Technical Program Committee (TPC) member for numerous international conferences, and editor-in-chief, associate editor, editorial board member, and guest editor for a number of scientific journals. His recent research interests cover human-centric ubiquitous computing, human–computer interaction, behavior and cognitive informatics, big data, personal analytics and individual modeling, social networking and cyber security, e-learning and learning analytics, cyber-enabled health care, and computing for well-being. He is a member of the Institute of Electrical and Electronics Engineers (IEEE), the IEEE Computer Society (IEEE CS), and the Association for Computing Machinery (ACM), New York, USA; Institute of Electronics, Information and Communication Engineers (IEICE), Tokyo, Japan; Information Processing Society of Japan (IPSJ), Tokyo, Japan; Japanese Society for Artificial Intelligence (JSAI), Tokyo, Japan; and China Computer Federation (CCF); People's Republic of China.

Hongbo Zhu is a professor and the former vice president of Nanjing University of Posts and Telecommunications (NUPT), Nanjing, People's Republic of China. At present, he is serving in the following positions: director of the Research Institute of Internet of Things

(IoT), NUPT; fellow of the China Institute of Communication (CIC), Bejing, People's Republic of China; chairman of the IoT Committee, CIC; vice chairman of the Academic Committee, CIC; fellow of the Chinese Institute of Electronics (CIE), Bejing, People's Republic of China; chairman of the Communication Society, CIE; vice chairman of the Education Committee, CIE; member of the Information Division of the Science and Technology Committee, the Ministry of Education of China, People's Republic of China; director of the Ministerial Engineering Research Center on Ubiquitous Network Health Service System, the Ministry of Education of People's Republic of China; standing deputy director of the Ministerial Key Laboratory on Wideband Wireless Communications and Sensor Networks, the Ministry of Education of People's Republic of China; director of Jiangsu Key Laboratory on Wireless Communications, Jiangsu, People's Republic of China; director of Jiangsu Collaborative Innovation Center on IoT and Its Applications, Jiangsu, People's Republic of China; and academic leader of the Jiangsu Key Discipline on Communications and Information Systems, Jiangsu, People's Republic of China.

He has authored or coauthored more than 500 papers and 3 books on information and communication area, such as *IEEE Journal on Selected Areas in Communications, IEEE Transactions on Wireless Communications, IEEE Transactions on Antennas and Propagation,* and *IEEE Transactions on Microwave Theory and Techniques.*

PART I
ARCHITECTURE

1

DEVICE-TO-DEVICE COMMUNICATIONS TECHNOLOGIES

1.1 Introduction

Recently, the number of mobile subscriptions continues to grow along with an explosive increase in the mobile data traffic demand. Thereby, it creates significant network capacity shortage concerns for mobile network operators. In order to address this problem, it is essential to increase the network capacity at a low additional cost. Device-to-device (D2D) communication is believed to be a promising future mobile communication technology capable of creating various new mobile service opportunities and offload traffic from the cellular infrastructure. Efforts have been made by wireless engineers to meet this sociotechnological trend: Qualcomm has pioneered a mobile communication system known as FlashLinQ wherein *wireless sense* is implemented to enable proximity-aware communication among devices [1].

Basically, D2D communications commonly refer to the technologies that enable devices to communicate directly without an infrastructure of access points or base stations (BSs), and the involvement of wireless operators. A surge of interest in supporting direct D2D communications is motivated by several factors: cost, efficiency, traffic offloading, and especially the popularity of proximity-based services driven largely by mobile social networking applications.

One of the key aspects of D2D communications is the set of spectrum bands in which D2D communications takes place. As shown in Figure 1.1, D2D can be divided into licensed D2D and unlicensed D2D communications [2].

Unlicensed D2D: Here the D2D links exploit unlicensed spectrum. Instead of widely used cellular interface, using unlicensed spectrum

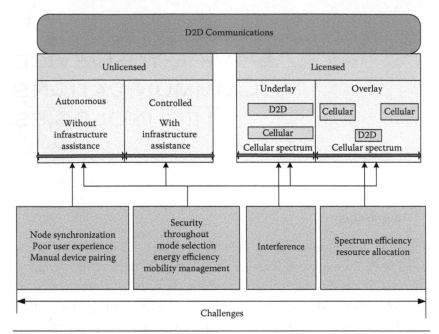

Figure 1.1   Classification of D2D technology.

requires an extra interface and usually adopts other wireless technologies such as Wi-Fi Direct, ZigBee, or Bluetooth. Some of the work on unlicensed D2D [3–5] suggest to give the control of the second interface/technology to the cellular network (i.e., controlled). In contrast, other schemes propose to leave the D2D communications to the individual users (i.e., autonomous) [6]. On one hand, in unlicensed D2D, there is no interference between D2D and cellular users, and users can have simultaneous D2D and cellular communications; on the other hand, unlicensed D2D may suffer from the uncontrolled nature of unlicensed spectrum, especially in an autonomous mode. Especially, unlicensed spectrum is reserved internationally for the use of radio frequency (RF) energy for industrial, scientific and medical, so-called ISM band, and abundant instruments and applications appeared in these bands including radio-frequency process heating, microwave ovens, and medical diathermy machines, and so on. Therefore, there exists a significant interference between mobile devices, unlicensed D2D, and ISM instruments.

For the unlicensed D2D communications with autonomous mode (i.e., without infrastructure assistance), most of the mobile users can only realize D2D function by Wi-Fi or Bluetooth, which is not an integral part of the cellular networks, and thus might cause inconvenient customer usage experience. For example, both Bluetooth and Wi-Fi require manual pairing between two devices. The distance of Wi-Fi Direct is claimed to be 656 in., which means that dozens of devices within the range may be on the list. This process will make the user quite cumbersome compared to making a phone call. Furthermore, since most of the existing D2D communications technologies work on the crowded 2.4 GHz unlicensed band, the interference is uncontrollable. For the similar reason, traditional D2D technologies cannot provide security and quality of service (QoS) guarantee as the cellular networks, and so on. In brief, autonomous unlicensed D2D communications is usually motivated by reducing the overhead of cellular networks. It does not require any change at the BS and can be deployed easily. However, it also has some problems such as node synchronization, poor user experience, and manual device pairing.

The controlled unlicensed D2D communications intentionally utilize the advanced management features of cellular infrastructure to improve the throughput, power efficiency, security, and reliability of D2D communications by allocating appropriate resource (avoid using the ISM bands that are currently occupied by other D2D), facilitating peer and service discovery, and selecting proper communications mode (cellular or D2D).

Licensed D2D: This category exploits the cellular spectrum for both D2D and cellular links. The motivation for choosing licensed communication is usually the high control over cellular (i.e., licensed) spectrum. Some researchers (see e.g., [7]) consider that the interference in the unlicensed spectrum is uncontrollable that imposes constraints for QoS provisioning. Under licensed D2D communication, user equipments (UEs) can reuse uplink/downlink resources in the same cell. Therefore, it is important to design the D2D mechanism in a manner that D2D users do not disrupt the cellular services. Interference management is usually addressed by power and resource allocation schemes. Ideally, the transmission power should be properly regulated, so that the D2D transmitter does not interfere with the cellular UE communication while maintaining minimum signal to interference plus noise ratio (SINR) requirement of the D2D receiver [8].

Interference can also be efficiently managed if the D2D users communicate over resource blocks (RBs) that are not used by nearby interfering cellular UEs.

As shown in Figure 1.1, licensed communication can be further divided into underlay and overlay paradigms. In underlay D2D communication, cellular and D2D communications share the same radio resources. Underlay D2D communications can improve the spectrum efficiency of cellular networks by reusing spectrum resources, but has the significant issue of interference between D2D devices and cellular devices. In contrast, D2D links in overlay communication are given dedicated cellular resources. Allocating dedicated cellular resources to D2D users can avoid the interference, but may waste the spectrum resource.

Irrespective of the unlicensed or licensed D2D, there are some common challenges, such as power efficiency, throughout improvement, security, and mobility management.

The chapter is organized as follows: Section 1.2 describes the basic technologies about unlicensed D2D communication and summarizes the fundamental issues and potential solutions. Section 1.3 describes the basic framework of licensed D2D communications and introduces some key problems and solutions. Finally, the conclusion is briefly provided.

1.2 Unlicensed D2D Communications

This section first describes the popular unlicensed D2D communications technologies, especially Wi-Fi Direct, summarizes the formation of two categories of unlicensed D2D communications: autonomous and network-assisted D2D communications, and then discusses several typical problems of Wi-Fi Direct technology, including power saving and security, and some solutions to solving these problems.

1.2.1 A Technical Overview on Wi-Fi Direct

Nowadays, IEEE 802.11 standard has become one of the most common ways to access the Internet. However, to continue with its striking success, the Wi-Fi technology needs to evolve and embrace a larger set of use cases. Given the wide adoption of Wi-Fi in many kinds of devices, a natural way to progress is to target D2D connectivity, that is, without requiring the presence of an access point (AP). This is the purpose of the Wi-Fi Direct technology recently developed by the Wi-Fi Alliance, which is one of the most popular technologies of unlicensed D2D communications.

In a typical Wi-Fi network, clients discover and associate to wireless local area networks (WLANs), which are created and announced by APs. In this way, a device unambiguously behaves either as an access point (AP) or as a client, each of these roles involving a different set of functionality. A major novelty of Wi-Fi Direct is that these roles are specified as dynamic, and hence a Wi-Fi Direct device has to implement both the role of a client and the role of an AP (sometimes referred to as SoftAP). These roles are therefore logical roles that could even be executed simultaneously by the same device, for instance, by using different frequencies (if the device has multiple physical radios) or by time-sharing the channel through virtualization techniques. In order to establish a communication, then, peer-to-peer (P2P) devices have to agree on the role that each device will assume. Wi-Fi Direct devices, formally known as P2P devices, communicate by establishing P2P groups, which are functionally equivalent to traditional Wi-Fi infrastructure networks. The device implementing AP-like functionality in the P2P group is referred to as the P2P group owner (P2P GO), and devices acting as clients are known as P2P clients. Given that these roles are not static, when two P2P devices discover each other, they negotiate their roles (P2P client and P2P GO) to establish a P2P group. Once the P2P group is established, other P2P clients can join the group as in a traditional Wi-Fi network. Legacy clients can also communicate with the P2P GO, as long as they are not 802.11b only devices, and support the required security mechanisms (discussed in subsection 1.2.3.1.2). In this way, legacy devices do not formally belong to the P2P group and do not support the enhanced functionalities defined in Wi-Fi Direct, but they simply *see* the P2P GO as a traditional AP [9].

As shown in Figure 1.2, the logical nature of the P2P role in Wi-Fi Direct supports different architectural deployments. The figure represents a scenario with two P2P groups. The first group is created by a mobile phone sharing its 3G connection with two mobile phones;

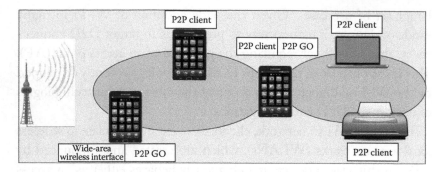

Figure 1.2 The logical nature of P2P role in Wi-Fi Direct group. (From Camps-Mur, D., Garcia-Saavedra, A., and Serrano, P. *IEEE Wirel. Commun.*, 20, 96–104, 2013.)

for this first group, the phone is acting as P2P GO, whereas the other phones behave as P2P clients. In order to extend the network, one of the mobile phones establishes a second P2P Group with a printer and a laptop; for this second group, the mobile phone acts as a P2P GO. In order to act both as P2P client and P2P GO the mobile phone will typically alternate between the two roles by time-sharing the Wi-Fi interface.

Table 1.1 gives some basic concepts/notations/terms and their meanings used in this chapter.

1.2.2 Formations of Unlicensed Device-to-Device Communication

In this part, we mainly illustrate the process of how to build the autonomous communications and a network-assisted communications in unlicensed D2D.

Table 1.1 Definitions about Wi-Fi Direct Used in this Chapter

NOTATIONS/TERMS	MEANINGS
Cluster	A group of mobiles that agree to share their connectivity over Wi-Fi interface
Cluster members/Group clients	All the mobiles that belong to the same cluster
Cluster head/Group owner (GO)	The cluster member with the highest channel quality and acts as the relay to the network
UE	A mobile that is capable of both cellular and Wi-Fi communications
Social channels	Channels 1, 6, and 11 of IEEE 802.11 that are used for device discovery purposes in Wi-Fi Direct

1.2.2.1 Autonomous Device-to-Device Communication   There are several ways in which two devices can establish a P2P group, depending on, for example, if they have to negotiate the role of P2P GO, or if there is some preshared security information available. Here we first describe the most complex case, which is named the standard case, to afterward highlight a couple of simplified cases which are named the autonomous and persistent cases [9].

Standard—In this case the P2P devices have to first discover each other, and then negotiate which device will act as P2P GO. Wi-Fi Direct devices usually start by performing a traditional Wi-Fi scan (active or passive), by means of which they can discover existent P2P groups and Wi-Fi networks. After this scan, a discovery algorithm is executed. Specifically, a P2P device selects one of the so-called social channels, namely channels 1, 6, or 11 in the 2.4 GHz band, as its listen channel. Then, it alternates between two states: a search state, in which the device performs active scanning by sending probe requests in each of the social channels; and a listen state, in which the device listens for probe requests in its listen channel to respond with probe responses. The amount of time that a P2P device spends on each state is randomly distributed, typically between 100 ms and 300 ms, but it is up to each implementation to decide on the actual mechanism to, for example, trade-off discovery time with energy savings by interleaving sleeping cycles in the discovery process. An example operation of this discovery algorithm is illustrated in Figure 1.3. Once the two P2P devices have found each other, they start the GO negotiation phase. This is implemented using a three-way handshake, namely GO negotiation request/response/confirmation, whereby the two devices agree on which device will act as P2P GO, and on the channel where the group will operate, which can be in the 2.4 GHz or 5 GHz bands. In order to agree on the device that will act as P2P GO, P2P devices send a numerical parameter, the GO intent value, within the three-way handshake, and the device declaring the highest value becomes the P2P GO. To prevent conflicts when two devices declare the same GO intent, a tiebreaker bit is included in the GO negotiation request, which is randomly set every time a GO negotiation request is sent. Once the devices have discovered each other and agreed on the respective roles, the next phase is the establishment of a secure communication using Wi-Fi protected setup, which we denote as WPS provisioning phase and described later, and finally a

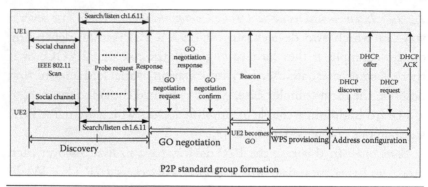

Figure 1.3 Illustration of P2P standard group formation. (From Camps-Mur, D., Garcia-Saavedra, A., and Serrano, P. *IEEE Wirel. Commun.*, 20, 96–104, 2013.)

dynamic host configuration protocol (DHCP) exchange to set up the IP configuration (the address configuration phase in Figure 1.3).

Autonomous—A P2P device may autonomously create a P2P group, where it immediately becomes the P2P GO, by sitting on a channel and starting to beacon. Other devices can discover the established group using traditional scanning mechanisms, and then directly proceed with the WPS provisioning and address configuration phases. Compared to the previous case, then, the discovery phase is simplified in this case as the device establishing the group does not alternate between states, and indeed no GO negotiation phase is required. As shown in Figure 1.4, an exemplary frame exchange for this case is illustrated.

Persistent—During the formation process, P2P devices can declare a group as persistent, by using a flag in the P2P capabilities attribute present in Beacon frames, probe responses and GO negotiation frames. In this way, the devices forming the group store network credentials

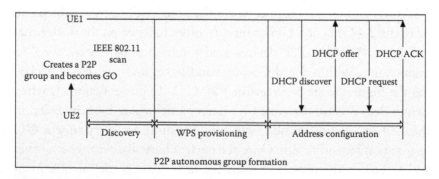

Figure 1.4 P2P autonomous group formation. (From Camps-Mur, D., Garcia-Saavedra, A., and Serrano, P. *IEEE Wirel. Commun.*, 20, 96–104, 2013.)

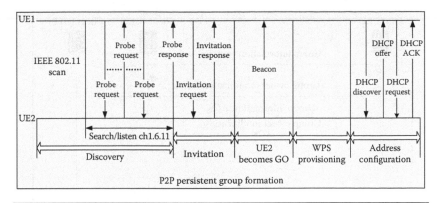

Figure 1.5 P2P persistent group formation. (From Camps-Mur, D., Garcia-Saavedra, A., and Serrano, P. *IEEE Wirel. Commun.*, 20, 96–104, 2013.)

and the assigned P2P GO and client roles for subsequent reinstantiations of the P2P group. Specifically, after the discovery phase, if a P2P device recognizes to have formed a persistent group with the corresponding peer in the past, any of the two P2P devices can use the invitation procedure (a two-way handshake) to quickly reinstantiate the group. This is illustrated in Figure 1.5, where the standard case is assumed as baseline, and the GO negotiation phase is replaced by the invitation exchange, and the WPS provisioning phase is significantly reduced because the stored network credentials can be reused.

As analyzed in Reference 5, long term evolution (LTE) cellular networks can benefit from D2D communications between mobile devices in order to boost energy efficiency, make channel utilization opportunistic, and achieve high fairness while improving network throughput. The basic idea presented in Reference 5 is that mobile users form groups, namely clusters, in which a particular member, the cluster head, is opportunistically selected based on its cellular channel quality, and is responsible for relaying the aggregate traffic of the entire cluster. The cluster head changes over time, as it follows a channel opportunistic selection scheme, which guarantees an optimal utilization of LTE resources.

GO transfer—One of the distinguished features of Wi-Fi Direct is that these roles are specified as dynamic, and hence a Wi-Fi Direct device has to implement both the roles of client and GO. However, in the original Wi-Fi Direct specification, the group ownership cannot be transferred dynamically. In some special scenario, dynamic switching of

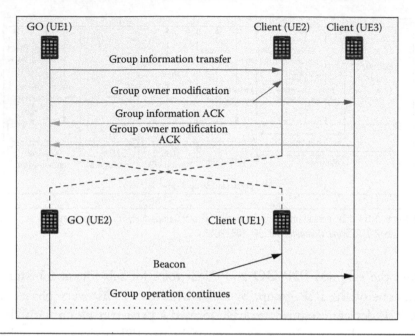

Figure 1.6   Switching of logical roles (i.e., GO and Client) in Wi-Fi Direct. (From Camps-Mur, D., Garcia-Saavedra, A., and Serrano, P. *IEEE Wirel. Commun.*, 20, 96–104, 2013.)

the roles can bring extra benefit. For instance, GO may transfer, when the BS (e.g., eNB) detects that another group member has a better cellular channel quality than the current GO. Two messages are to technically define the support GO transfer in Wi-Fi Direct [9]. As shown in Figure 1.6, first, the GO sends the group information transfer message to the provisioned GO. This message contains the updated list of members and their power-saving related parameters. Second, the GO sends the GO modification broadcast message. Each group client should individually acknowledge this message before the GO transfer is completed.

1.2.2.2 Network-Assisted Device-to-Device Communication (Controlled)   Current WLAN technologies running on the unlicensed bands can be made to cause no interference to cellular networks. But while this makes Wi-Fi Direct a great choice for the network infrastructure, this may not always be the case for the client. Wi-Fi Direct lacks a fast and resource efficient way of notifying clients when/if they are in D2D range. Thus, if a client is searching for a particular user who is out of range for a longer period of time, it will suffer significant battery drain. This is where the LTE network can be of help.

If clients are continuously connected to the LTE network, it knows which cell(s) they are associated with, which tracking area(s) they are in, and their locations within a few meters (if location services are enabled). Therefore, the network can quickly and without significant overhead determine if/when clients are potentially within D2D range and inform them when this is the case.

Although D2D communications have been widely studied within the past few years, the majority of the literature is confined to new theoretical proposals and did not consider implementation challenges. In fact, the implementation feasibility of D2D communications and its challenges are still a relevant research question. In Reference 6, the authors introduced a protocol that focuses on D2D communications using LTE and Wi-Fi Direct technologies. This paper also shows that currently available Wi-Fi Direct features permits to deploy the D2D paradigm on top of the LTE cellular infrastructure, without requiring any fundamental change in LTE protocols. The proposed architecture requires LTE mobiles to form Wi-Fi clusters, and the cluster traffic flows through a cluster head opportunistically and timely selected. The next subsection explains how to implement a D2D system to LTE system using Wi-Fi Direct. It is required two steps to implement a D2D system to LTE system using Wi-Fi Direct: Cluster Registration in LTE and bearer establishment [10].

1. Cluster Registration in LTE
   Once a cluster is formed over Wi-Fi Direct, it should register at the LTE network. Cluster registration procedure is shown in Figure 1.7. This procedure consists of two phases: (a) cluster notification and (b) cluster verification.
   a. Cluster notification: After cluster formation over Wi-Fi Direct, the cluster head must notify this event to the evolved node base station (eNB). The cluster head notifies the eNB by sending the cluster radio resource control (RRC) connection management message (on signaling radio bearer [SRB]) with the request cause set to connection initiation. Next, the cluster head sends the RRC connection setup complete to finish the RRC setup. Here, the dedicated network administration system (NAS) information field is extended to include the EPS mobile identities of all cluster members.

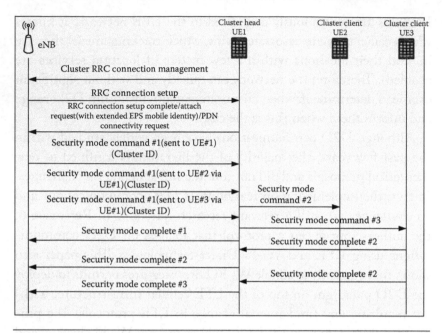

Figure 1.7   Process of cluster registration in LTE. (From Asadi, A. and Mancuso, V. Wi-Fi Direct and LTE D2D in action. In *IFIP Wireless Days*, Valencia, Spain, November 13–15, 2013.)

b. Cluster verification: Once the RRC connection is established, the eNB sends a security mode command message to each cluster member. Note that the security mode command for all clients is received by the cluster head and forwarded to the corresponding client over Wi-Fi. The cluster head also collects the client's responses to the eNB security mode command over Wi-Fi and forward them to the eNB. By forcing the security verification to pass through the cluster head, the eNB ensures that all the cluster clients are already members of the cluster over Wi-Fi.

2. Bearer Establishment

After cluster registration, the cluster head should initiate a cluster bearer establishment procedure. The difference between cluster bearer and UE bearer is in resource provisioning. The allocated resources for a cluster bearer are equivalent to the aggregate of resources allocated to all cluster members. LTE standard defines two types of bearers, namely default and dedicated, to support services with different QoS. The default bearer is established once a UE attaches to the network and it remains until the UE leaves the network. On the other hand, the dedicated bearer is established for services with specific QoS requirement and it remains active for life time of the service. For brevity, we suffice to elaborate on the default bearer establishment. The procedure of dedicated bear establishment requires minor changes in the address field in order to accommodate all cluster members. As shown in Figure 1.8, the procedure for default bearer establishment is depicted, and it consists of three steps: (a) bearer request; (b) bearer request response; and (c) bearer request confirmation.

a. Bearer request: After cluster registration is completed, the eNB sends the attach request to the mobility management entity (MME). The MME determines the international mobile subscriber identity (IMSI) of each cluster member from the information provided in EPS mobile identity fields of the attach request. In case the IMSI of a member cannot be identified, the MME explicitly asks for it. Next, the MME sends a create session request to the serving gateway (S-GW), which contains information such as IMSI of the cluster members and requested packet data network (PDN) connectivity. The S-GW updates its EPS

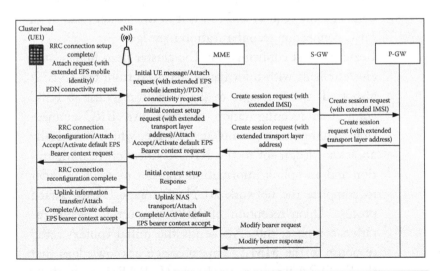

Figure 1.8 Signaling required for default cluster bearer establishment. (From Asadi, A. and Mancuso, V. Wi-Fi Direct and LTE D2D in action. In *IFIP Wireless Days*, Valencia, Spain, November 13–15, 2013.)

bearer table and forward the create session request message to the PDN gateway (P-GW). The P-GW updates its EPS bearer context Table and generates a charging profile for every member who does not have one yet.

b. Bearer request response: The P-GW responds to the S-GW request with create session response message. In this message, the address (assigned by P-GW) and QoS parameters (assigned by policy and charging rules function [PCRF]) fields are extended to accommodate all cluster members. Naturally, QoS parameter of the cluster bearer is equivalent to the aggregate of members' QoS. Next, the S-GW forward the create session response message to the MME, which triggers the initial context setup request sent from the MME to the eNB. This message provides the eNB with settings such as the IP address and the QoS parameters of each cluster members. Again, the IP address and QoS parameters fields of the initial context setup request message are extended to accommodate all cluster members. Note that this reduces the signaling overhead compared to standard LTE operation because the network does not need to send this information to each UE separately. Finally, the eNB extracts the attach accept message from the initial context setup request and sends it to the cluster head in an RRC connection reconfiguration message.

c. Bearer request confirmation: The cluster head updates the cluster clients with information received from the eNB. It also sends two messages to the eNB in response to RRC connection reconfiguration message. An RRC connection reconfiguration complete message, which is basically an acknowledgment to the RRC connection reconfiguration and an uplink information transfer message in order to complete the network attached storage (NAS) attach process. Upon reception of RRC connection reconfiguration complete, the eNB sends the initial context setup response to the MME. This message acknowledges that the E-UTRAN radio access bearer (E-RAB) is successfully setup for the default bearer. It also provides an IP address for communication between the eNB and S-GW for downlink data transfer. After the eNB received the uplink information transfer message, it sends the attach complete message to the MME. The attach complete and active default EPS bearer context accept messages trigger the MME to send the modify bearer request to S-GW. This message mainly serves as an acknowledgment. Finally, the S-GW completes the process by sending modify bearer respond to the MME. As concerns IP addressing, in LTE, each active UE has at least one default bearer and each default bearer has a unique IP address. Therefore, if a cluster member had bearer(s) before cluster formation, the PDN-GW keeps the existing IP address(es) associated to the default bearer(s). Once the cluster bearer is activated, the PDN-GW automatically terminates the old default bearer(s).

1.2.3 Key Problems

In this subsection, we summarize some key issues related to unlicensed D2D communications and some solutions to solve these problems.

1.2.3.1 Issues in Autonomous Device-to-Device Communication

1.2.3.1.1 Power Saving
1. Existing power-saving technologies

   Using Wi-Fi Direct, battery-constrained devices may typically act as P2P GO (soft-AP), and therefore energy efficiency is of capital importance. However, power-saving mechanisms in current Wi-Fi networks are not defined for APs but only for clients. Notice that with Wi-Fi Direct, a P2P client can benefit from the existing Wi-Fi power-saving protocols, that is, legacy power save mode or U-APSD. In order to support energy savings for the AP, Wi-Fi Direct defines two new power saving mechanisms: the opportunistic power save protocol and the notice of absence (NoA) protocol [9].

   Opportunistic Power Save
   The basic idea of opportunistic power save is to leverage the sleeping periods of P2P clients. The mechanism assumes the existence of a legacy power-saving protocol, and works as follows.

The P2P GO advertises a time window, denoted as CTWindow, within each beacon and probe response frames. This window specifies the minimum amount of time after the reception of a Beacon during which the P2P GO will stay awake and therefore P2P Clients in power saving can send their frames. If after the CTWindow the P2P GO determines that all connected clients are in doze state, either because they announced a switch to that state by sending a frame with the power management (PM) bit set to 1, or because they were already in the doze state during the previous beacon interval, the P2P GO can enter sleep mode until the next Beacon is scheduled; otherwise, if a P2P client leaves the power-saving mode (which is announced by sending a frame with the PM bit set to 0) the P2P GO is forced to stay awake until all P2P Clients return to power-saving mode. Figure 1.9a provides an example of the operation of the opportunistic power save protocol for a scenario consisting of one P2P GO and one P2P client.

Notice that, using this mechanism, a P2P GO does not have the final decision on whether to switch to sleep mode or not, as this depends on the activity of the associated P2P clients.

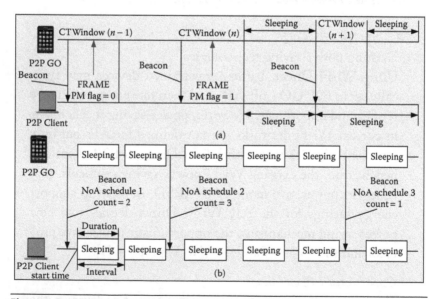

Figure 1.9 Operations of power saving. (From Camps-Mur, D., Garcia-Saavedra, A., and Serrano, P. *IEEE Wirel. Commun.*, 20, 96–104, 2013.)

Notice of Absence

The NoA protocol allows a P2P GO to announce time intervals, referred to as absence periods, where P2P clients are not allowed to access the channel, regardless of whether they are in power save or in active mode. In this way, a P2P GO can autonomously decide to power down its radio to save energy. Like in the opportunistic power save protocol, in the case of NoA the P2P GO defines absence periods with a signaling element included in Beacon frames and probe responses. In particular, a P2P GO defines a NoA schedule using four parameters:

- Duration that specifies the length of each absence period
- Interval that specifies the time between consecutive absence periods
- Start time that specifies the start time of the first absence period after the current Beacon frame
- Count that specifies how many absence periods will be scheduled during the current NoA schedule

A P2P GO can either cancel or update the current NoA schedule at any time by respectively omitting or modifying the signaling element. P2P clients always adhere to the most recently received NoA schedule. Figure 1.9b depicts an example operation of the NoA protocol.

2. Problems of existing Wi-Fi Direct Power Management

For periodical data transmissions, it is generally expected that each packet is sequentially transmitted with minimal delay. When opportunistic power saving is used, GO will doze if a periodical packet is due for transmission after the CTWindow. This may result in high end-to-end delay. If a packet periodically arrives during the CTWindow, PM configuration of 0 will keep the GO in the active state, which is a factor for energy waste. To avoid this, PM configuration of 1 can be used. However, this will in turn cause the GO to always doze after CTWindow, inducing transmission delay for packets that arrive after CTWindow. In case of typical bulk transmissions, there is a clear indicator for configuring the PM to 0 or 1: transmission start and end. However, periodical data do not have clear indication for configuring the PM value.

To handle periodical data transmissions, NoA mode can be utilized to provide efficient duty cycling. However, this is valid only if the interval of NoA and periodic data transmission is exactly synchronized. If the interval of NoA is smaller than the data transmission interval, devices wake up too frequently, wasting more energy. If the data interval is smaller, then NoA cannot cope with the transmission rate of the periodical data, inducing transmission delay [11].

1.2.3.1.2 Security Despite all the benefits of D2D communications, security is one of the major concerns that need to be well addressed before D2D technique gets widely accepted and implemented. It is well known that due to the broadcast nature of wireless channels, wireless communication such as Wi-Fi and Bluetooth is vulnerable to a variety of attacks that challenges the three basic principles of security—confidentiality, integrity, and availability. Some common attack vectors include surreptitious eavesdropping, message modification, and node impersonation. For example, by stealthy listening to the communication between two devices, an attacker can gain critical or privacy information, such as trade secrets or identity-related information. Thus, the D2D communications between devices need to be properly secured. Specifically, how to establish a shared secret between devices is one of the main challenges for secure D2D communications.

Reference 12 investigated the security requirements and challenges for D2D communications, and presented a secure and efficient key agreement protocol, which enables two mobile devices to establish a shared secret key for D2D communications without prior knowledge. Their approach is based on the Diffie–Hellman key agreement protocol and commitment schemes. Compared to previous work, their proposed protocol introduces less communication and computation overhead. The authors present the design details and security analysis of the proposed protocol. They also integrate their proposed protocol into the existing Wi-Fi Direct protocol, and implement it using Android smartphones.

Commitment Schemes:

A commitment scheme allows one user to commit to a chosen value or statement while keeping it hidden to others, with the ability to reveal the commitment value latter. In addition, the commitment schemes contain two important parts: commit and open.

Commit: $(c; d) \leftarrow m$ transforms a value m into a commitment/open pair $(c; d)$. The commit value c reveals no information of m, but with decommit value d together with $(c; d)$ will reveal m.

Open: $m \rightarrow (c; d)$ output original value m if (c, d) is the commitment/open pare generated by commit(m).

Protocol Design:

The key agreement protocol in Reference 12 is based on the traditional Diffie–Hellman key agreement protocol and a commitment scheme. In their protocol, two mobile users A and B, respectively, generate k-bit random strings, N_A and N_B and $N_A \oplus N_B$ as the short authentication string for mutual authentication.

Figure 1.10 shows the message flow of the proposed protocol. At the initial stage, users A and B select their Diffie–Hellman parameters a and b, then compute g^a and g^b. A and B randomly generate their k-bit strings N_A and N_B. $m_A = \text{ID}_A \, g^a N_A$ and $m_B = \text{ID}_B \, g^b N_B$ are formed by concatenation, in which ID_A and ID_B are human-readable identifiers for users A and B, such as names or e-mail addresses. User A also needs to calculate the commitment/opening $(c; d)$ for m_A.

After the initial stage, user A and user B perform the following message exchange over their D2D communications channel. User A sends the c, the commitment value of m_A to user B; after receiving c, user B sends m_B to user A. In return, user A sends the decommit value d to user B. User B opens the commitment and gets $m_A = \text{ID}_A g^a N_A$. In the final stage, users A and B generate the k-bits authentication string by $S_A = N_A \oplus N'_B$ and $S_B = N'_A \oplus N_B$, in which N'_B and N'_A are derived from messages received by A and B. Then users A and B verify if $S_A = S_B$ via trusted channel (visual or verbal comparison). If the authentication strings match, A and B accept each other's Diffie–Hellman parameters and calculate the shared secret key, $K = g^{ab} \bmod p$. The reason for comparing authentication string before generating Diffie–Hellman secret key is that

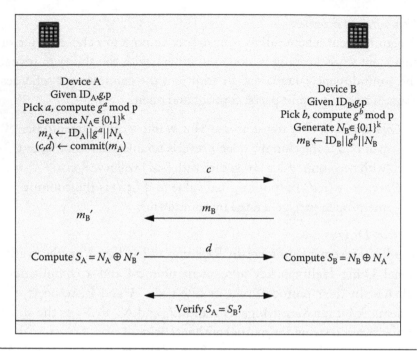

Figure 1.10 Secure key exchange protocol. (From Shen, W., Hong, W., Cao, X., Yin, B., Shila, D.M., and Cheng, Y. Secure key establishment for Device-to-Device communications. In *Proceedings of Global Communications Conference* (*GLOBECOM*), Austin, TX, December 8–12, 2014.)

if the strings do not match, both users can save the computation for secret key generation.

Wi-Fi Direct protocol enables two devices to establish a D2D connection using Wi-Fi frequency without the help of access points. Figure 1.10 shows the procedure for a D2D connection establishment using Wi-Fi Direct. First, two devices perform the channel probing and discover each other. Then the two devices will go through a three-way handshake to determine the group owner (works as an access point) for this D2D connection. After the devices have agreed on their respective roles, a Dynamic Host Configuration Protocol (DHCP) exchange will be conducted to set up the IP addresses for both devices. Thus, the D2D connection between these two devices has been established. We add our proposed key agreement protocol on top of the existing Wi-Fi Direct protocol, as shown in Figure 1.11. After the address configuring phase, the two devices will go through our proposed key agreement protocol as well as the mutual authentication

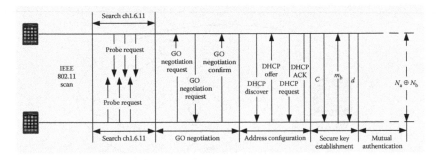

Figure 1.11 Secure Wi-Fi Direct protocol. (From Shen, W., Hong, W., Cao, X., Yin, B., Shila, D.M., and Cheng, Y. Secure key establishment for Device-to-Device communications. In *Proceedings of Global Communications Conference (GLOBECOM)*, Austin, TX, December 8–12, 2014.)

process to agree on a shared secrete key. As long as the two devices have agreed on the authentication message, they can subsequently use their shared secret key for future communication.

1.2.3.2 Network-Assisted Device-to-Device Communication   In order to facilitate network-assisted D2D on ISM band, the main challenges are related to the coexistence of D2D and the popular IEEE 802.11 WLAN in the same frequency band and geographical area, arising from the intersystem interference and potential vicious competition. Although the works are related, schemes can improve the fairness and efficiency of ISM band resource usage to some extent, some important issues are still open that are as follows: (1) The bandwidth of WLAN channels, that is, 22 MHz, is too wide for any single D2D pair usage. For more efficient usage, a D2D winner in resource contention should share the seized 22 MHz resource with other D2D peers; (2) too many simultaneous channel-contending attempts among D2D pairs will greatly lower the resource utilization efficiency; and (3) the high density of D2D devices (e.g., in some hotspots) may cause an unfair competition with their WLAN rivals. To address these problems, we proposed a groupwise channel sensing and resource preallocation scheme for D2D devices, where D2D pairs with different QoS/bandwidth requirements are grouped and prescheduled to approximately fill in an overall flexible RB, and a representative D2D pair is appointed in each group for contending resource to avoid intragroup contention [13].

As shown in Figure 1.12, a typical D2D system operating on ISM band consists of a centralized D2D controller (e.g., AP or eNB) and

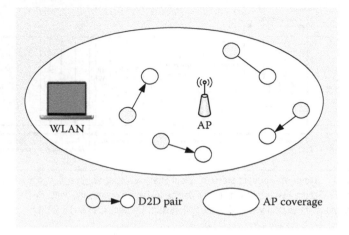

Figure 1.12   A typical D2D system operating on ISM band. (From Zhou, B., Ma, S., Xu, J., and Li, Z. Group-wise channel sensing and resource pre-allocation for LTE D2D on ISM band. In *Proceedings of Wireless Communications and Networking Conference* (*WCNC*), Shanghai, China, April 7–10, 2013.)

some D2D communicating pairs. For every D2D pair, the transmitter can communicate directly with the recipient using the resource on ISM band. At the same time, some WLAN stations are also deployed in the same geographic area acting as the potential interference sources to D2D communication.

As shown in Figure 1.13, first, a D2D sender (denoted as gray point in Figure 1.12) willing to transmit a packet will first sense the medium, if the medium is busy then it defers. If the medium is free for a specified time (called DIFS, distributed inter frame space, in IEEE 802.11), it is allowed to transmit a short control packet called request to send (RTS), which includes the source address, the destination address, and the duration of the following D2D transaction (i.e., the D2D packet and the corresponding ACK); a D2D recipient (denoted as white point) will respond (if the medium is free) with a response control packet called clear to send (CTS), which will include the same duration information. Then, all the WLAN stations received either RTS and/or CTS will set their virtual carrier sense indicator (called network allocation vector [NAV]) and keep silent for the given duration indicated by RTS or CTS, following the IEEE 802.11 DCF protocol. By this means, the medium can be reserved for D2D communications without intersystem interference.

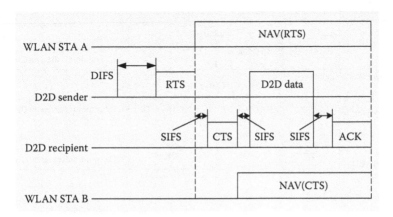

Figure 1.13 Illustration of the proposed groupwise channel sensing (only include single D2D pair and a WLAN device). (From Zhou, B., Ma, S., Xu, J., and Li, Z. Group-wise channel sensing and resource pre-allocation for LTE D2D on ISM band. In *Proceedings of Wireless Communications and Networking Conference* (*WCNC*), Shanghai, China, April 7–10, 2013.)

This mechanism can protect D2D communications on ISM band form the interference caused by its WLAN rivals. The 22 MHz WLAN channel bandwidth is too wide for a single D2D pair exclusive usage. To improve resource efficiency, it is necessary to let multiple D2D pairs share the seized channel with the help of a centralized D2D controller.

As in Figure 1.14, Reference 13 proposed a new scheme for multi D2D pairs and WLAN devices to share the 22 MHZ spectrum. First, the spectrum is divided into three subspectrums(CH#1, CH#2, CH#3). All D2D pairs will listen on these subspectrums in turn in a slot. The sender of D2D pair1 will sense the medium and it finds that CH#3 is free. The sender will transmit a short control packet called RTS to all devices. The receiver of D2D pair1 will transmit a short control packet called RTS to respond. Then, all the WLAN stations received either RTS and/or CTS will set their virtual carrier sense indicator (called NAV, for network allocation vector) and will keep silent for the given duration indicated by RTS or CTS on CH#3. Other D2D pairs will communicate on CH#3 in turn.

References 14 and 15 described a novel architecture to improve the throughput of video transmission in cellular networks, based on caching of popular video files in cellphones and BS-controlled D2D communications. The architecture exploits the large storage available on

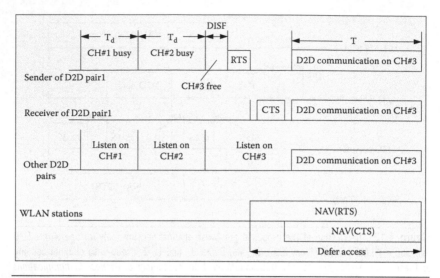

Figure 1.14 Illustration of the proposed groupwise channel sensing (include multi D2D pairs and WLAN devices). (From Zhou, B., Ma, S., Xu, J., and Li, Z. Group-wise channel sensing and resource pre-allocation for LTE D2D on ISM band. In *Proceedings of Wireless Communications and Networking Conference (WCNC)*, Shanghai, China, April 7–10, 2013.)

modern smartphones to cache video files that might be requested by other users. BSs keep track of the cache content and direct requests to the nearest smartphone that has the desired file, which is then transmitted via a D2D link. Since the distance between the requesting user and the smartphone with the stored file will be small in most cases, multiple D2D links can be operated on the same time/frequency resources within one cell. This in turn leads to a dramatic increase in spectral efficiency.

The stored files are transmitted, upon request, to a user requesting a particular file. The transmission occurs by D2D communications; since the distance between transmitting and receiving device is much shorter than between device and BS, multiple D2D links can be operated on the same time/frequency resources within one cell. This in turn leads to a dramatic increase in spectral efficiency. The BS keeps track of which phone has which files stored. Thus, when a user requests a certain video file, the BS can direct it to the nearest smartphone that has the file stored, which is then transmitted via a D2D link, and can optimize the frequency reuse between the devices. According to their simulations, their proposal improves the video throughput by one or two orders of magnitude.

1.3 Licensed Device-to-Device Communication

LTE Direct is a new and innovative direct D2D technology that enables discovering thousands of devices and their services in the proximity, in a privacy sensitive and battery efficient way. This allows the discovery to be *Always ON* and autonomous, without drastically affecting the device battery life. LTE Direct uses licensed spectrum, allowing mobile operators to employ it as a way to offer a range of differentiated applications and services to users. It relies on the LTE physical layer to provide a scalable and universal framework for discovering and connecting proximate peers [16,17].

1.3.1 LTE Direct Architecture

3GPP technology has the opportunity to become the platform of choice to enable D2D communication. The proximity services (ProSe) can be divided into two parts: proximity discovery and direct communication [19]. With proximity discovery, users can discover other users that are in proximity. Discovery mechanisms can be network or user assisted. From an architectural viewpoint, the solution for ProSe involves UEs, the radio access network (RAN), core network, and application servers. Although different architecture alternatives are still being evaluated in 3GPP, there are some common characteristics that can be identified. As shown in Figure 1.15, in the core, a new network function, so-called the proximity function or proximity server is proposed for addition in order to provide the proximity services. The proximity server would provide the connection between application servers and the mobile network. It could identify proximity between UEs and inform the application servers of the opportunities. D2D sessions could be initiated from the proximity servers by sending an initiation request to the MME. Then the MME initiates the D2D radio bearer setup. It is also responsible for transmitting the IP address toward D2D-terminating devices. The key functional elements for this process are the following:

1. The MME, in addition to being the control node responsible for signaling issues related to mobility (tracking and paging) and security access for the evolved universal terrestrial radio access network (E-UTRAN), it will cache a copy of the user's profile related to ProSe after being authenticated

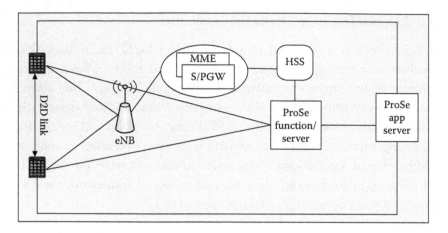

Figure 1.15 Functional block of D2D inside LTE-A SAE (System architecture evolution). (From Doumiati, S., Artail, H., and Gutierrez-Estevez, D.M. *Procedia Computer Science*, 34, 87–94, 2014.)

by the home subscriber server (HSS), and informs the eNB about the user's permission.

2. The HSS is a data repository for subscribers' profiles that authenticates/authorizes user access to the system, and more specifically will check whether the requesting users are ProSe subscribers or not.

3. The ProSe function generates the IDs of the ProSe users after being authorized by the HSS and handles these IDs along with their corresponding application layer user IDs. It also stores a list of authorized applications IDs to use EPC-level ProSe discovery. The ProSe function also plays the role of location services client (service logic processing [SLP] agent) to communicate with the SLP and be aware of the UEs' locations to determine their proximity.

4. The ProSe application server contains the applications offering services based on the corresponding application programming interfaces (APIs) for ProSe, provided by the 3GPP operator in the service agreement. It is the entity on the service network from which the user downloads the apps. It also stores the identities of the ProSe users, as defined at the network level, and maps these identities to the application layer user identities, which identify specific users within an application. Moreover, the ProSe function ID corresponding to each user can also be saved there.

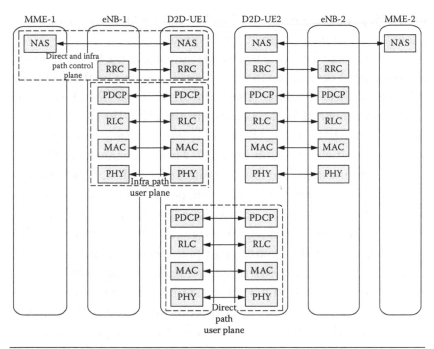

Figure 1.16 Protocol stack for LTE Direct communication. (From Raghothaman, B., Deng, E., Pragada, R., Sternberg, G., Deng, T., and Vanganuru, K. Architecture and protocols for LTE-based device to device communication. In *Proceedings of 2013 International Conference on Computing, Networking and Communications* (*ICNC*), San Diego, CA, January 28–31, 2013.)

Figure 1.16 shows a proposed protocol stack for LTE Direct communication. One or more separate data radio bearers are setup for transmission of user plane data over the direct path. The PHY, MAC, RLC, and packet data convergence protocol (PDCP) layers for these bearers are terminated at the UEs. Each UE is also simultaneously connected to an eNB. The control plane protocols, namely RRC and NAS, are terminated between each UE and the corresponding eNB and MME respectively.

The NAS procedures for service request are altered to include D2D aspects, as described in Reference 20. Changes to the RRC protocol are necessary for the establishment and management of the direct path data radio bearer, and these are discussed in Reference 21. In contrast to current LTE systems, where the radio bearer is terminated at the UE and the eNB, the end points of the direct path communication are at the two UEs. Hence, the RRC configuration provided by the eNB to both UEs should be compatible with each other. Radio-link monitoring, measurement, and handover procedures also need to be altered to accommodate direct path aspects.

The direct path data radio bearer is terminated at the UEs. However, the resource allocation functionality is still retained by the eNB. For this reason, there should be a separate MAC entity for D2D in the UE, and a similar MAC D2D entity in the eNB. The functionalities of the D2D MAC in the UE include data transfer, HARQ, BSR reporting, multiplexing of multiple D2D logical channels, and logical channel prioritization. The functionalities of the MAC D2D entity in the eNB include resource allocation and per-TTI scheduling. Varying degrees of eNB control can be envisioned in the scheduling process, broadly divided into two categories—(a) Full eNB control of per-TTI scheduling or (b) eNB resource grant plus UE scheduling.

The direct path may have a simplified RLC, perhaps with no RLC retransmissions. In the PDCP functionality, UE support of simultaneous separate ciphering for infrastructure and direct path bearers is required, as well as altered procedures for PDCP sequence preservation during mobility events between infrastructure and direct path, as described in Reference 20.

1.3.2 Underlay and Overlay Device-to-Device Communication

The literature under this category, which contains the majority of the available work, proposes to use the cellular spectrum for both D2D and cellular links. The motivation for choosing inband communication is usually the high control over cellular (i.e., licensed) spectrum. Some researchers consider that the interference in the unlicensed spectrum is uncontrollable, which imposes constraints for QoS provisioning. Inband communication can be further divided into underlay and overlay categories. In underlay D2D communication, cellular and D2D communications share the same radio resources. In contrast, D2D links in overlay communication are given dedicated cellular resources. Inband D2D can improve the spectrum efficiency of cellular networks by reusing spectrum resources (i.e., underlay) or allocating dedicated cellular resources to D2D users that accommodates direct connection between the transmitter and the receiver (i.e., overlay). The distinguished disadvantage of inband D2D is the interference caused by D2D users to cellular communications and vice versa. This interference can be mitigated by introducing high-complexity resource allocation methods, which increase the computational overhead of the BS or D2D users.

1.3.3 Key Challenges

1.3.3.1 Resource Allocation

1.3.3.1.1 Overlay Resource Allocation In Reference 22, authors analyze the resource sharing in a D2D communication overlaying cellular system, in which both cellular traffic and D2D traffic use the same resources, and the system aims to optimize the total throughput over the shared resources while fulfilling possible spectral efficiency restrictions and power constraints.

Figure 1.17 illustrates the considered scenario, where g_i is the channel response between the BS and UE_i, and g_{ij} is the channel response between UE_i and UE_j. The D2D pair can communicate directly with coordination from the BS. The channel response can include the path loss, the shadow, and the fast fading effects.

The sharing of resources between D2D and cellular connections is determined by the BS. If D2D users are assigned resources that are orthogonal to those occupied by the cellular user, they cause no interference to each other and the analysis is simpler. Figure 1.18 illustrates the resource allocation of the orthogonal sharing mode. Here, an overlay

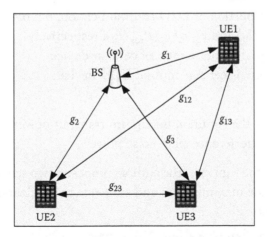

Figure 1.17 D2D communications as an underlay network to a cellular network. (From Yu, C.H., Doppler, K., Ribeiro, C.B., and Tirkkonen, O. *IEEE Transactions Wireless Communications*, 10, 2752–2763, 2011.)

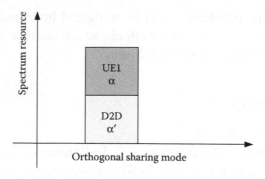

Figure 1.18   Resource allocation in overlay D2D modes. (From Yu, C.H., Doppler, K., Ribeiro, C.B., and Tirkkonen, O. *IEEE Transactions Wireless Communications*, 10, 2752–2763, 2011.)

mode is investigated: D2D communication gets part of the resources and leaves the remaining part of resources to the cellular user. There is no interference between cellular and D2D communication. The resources allocated to D2D and cellular connections are to be optimized.

With orthogonal resource sharing, the sum rate expression with power/energy constraints is Equation 1.1:

$$R_{\text{OS-P}}(\alpha) = \alpha\log_2(1+\gamma_1) + \alpha'\log_2(1+\gamma_{23}) \tag{1.1}$$

where:

$R_{\text{OS-P}}(\alpha)$ is the sum rate with maximum power constraint

α is the proportion of cellular communication accounted for the entire band

α' is the proportion of D2D communication, $0 \le \alpha \le 1$, $\alpha' = 1 - \alpha$

$\gamma_1 = (g_1 P_{\text{max}}/I_0)$ and $\gamma_{23} = (g_{23}P_{\text{max}}/I_0)$, respectively

γ_1 is the signal-to-noise ratio of cellular device

γ_{23} is the signal-to-noise ratio of D2D device

where:

P_{max} denotes the common maximum transmit power

I_0 denotes interference-plus-noise power

To maximize the sum rate, the authors proposed two schemes for this: greedy sum-rate maximization and sum-rate maximization subject to rate constraints

1. *Greedy Sum–Rate Maximization*: To maximize the sum rate, Reference 22 allocates all resources to the type of communication dominating the performance, resulting in Equation 1.2:

$$\alpha = \begin{cases} 1 & \text{if } \gamma_1 \geq \gamma_{23} \\ 0 & \text{if } \gamma_1 \leq \gamma_{23} \end{cases} \tag{1.2}$$

where:

γ_1 is the signal-to-noise ratio of cellular device

γ_{23} is the signal-to-noise ratio of D2D device

The optimized sum rate is Equation 1.3:

$$R_{\text{OS-P}}(\alpha) = \begin{cases} \log_2(1+\gamma_1) & \text{if } \gamma_1 \geq \gamma_{23} \\ \log_2(1+\gamma_{23}) & \text{if } \gamma_1 \leq \gamma_{23} \end{cases} \tag{1.3}$$

Note that the optimized $R_{\text{OS-P}}(\alpha)$ is simply a subset of the corresponding case for nonorthogonal sharing.

2. *Sum-rate Maximization Subject to Rate Constraints*: With a cellular service guarantee, we need to avoid allocating all the resources to D2D users. When the D2D link is stronger than the cellular link, we allocate enough resources to the cellular user to fulfill the cellular service guarantee and give the remaining resources to the D2D users to maximize the sum rate. Therefore, α can be obtained by Equation 1.4:

$$\alpha = \begin{cases} 1 & \text{if } \gamma_1 \geq \gamma_{23} \\ \min\left[\dfrac{\gamma 1}{\log_2\left(1+\min(\gamma_1,\gamma_b)\right)}, 1 \right] & \text{if } \gamma_1 \geq \gamma_{23} \end{cases} \tag{1.4}$$

where:

γ_1 is the signal-to-noise ratio of cellular device

γ_{23} is the signal-to-noise ratio of D2D device

γ_b is the maximum signal-to-noise ratio

The sum rate is Equation 1.5:

$$R_{\text{OS-P}}(\alpha) = \alpha\log_2\left[1+\min(\gamma_1,\gamma_b) \right] \\ + \alpha'\log_2\left[1+\min(\gamma_{23},\gamma_b) \right] \tag{1.5}$$

1.3.3.1.2 Underlay Resource Allocation (Interference Management) In underlay D2D communication, cellular and D2D communications

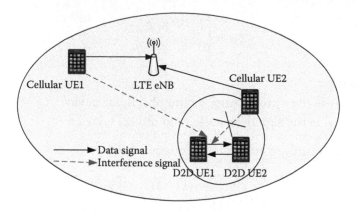

Figure 1.19   The interference in underlay D2D communications.

share the same radio resources. By exploiting the spatial diversity, underlay D2D communication is able to increase the cellular spectrum efficiency. Interference between the cellular and D2D communications is the most important issue in underlay D2D communications. Until now most researches of D2D transmission focus on the scenario that D2D users reuse the uplink spectrum of the LTE system. Since D2D works autonomously and fully share frequency resources with cellular users, sometimes there is near-far interference from cellular users to D2D pairs. The addressed problems are illustrated in Figure 1.19, where cellular UE1 may impose serious near-far interference to D2D transmission if they are sharing the same resource.

In Reference 23, a new interference cancelation scheme is designed based on the location of users. The authors propose to allocate a dedicated control channel for D2D users. Cellular users listen to this channel and measure the SINR. If the SINR is higher than a predefined threshold, a report is sent to the eNB. Accordingly, the eNB stops scheduling cellular users on the RBs that are currently occupied by D2D users. The eNB also sends broadcast information regarding the location of the users and their allocated RBs. Hence, D2D users can avoid using RBs that interfere with cellular users.

The work in Reference 24 proposes a new interference management in which the interference is not controlled by limiting D2D transmission power as in the conventional D2D interference management mechanisms. The proposed scheme defines an interference limited area in which no cellular users can occupy the same resources as the D2D pair.

Therefore, the interference between the D2D pair and cellular users is avoided. The disadvantage of this approach is reducing multiuser diversity because physical separation limits the scheduling alternatives for the BS.

In Reference 25, the authors propose a novel scheme to realize UL resources sharing meanwhile avoiding near-far interference to D2D transmission in a hybrid network. By monitoring the common control channel (CCCH), eNB will identify the *near-far-risk* cellular users, and broadcast the information on their allocated resources. Based on this knowledge, D2D devices can proactively perform radio resource management (RRM) to avoid the near-far interference from cellular UEs. Different from the previous work, by utilizing the proposed scheme, the cell radio network temporary identifier (C-RNTI) of D2D UEs is not always necessary and the interference from the neighboring cell can be suppressed as well.

Considering the spectrum efficiency of D2D communications underlaying an LTE-advanced network, a resource allocation scheme is proposed in Reference 26 to minimize the transmission length of D2D links (i.e., maximize the spectrum utilization) by allowing multiple D2D transmissions in the same resource block (RB) of a cellular user. The proposed scheme jointly allocates the physical RBs in both time domain (i.e., time slots) and the frequency domain (i.e., channels) and performs power control for each D2D link subject to interference constraints for cellular users and traffic demands of D2D links. The problem can be formulated as a mixed integer programming (MIP), which is an NP-complete problem. It implies that there is no known polynomial-time algorithm for finding all feasible schedules and the corresponding transmission power vectors. Therefore, the authors propose a column generation method to solve the resource allocation problem with low complexity. The main idea is to find the maximum of active D2D links that can simultaneously transmit data in each time slot while the feasible access pattern constraints are satisfied.

In Reference 27, the authors introduce a reverse iterative combinatorial auction as the allocation mechanism. In the auction, all the spectrum resources are considered as a set of resource units, which as bidders compete to obtain business while the packages of the D2D pairs are auctioned off as goods in each auction round.

A model of a single cell with multiple users is considered, in which, as shown in Figure 1.20, UEs with data signals between each other are in the D2D communication mode, whereas UEs that transmit

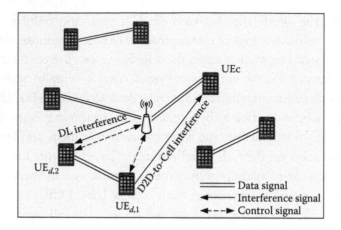

Figure 1.20   System model of D2D communication underlaying cellular networks with downlink resource sharing. (From Xu, C. et al. *IEEE J. Sel. Area. Comm.*, 31, 348–358, 2013.)

data signals with the BS keep in the traditional cellular mode. Each user is equipped with a single omnidirectional antenna. The locations of cellular users and D2D pairs are randomly set and traversing the whole cell. For simplicity and clarity, we illustrate cochannel interference scenario involving three UEs (UE_c, $UE_{d,1}$, and $UE_{d,2}$), and omit the interference and control signal signs among others. UE_c is a traditional cellular user that is distributed uniformly in the cell. $UE_{d,1}$ and $UE_{d,2}$ are close enough to satisfy the distance constraints of D2D communication, and at the same time they also have communicating demands. One member of the D2D pair $UE_{d,1}$ is distributed uniformly in the cell, and the position of the other member $UE_{d,2}$ follows a uniform distribution inside a region at most L from $UE_{d,1}$.

The authors define D as a package of variables representing the index of D2D pairs that share the same resources.

The authors assume that the total pairs can form N such packages. Thus, if the members of the k-th (k = 1, 2,..., N) D2D user package share resources with cellular user c, then the channel rate is given by Equation 1.6:

$$V_c(k) = R_{ck} + \sum_{d \in D_k} R_d^k \qquad (1.6)$$

where:

R_{ck} denotes the channel rates of UE_c

R_d^k denotes the channel rates of D2D pair d (where d denotes a D2D pair).

In the proposed reverse iterative combinatorial auction (I-CA) mechanism, the authors consider spectrum resources occupied by cellular user c as one of the bidders who submit bids to compete for the packages of D2D pairs, in order to maximize the channel rate. It is obvious that there would be a gain of channel rate owing to D2D communication as long as the contribution to data signals from D2D is larger than that to interference signals. Considering the constraint of a positive value, the authors define the performance gain by Equation 1.7, which is the private valuation of bidder c for the package of D2D pair D_k:

$$v_c(k) = \max[V_c(k) - V_c, 0] \qquad (1.7)$$

Here, V_c denotes the channel rate of UE_c without cochannel interference.

In the auction, the cellular resource denoted by c obtains a gain by getting a package of D2D communications. However, there exists some cost such as control signals transmission and information feedback during the access process. The authors define the cost as a pay price. The price to be paid by the bidder c for the package D_k is called pay price denoted by $P_c(k)$. The unit price of item $d(\forall k, d \in D_k)$ can be denoted by $P_c(d)$.

In addition, the scheme is to maximize the overall gain of the system, and the system gain can be obtained by Equation 1.8:

$$\text{Max} \sum_{c=1}^{C} \sum_{k=1}^{N} x_c(k) v_c(k) \qquad (1.8)$$

where $x_c(k)$ denotes that if the bidder c choose the D_k package and if the bidder c choose the D_k package, $x_c(k) = 1$, otherwise $x_c(k) = 0$.

In Reference 27, the authors proposed the specific algorithm for resource allocation to maximize the overall gain. In addition, a detailed nonmonotonic descending price auction algorithm is explained depending on the utility function that accounts for the channel gain from D2D and the costs for the system. Further, the authors prove that the proposed auction-based scheme is cheat-proof, and converges in a finite number of iteration rounds. The authors explain nonmonotonicity in the price update process and show lower complexity compared to a traditional combinatorial allocation. The simulation results show that the system sum rate goes up with both the number of D2D pairs and the number of resource units increasing. The proposed auction algorithm is much superior to the random allocation, and provides high system efficiency, which is stable over different parameters of users and resources.

1.3.3.2 Transmission Mode Selection   The transmission mode and communication establishment are two key issues in the initial procedure for the D2D underlaying cellular network. In a hybrid system, a UE can operate in two transmission modes, namely cellular mode and D2D mode. In this book, the D2D mode indicates the direct communication between a pair of the D2D devices, and the cellular mode indicates the communication via the BS as in Reference 28. The mode selection scheme makes a decision on whether the device operates in the D2D mode or cellular mode.

In some literatures a criterion of transmission mode selection, which is uniquely dependent on the distance between two UEs is suggested. When the BS detects that two mobile stations are in sufficient proximity to each other, so that a clear radio signal can be maintained, it will initiate a D2D communication otherwise a cellular transmission is held. However, such distance-based criterion does not consider the interference to the cellular users such that the total system performance may be degraded. In Reference 29, the author thinks that resource control in network-assisted D2D communications mainly includes three actions: mode selection, power control, and resource allocation. Based on the network model in Reference 29, the resource control actions, including mode selection, resource allocation, and power control, can be optimized. As an initial attempt to address this problem, an optimization framework was introduced. Specifically, when a D2D pair wants to set up a connection between them, the BS first evaluates its performance such as average delay and dropping probability under D2D RM, cellular RM, and hybrid RM, respectively, assuming that the data arrival pattern of all the connections are known. Although the hybrid RM generally achieves better performance than the D2D RM and cellular RM due to its ability to exploit the channel variation opportunity, it involves larger computation complexity and signaling overhead, since more than one route exists for the D2D connection.

Therefore, the selected RM should consider the trade-off between performance and complexity [30].

In order to save the transmission power of the device, the mode selection scheme to minimize the transmission power has been proposed in Reference 31. Also, the mode selection scheme to maximize the system capacity has been investigated to enhance the cell throughput in Reference 22. Therefore, it is necessary to jointly consider those two aspects, the transmission power and the system capacity in D2D communication underlaying cellular networks. In Reference 29, the authors define the utility function as power efficiency in order to jointly consider the system capacity and the transmission power. The power efficiency is defined as the system capacity per total power in the paper. The joint mode selection and power allocation scheme is proposed to maximize the utility function. In the proposed scheme, the authors calculate the optimal power with respect to the maximal power efficiency for each of all the possible modes of each device.

The aforementioned studies do not cover multiclass service requirements and assume that one D2D user can only share resource with one cellular user. Such assumptions will limit scheduling flexibility and cannot meet throughput demands of high-rate services. In Reference 32, the authors focus on a mode selection scheme considering both interference and QoS requirements of different users. A joint mode selection and resource allocation (JMSRA) scheme is proposed, which gives cellular user and D2D user the same priority in scheduling and allocates resource according to service demands. Their scheme provides three D2D modes, that is, underlay D2D mode, dedicated D2D mode, and cellular mode, and allows D2D users to share any continuous resources of cellular users or transmit on dedicated resources. Unlike the previous studies, the authors estimate the transmission mode by channel states in the initialization phase, but dynamically adjust the transmission mode according to the service history. In this way, the scheduling scheme will result in higher spectrum efficiency and QoS satisfaction degree.

The authors consider a multicell deployment in LTE uplink system, where cellular users are uniformly distributed. As shown in Figure 1.21, devices can communicate directly as a D2D link or via BS as a cellular link, for example, UE1 and UE2 are both in cellular

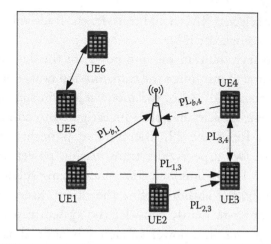

Figure 1.21 The communication modes and interference scenario between devices. (From Jung, M., Hwang, K., and Choi, S. Joint mode selection and power allocation scheme for power-efficient device-to-device (D2D) communication. In *Proceedings of Vehicular Technology Conference* (*VTC Spring*), Yokohama, Japan, May 6–9, 2012.)

mode and UE 3, 4 and 5, 6 communicate with a direct link. In the uplink system, a cellular link suffers from the interference at BS that comes both from cellular users in adjacent cells and D2D users who share the same resources, while a D2D link suffers from the interference at the device, which is caused by cellular user and other D2D users on the same RB. Figure 1.21 also depicts the interference between devices. Since cellular user 1, 2 and D2D user 5, 6 are both in dedicated mode and their resources are orthogonal, no interference exists between them. However, if D2D user 3 chooses underlay mode and share the same resources with user 1, its transmit power will be an interference to the cellular link. At the same time, D2D user 3 will suffer great interference from cellular user 1 if the isolation between them is not enough.

In Reference 32, the authors study the interference and SINR calculations of cellular and underlay D2D users. The authors assume that there are N users in the system, and some of them have a probability for a direct link. According to Figure 1.21, when user 1 chooses the cellular mode, its SINR can be expressed by Equation 1.9:

$$\beta_{c,ik} = \frac{p_{ik}/\mathrm{PL}_{b,ik}}{\sum_{j=1, j\neq i}^{N} p_{jk}/\mathrm{PL}_{b,jk} + \mathrm{IoT}_b + n_b} \tag{1.9}$$

where $\beta_{c,ik}$ is the SINR for user i with cellular link on RB k, $p_{ik}(p_{jk})$ is the transmitting power of device $i(j)$ on RB k, $PL_{b,ik}(PL_{b,jk})$ is total path loss between device $i(j)$ and BS on RB k (including fast fading), IoT_b is the interference from adjacent cells and n_b is the noise power at the receiver of BS. Similarly, SINR of user 3 with underlay D2D mode is given by Equation 1.10 ($PL_{u,iik}$ means the inner path loss of direct link):

$$\beta_{d,ik} = \frac{p_{ik}/PL_{u,iik}}{\sum_{j=1,j\neq i}^{N} p_{jk}/PL_{u,ijk} + IoT_d + n_u} \qquad (1.10)$$

where $\beta_{d,ik}$ is the SINR for user i with direct link on RB k.

To give a general expression, m_i is introduced to denote the communication mode of user i, then the SINR of user i is obtained by Equation 1.11:

$$\beta_{ik} = (1-m_i)*\beta_{c,ik} + m_i*\beta_{d,ik} \qquad (1.11)$$

Since β_{ik} reflects current channel state and decides modulation and coding scheme (MCS), it will affect the transmit rate directly. The authors define a function as $F(\beta_{ik},\varepsilon) = r_{ik}$ to express the mapping from β_{ik} to transmit rate r_{ik}. Here ε denotes the bit error ratio (BER), which is limited by QoS requirement.

Since the single carrier FDMA (SC-FDMA) has been defined for the LTE uplink, designers must devise scheduling algorithms with contiguous RB allocation constraint. Using these equations and constraints, the optimal mode selection vector $m = \{m_1, m_2, ..., m_N\}$ and resource allocation matrix $P = \{p_{ik}, 1 \le i \le N \text{ and } 0 \le k \le L\}$ can be found, in which p_{ik} denotes the transmitting power of user i on RB k.

$$argmax \sum_{i=1}^{N} \sum_{k=1}^{L} r_{ik} \qquad (1.12)$$

subject to

$$F(\beta_{ik}, \varepsilon_{i,k}) = r_{ik} \qquad (1.13)$$

The target of problem (1.12) is to maximize the throughput of the hybrid network. Equation 1.13 reflects the mapping from SINR to transmit rate, which can be done by LTE link-level simulation.

By maximizing the problem (1.12), the optimal mode selection can be found. However, the problem (1.12) is a nonconvex optimization problem, thus it is difficult to find the optimal result. Reference 32 proposed a joint mode selection and resource allocation scheme for this problem and discussed the algorithm to design the suboptimal scheduling method with system and QoS constraints.

Finally, numerical results show that the proposed JMSRA scheme achieves higher system capacity than both pure cellular and force D2D schemes, especially when the maximum distance between D2D pairs or system load increases. Simulation also proves that their scheme can obtain higher service satisfaction degree in the hybrid network.

1.4 Conclusion

In this chapter, the communication spectrum of D2D transmission is divided into two major groups, namely, inband and outband. The works under inband D2D were further divided into underlay and overlay. Outband D2D-related literature was also subcategorized as controlled and autonomous. The major issue faced in underlay D2D communication is the power control and interference management between D2D and cellular users. Overlay D2D communication does not have the interference issue because D2D and cellular resources do not overlap. However, this approach allocates dedicated cellular resources to D2D users and has lower spectral efficiency than underlay. In outband D2D, there is no interference and power control issue between D2D and cellular users. Nevertheless, the interference level of the unlicensed spectrum is uncontrollable, hence, QoS guaranteeing in highly saturated wireless areas is a challenging task.

References

1. Lei, L., Z. Zhong, C. Lin, and X. Shen. Operator controlled device-to-device communications in LTE-advanced networks. *IEEE Wireless Communications*, 2012; 19(3): 96–104.
2. Asadi, A., Q. Wang, and V. Mancuso. A survey on device-to-device communication in cellular networks. *IEEE Communications Surveys & Tutorials*, 2014; 16(4): 1801–1819.
3. Golrezaei, N., A.G. Dimakis, and A.F. Molisch. Device-to-device collaboration through distributed storage. In: *Proceedings of Global*

Communications Conference (*GLOBECOM*), Anaheim, CA, December 3–7, 2012.

4. Asadi, A. and V. Mancuso. Energy efficient opportunistic uplink packet forwarding in hybrid wireless networks. In: *Proceedings of the Fourth International Conference on Future Energy Systems* (*ACM e-Energy*), Berkeley, CA, May 22–24, 2013.

5. Asadi, A. and V. Mancuso. On the compound impact of opportunistic scheduling and D2D communications in cellular networks. In: *Proceedings of the 16th ACM International Conference on Modeling, Analysis & Simulation of Wireless and Mobile Systems* (*MSWiM*), Barcelona, Spain, November 3–8, 2013.

6. Wang, Q. and B. Rengarajan. Recouping opportunistic gain in dense base station layouts through energy-aware user cooperation. In: *Proceedings of the 14th IEEE International Symposium on a World of Wireless, Mobile and Multimedia Networks* (*WoWMoM*), Madrid, Spain, June 4–7, 2013.

7. Akkarajitsakul, K., P. Phunchongharn, E. Hossain, and V.K. Bhargava. Mode selection for energy-efficient D2D communications in LTE-advanced networks: A coalitional game approach. In: *Proceedings of 2012 IEEE International Conference on Communication Systems* (*ICCS*), Singapore, November 21–23, 2012.

8. Lin, X., J. Andrews, A. Ghosh, and R. Ratasuk. An overview of 3GPP device-to-device proximity services. *IEEE Communications Magazine*, 2014; 52(4): 40–48.

9. Camps-Mur, D., A. Garcia-Saavedra, and P. Serrano. Device-to-device communications with WiFi direct: Overview and experimentation. *IEEE Wireless Communications*, 2013; 20(3): 96–104.

10. Asadi, A. and V. Mancuso. WiFi Direct and LTE D2D in action. In: *IFIP Wireless Days*, Valencia, Spain, November 13–15, 2013.

11. Lim K.W., W.S. Jung, H. Kim, J. Han, and Y.B. Ko. Enhanced power management for Wi-Fi direct. In: *Proceedings of Wireless Communications and Networking Conference* (*WCNC*), Shanghai, China, April 7–10, 2013.

12. Shen, W., W. Hong, X. Cao, B. Yin, D.M. Shila, and Y. Cheng. Secure key establishment for device-to-device communications. In: *Proceedings of Global Communications Conference* (*GLOBECOM*), Austin, TX, December 8–12, 2014.

13. Zhou, B., S. Ma, J. Xu, and Z. Li. Group-wise channel sensing and resource pre-allocation for LTE D2D on ISM band. In: *Proceedings of Wireless Communications and Networking Conference* (*WCNC*), Shanghai, China, April 7–10, 2013.

14. Golrezaei, N., P. Mansourifard, A.F. Molisch, and A.G. Dimakis. Base-station assisted device-to-device communications for high-throughput wireless video networks. *IEEE Transactions on Wireless Communications*, 2014; 13(7): 3665–3676.

15. Golrezaei, N., A.G. Dimakis, and A.F. Molisch. Device-to-device collaboration through distributed storage. In: *Proceedings of Global Communications Conference* (*GLOBECOM*), Anaheim, CA, December 3–7, 2012.

16. LTE Direct always-on Device-to-Device proximal discovery. Technical report, Qualcomm Research, 2014. Available online: https://www.qualcomm.com/media/documents/files/lte-direct-always-on-device-to-device-proximal-discovery.pdf

17. LTE Direct Trial, White Paper, Qualcomm Research, 2015. Available online: https://www.qualcomm.com/media/documents/files/lte-direct-trial-white-paper.pdf

18. Doumiati, S., H. Artail, and D.M. Gutierrez-Estevez. A framework for LTE-A proximity-based device-to-device service registration and discovery. *Procedia Computer Science*, 2014; 34: 87–94.

19. Mumtaz, S., K.M.S. Huq, and J. Rodriguez. Direct mobile-to-mobile communication: Paradigm for 5G. *IEEE Wireless Communications*, 2014; 21(5): 14–23.

20. Raghothaman, B., E. Deng, R. Pragada, G. Sternberg, T. Deng, and K. Vanganuru. Architecture and protocols for LTE-based device to device communication. In: *Proceedings of 2013 International Conference on Computing, Networking and Communications (ICNC)*, San Diego, CA, January 28–31, 2013.

21. Fodor, G. et al. Design aspects of network assisted device-to-device communications. *IEEE Communications Magazine*, 2012; 50(3): 170–177.

22. Yu C.H., K. Doppler, C.B. Ribeiro, and O. Tirkkonen. Resource sharing optimization for device-to-device communication underlaying cellular networks. *IEEE Transactions Wireless Communications*, 2011; 10(8): 2752–2763.

23. Janis, P., V. Koivunen, C. Ribeiro, J. Korhonen, K. Doppler, and K. Hugl. Interference-aware resource allocation for device-to-device radio underlaying cellular networks. In: *Proceedings of Vehicular Technology Conference (VTC Spring)*, Barcelona, Spain, April 26–29, 2009.

24. Min, H., J. Lee, S. Park, and D. Hong. Capacity enhancement using an interference limited area for device-to-device uplink underlaying cellular networks. *IEEE Transactions Wireless Communications*, 2011; 10(12): 3995–4000.

25. Xu, S., H. Wang, T. Chen, Q. Huang, and T. Peng. Effective interference cancellation scheme for device-to-device communication underlaying cellular networks. In: *Proceedings of Vehicular Technology Conference Fall (VTC 2010-Fall)*, Ottawa, Canada, September 6–9, 2010.

26. Phunchongharn, P., E. Hossain, and D.I. Kim. Resource allocation for device-to-device communications underlaying LTE-advanced networks. *IEEE Wireless Communications*, 2013; 20(4): 91–100.

27. Xu, C. et al. Efficiency resource allocation for device-to-device underlay communication systems: A reverse iterative combinatorial auction based approach. *IEEE Journal Selected Areas in Communications*, 2013; 31(9): 348–358.

28. Min, H., W. Seo, J. Lee, S. Park, and D. Hong. Reliability improvement using receive mode selection in the device-to-device uplink period underlaying cellular networks. *IEEE Transactions Wireless Communications*, 2011; 10(2): 413–418.
29. Jung, M., K. Hwang, and S. Choi. Joint mode selection and power allocation scheme for power-efficient device-to-device (D2D) communication. In: *Proceedings of Vehicular Technology Conference (VTC Spring)*, Yokohama, Japan, May 6–9, 2012.
30. Kuang, Y., X. Shen, C. Lin, and Z. Zhong. Resource control in network assisted device-to-device communications: Solutions and challenges. *IEEE Communications Magazine*, 2014; 52(6): 108–117.
31. Hakola, S., T. Chen, J. Lehtomaki, and T. Koskela. Device-to-device (D2D) communication in cellular network-performance analysis of optimum and practical communication mode selection. In: *Proceedings of Wireless Communications and Networking Conference (WCNC)*, Sydney, Australia, April 18–21, 2010.
32. Wen, S., X. Zhu, X. Zhang, and D. Yang. QoS-aware mode selection and resource allocation scheme for device-to-device (D2D) communication in cellular networks. In: *Proceedings of 2013 IEEE International Conference on Communications Workshops*, Budapest, Hungary, June 9–13, 2013.

2

PEER AND SERVICE DISCOVERY IN PROXIMITY SERVICE

2.1 Introduction

Proximity services intentionally exploit the geographical position of mobile devices, particularly the geographical proximity and social relationships of mobile users, to enable various applications, including commercial advertisement, communication offloading, and mobile social networking in proximity [1]. Existing technologies used to serve the proximity awareness can be broadly divided into over-the-top (OTT), and device-to-device (D2D) (peer-to-peer [P2P]) solutions. In the OTT model, the centralized server (usually located in the cloud) receives periodic location updates from user mobile devices (using GPS), and determines proximity based on location updates and interests. The constant location updates not only result in significant battery impact because of GPS power consumption and the periodic establishment of cellular connections, but also causes serious privacy problem. Moreover, OTT approaches may incur undesired network overheads and latency for discovery and communication. Different from OTT, D2D schemes forego centralized processing in identifying relevancy matches, instead of autonomously determining relevance at the device level by transmitting and monitoring for relevant attributes. This approach offers crucial privacy benefits. In addition, keeping discovery on the device rather than in the cloud will allow for user-level controls over what are shared.

Peer and service discovery is a major issue in D2D-based proximity services, since before two devices can directly communicate and exchange contents with one another, they must first know (discover) that whether they are near each other and whether there is service that they need [2]. Basically, there are a priori device discovery and a posteriori device discovery. In a priori schemes, the network and/or the devices themselves detect D2D candidates before commencing a communication session between the devices. In a posteriori device discovery, the network infrastructure (e.g., an eNB) realizes that two communicating devices are in the proximity of one another, and thereby they could be D2D candidates when the communication session is already ongoing (in cellular mode) between the UEs (user equipments) [3].

From a cellular network perspective, peer discovery has a similar functionality as cell search in LTE by which the UE determines the time and frequency parameters that are necessary to demodulate the downlink and to determine the cell identity (and thereby the UE effectively *discovers*, i.e., detects the cell). Service discovery is defined as a process allowing networked entities to advertise their services as well as to query about services provided by other entities located in proximal area.

As shown in Figure 2.1, peer and service discovery may be broadly categorized into two types: discovery without network assistance (distributed approach) and discovery assisted by infrastructural network (centralized approach) [4]. In both the distributed and centralized approaches, peer and service discovery is made possible by one party transmitting a known synchronization or reference signal sequence (that may contain service description and is referred to as the beacon), and then another party can receive the signal and respond to it. Irrespective of the technology details, the fundamental problem of peer and service discovery is that the two peer devices have to meet in space, time, and frequency.

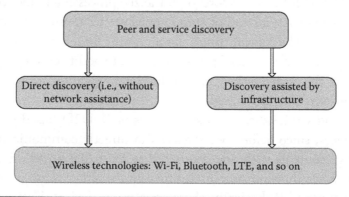

Figure 2.1 Category of peer and service discovery in proximity service.

The distributed approach of peer discovery does not need the involvement of the base station (BS), which is more flexible and scalable than the centralized one, since it operates under local-level requirements, and the complexity is transferred to the end users. It is also a suitable solution in case of out-of-coverage (no cellular network available) D2D communication. However, without network coordination, D2D devices have to blindly decode discovery signals in the procedures of peer and service discovery [5]: some randomized procedure of transmitting/searching beacons sent by D2D users are needed, which are typically time and energy consuming for the mobile device. Furthermore, if D2D communications utilizing licensed spectrum resources in modern cellular systems, this approach (i.e., LTE Direct) will lead to uncontrolled use of the licensed band and cause severe inferences to traditional cellular users.

The centralized approach can better cope with the synchronization problem, one of the most important open issues in the field of D2D communications, because the central node can provide devices with vital synchronization information. Moreover, with network assistance, a device can register its own service across the network, such that other devices may discover it. This will allow other devices to verify that there are other services on the network and after resolving their metadata it can be analyzed and checked what kind of service is being provided [6].

The chapter is organized as follows: Section 2.2 describes how to complete peer and service discovery in the case of no network assistance, based on two popular wireless technologies: Wi-Fi and Bluetooth. Section 2.3 summarizes the methods of peer and service discovery with network assistance. Finally, this chapter is briefly concluded.

2.2 Peer and Service Discovery without Network Assistance

2.2.1 Peer Discovery

In the case of direct discovery, mobile device should be able to search and find other devices fast in its vicinity without infrastructure support, without consuming too much battery power, and preferably without requiring clock synchronization, it requires UE devices to participate in the device discovery process by periodically transmitting/receiving discovery signals. This section focuses on the peer discovery in two main D2D technologies in unlicensed spectrum: Wi-Fi and Bluetooth [7]. First, the main challenges of peer discovery without network assistance are introduced. Then some typical existing approaches to solving those challenges are summarized. Finally, a scheme is illustrated that integrates two different wireless technologies Wi-Fi and Bluetooth, to efficiently conduct peer discovery.

2.2.1.1 Main Challenges   As shown in Figure 2.2, the design of direct peer discovery faces the main challenges of energy efficiency, fast discovery, asynchronous discovery, and scalability.

If energy were not a constraint, Wi-Fi peer discovery could be as simple as letting every device periodically beacon its presence to all other devices. For example, assume that we have device A and device B, and device B wants to discover device A. Without the help of other infrastructures, typically, device A transmits a beacon periodically to announce its existence. Hence, device B can simply scan to discover A. However, energy efficiency and fast discovery is a pair of contradictory to a certain extent. In order to have fast discovery, device A needs to transmit the beacon with shorter period and spend more energy, so that there is a trade-off between energy efficiency and fast discovery.

Instead, a device could adopt duty cycle, keeping the wireless interface in a sleep state most of the time, and periodically waking it up to execute the discovery process. Naturally, both devices must be awake to discover each other, so duty-cycling schemes should ensure that the wakeup times of two neighboring devices overlap. This is not hard to

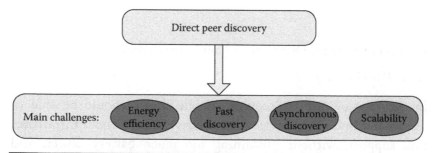

Figure 2.2   Main challenges in peer discovery without network assistance.

achieve with synchronization information, by which the devices can wake up at the same time. However, synchronization requires infrastructural support (4G/Wi-Fi BS), which is not available. In response, periodic schemes (asynchronous discovery) have been designed that ensure such overlap within a reasonable time bound while operating at low-duty cycles, requiring no synchronization information, which is provided by infrastructure. Compared to asynchronous discovery, synchronous schemes are obviously appealing as they are more efficient in terms of energy consumption, and they result in more reliable, faster discovery. However, without the support of the infrastructure, appropriate asynchronous discovery schemes have to be designed, so that devices can save energy as soon as possible [8]. In addition, scalability is also an important target of the discovery protocol. The protocol of peer discovery should work well in highly dense environment. If many devices want to discover or be discovered, all devices have to transmit beacons at the same time, which may cause traffic congestion, and lead to some devices that cannot be discovered [9].

2.2.1.2 Solution to the Challenges In this section, several peer discovery schemes are summarized, which can efficiently resolve the challenges mentioned above. The commonly used concepts in those peer discovery schemes are briefly defined including: time slot, discovery cycle, discovery latency, the worst-case discovery latency, duty cycle, symmetry, and asymmetry.

Time slot: Time is divided into intervals with equal size, and each time interval is called a time slot. The concept of time slot can reduce the difficulty of realizing the experiment and ensures that when the length of the time slot is greater than the total clock skew, it is possible to effectively overcome the impact of clock offset.

Discovery cycle: A discovery cycle is organized by n time slots. In a discovery cycle, nodes arrange their work status in accordance with a certain time-slot schedule.

Discovery latency: The length of the time from the moment that the two nodes are within communication range of each other and start to work to the instant that they first discover each other.

The worst-case discovery latency: The longest latency of one node to discover its entire neighborhood. It determines the minimum time that the two nodes within their communication range can definitely discover each other.

Duty cycle: Nodes periodically alternate from working or sleeping status. The ratio between the duration of one node's working status and the total time length is called duty cycle. The smaller the duty cycle, the lower is the energy the node consumes.

Symmetry: The same scheduling of time-slot sequence representing the same duty cycle is called symmetry.

Asymmetry: The different scheduling of time-slot sequence representing the different duty cycle is called asymmetry [10].

Energy efficient asynchronous neighbor-discovery schemes fall broadly into two categories: probabilistic and deterministic. Unfortunately, there is no clear winner. Instead, there is a trade-off between the average and worst-case discovery latency. Specifically, probabilistic approaches (e.g., Birthday protocol) transmit/receive or sleep with different probabilities and perform well for average discovery latency, but they exhibit long tails resulting in high upper bounds on neighbor-discovery time. On the contrary, deterministic protocols, which let devices wake up at specific time slots according to a deterministically designed schedule, improve on the worst-case discovery latency, but do so by sacrificing average discovery latency.

To address the issues above, searchlight [11], an asynchronous neighbor-discovery protocol was proposed to strike a balance between the two conflicting goals of low-power operation (energy efficiency) and small discovery latency (fast discovery), in which the deterministic and probabilistic methods are integrated to have both good average-case performance and the best worst-case bound for any given energy budget. In detail, searchlight utilizes a periodic slot-based discovery scheme, where a period consists of t contiguous slots as determined by the target duty cycle at which a device wants to operate and author focuses only on the symmetric operation. To save energy, the device sleeps in the majority of slots.

In every period, there are two active slots—an anchor (A) and a probe (P) slot (as shown in Figure 2.3, $t = 7$). There are three cases

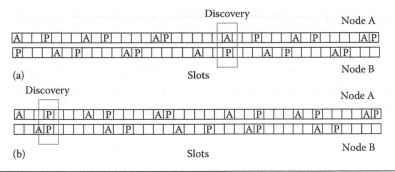

Figure 2.3   Overlap of anchor and probe slots with (a) sequential and (b) randomized probing, $t = 7$.

of successful discovery: (1) anchor–anchor overlap, (2) probe–anchor overlap, and (3) probe–probe overlap. Among them, in general, the condition of anchor–anchor overlap will not happen since t is large enough (to simplify, the t of cases in Figure 2.3 is small), so the author only focuses on another two conditions of overlap.

The position of the anchor slot is fixed at the first slot in a period. If the period is the same for two devices, the relative position of the anchor slot of one device will remain the same with respect to that of the other device. For example, as shown in Figure 2.3a, node B's anchor slot is always four slots after node A's anchor slot.

Searchlight is based on this key observation about the constant relative offset. Essentially, it introduces an additional probe slot in each period to systematically search for the other device's anchor slot, either sequentially or via a random permutation of the probed slots. The benefit of this probe slot is that it needs only probe half of the slots in the period, for example, the position of the probe slot can follow the pattern 1, 2, 3...$t/2$. By this schedule, it is enough to guarantee that the condition of probe–anchor overlap will occur.

In addition, the randomized case, called searchlight-R, introduces the potential of a probe slot overlapping with another probe slot (as shown in Figure 2.3b). The overlap of two probe slots (probe–probe overlap) can also result in a successful discovery. But in sequential probing, which is referred to as searchlight-S (as shown in Figure 2.3a), the probe slots of two nodes follow the same pattern, and hence they are often in sync with each other, greatly reducing the probability of a probe–probe overlap. To increase the probability of a probe–probe overlap, searchlight introduces a probabilistic component. In searchlight-R, instead of being restricted to the only pattern, nodes can randomly pick any probe slot pattern, that is, a permutation of values from 1 to $t/2$. This probabilistic approach essentially increases the possibility of discovery through a probe–probe overlap without changing the worst-case bound, so that it achieves combination of probabilistic and deterministic method.

For illustration purpose, the author has shown the slots to be aligned in previous figures. However, searchlight is designed as an asynchronous protocol, and hence does not assume or rely on the alignment of slot boundaries at different nodes. To ensure that an overlap between two active slots always leads to discovery, a beacon gets sent both at the beginning and end of an active slot and the node remains in listening mode in the intermediate period. A discovery is successful if the active slots of any two nodes overlap, at which point they can launch communication.

By adopting a systematic approach that has both deterministic and probabilistic components, searchlight achieves better average discovery performance and significantly improves worst-case discovery bounds for symmetric operation in comparison to probabilistic protocols for birthday and the current best deterministic protocols Uconnect and Disco.

First, at the worst-case of discovery latency for all protocols operating at 5% duty cycle, searchlight-R always achieves lower latency over all other protocols except for the birthday protocol. For 65% of the time, searchlight-R performs on par with the birthday protocol or slightly lags behind. Beyond that, the probabilistic nature of the birthday protocol leads to a long tail, and searchlight-R achieves the lowest latency. In comparison to Uconnect, searchlight-R achieves better latency all along. For both versions of searchlight, maximum discovery latency of Uconnect is around 20% more than searchlight. Searchlight-S always performs better than Uconnect and Disco but lags behind searchlight-R in the average case.

Next, at the average discovery latency of the protocols for different duty cycles, for all duty cycles, searchlight-R achieves the lowest average latency. Searchlight-R reduces average latency by at least 25% for Uconnect and 16% for the birthday protocol, whereas the performance of Disco ranks worst by a distance. Average latency for searchlight-S lies between that of searchlight-R and the birthday protocol for all duty cycles. The difference between the performance of searchlight-R and searchlight-S clearly demonstrates the advantage of incorporating randomization in moving the probe slot.

As future work, this research would like to integrate neighbor discovery with data transmission between discovered peers.

Reference 9 proposed an energy efficient, fast, and scalable group-based D2D discovery protocol. The key idea behind the protocol is to group devices in a neighborhood such that devices in a group will take turns to announce the existence of other devices in a group. Hence, a device can reduce the period of announcing its existence and have the advantages of energy efficiency and scalability. Further, the transmissions from other devices in the group still guarantee fast discovery. The protocol design is validated by analysis and simulations.

There are three main problems in this protocol design. First, how can devices distributed form a reliable group such that the beacon can be transmitted in sequence? Second, the devices may keep moving, and how can a group be maintained in a mobile environment? Third, forming a group may introduce additional energy cost, thus will this extra energy cost be exceedingly large that outweigh the benefits of forming a group?

For the first problem, the operations of synced distributed protocol are divided into two parts: the join operations and the group operations, which will be described next. The join operations determine how a device chooses to join a group and how a device is accepted in a group. The group operations determine how the group list (which is a circular list of device identities. When several devices form a group, a group list will be generated, and the beacon transmission will follow the order defined by the group list) is reliably passed around the group and how the group list is adjusted if one device is added to or removed from the group. In order to achieve the two operations, three windows are introduced in this protocol: join window, contact window, and update window, as shown in Figure 2.4. A join window is allocated such that a device can accept other devices into the group. A contact window is allocated such that a device can be contacted to start the operations after device discovery. An update window is allocated when the number of devices in a group list is larger than 1 such that a device can update the group list information. Finally, two information elements are appended to the beacon

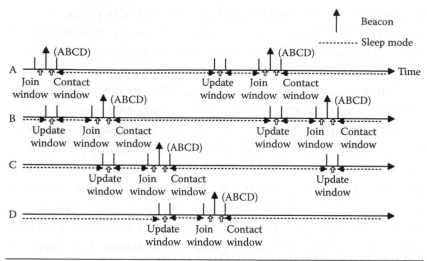

Figure 2.4 A beacon transmission pattern for synced distributed protocol.

body: the join state and the group list. A join state information element is added to prevent the racing condition, where two devices may join each other at the same time. The group list is added such that the existence of other devices can be broadcasted. Further, if a device receives a beacon, it can then use the group list, the timing synchronization function (TSF), and the beacon interval information elements to infer the synchronization timing of a group such as the next target beacon transmission time (TBTT). The group formation is composed of two operations: join and group operations.

Join Operation:

Suppose that a device D just enters the system, and there is a group (ABC) in the system. Device D will first scan T to discover the existing beacon transmissions. Assume that device D receives a beacon transmission from device C, then device D can calculate the starting time of the next join window, that is, the join window of device A. After that, the join process follows a three-way handshake. First, device D will schedule a join request transmission to device A. Second, if device A accepts the join request from device D, then a join response is sent to device D. Third, if a join response is received by device D, an ACK is transmitted to confirm the reception. Otherwise, if no join responses are received by device D at the end of join window, device D will scan another period T to find the beacon transmission.

Note that at most one device will be accepted into the group in a join window. Further, to alleviate the situation that multiple join requests may be transmitted to a device simultaneously, device D will first perform random back off at the beginning of the join window before the transmission of the join request.

Group Operation:

1. In a join window, if device A is accepted by device B, B adds A as its up-chain in the group list. Similarly, A views B as its down-chain and generates the group list correspondingly. Then, A schedules the starting time of the update window at next $TBTT + (N - 2) \cdot T$ and the next beacon transmission at next $TBTT + (N - 1) \cdot T$, where N is the number of devices in the group list of A.

2. At the TBTT of device A, A schedules the starting time of the next update window after $(N - 1) \cdot T$ and the next beacon transmission time after NT, where N is the number of devices in the group list of A.

3. In an update window, if device A receives a beacon from its up-chain device B, then

   a. A updates its group list and sends an ACK to B if A is included in the group list of B.

   b. A leaves the group if A is not included in the group list of B. If A does not receive a beacon from B, A removes its up-chain B from its group list. Note that for (1), device A must have scanned a beacon from the up-chain of device B, say device C, before accepted by device B. By inserting A between B and C in the group list, it guarantees that the group information can be passed sequentially after device A is added to the group. Also note that when a beacon is broadcasted, the down-chain device is responsible for replying an ACK. This additional feature is included because beacon transmission is a broadcast transmission, which is not reliable. By including the additional ACK mechanism, the beacon can be retransmitted if the ACK is not received and still remain as a broadcast transmission.

For the second problem that the devices may keep moving, enhanced techniques are introduced. In a mobile environment, devices in the group may frequently move out of the communication range of the group members. Hence, a group may frequently go to group reorganization procedure, and the group may break. This is not suitable for this protocol to have better energy saving and scalability. Motivated by this observation, the enhanced techniques are proposed. The idea is that if a device observes that the device, which transmits beacon right after it, may stop functioning because of no ACK reception, it transmits another beacon to increase the probability of passing the group list to another group member. Hence, this can reduce the frequency of the group reorganization.

Enhanced techniques: Each device maintains a variable called count with initial value 0.

4. At each scheduled TBTT, a device first randomly back off a period before it transmits a beacon. Also,
   a. If an ACK is received, the device follows (2). Further, the value of count is reset to 0.
   b. If a device does not receive the ACK from its down-chain, it removes its down-chain from the group list and increases count by 1. If count = 1, another beacon is scheduled at next group TBTT. If count > 1, the device leaves the group.

5. In an update window of a device A, suppose that A's up-chain is B, and B's up-chain is C. A will ask back the beacon transmission from B or C. Further, at the end of an update window,
   a. If A does not receive any beacons, or it only receives beacon from B, it follows (3).
   b. If A only receives beacon from C, then it treats C as an up-chain and follows (3).
   c. If A receives beacons from both B and C, then A updates the group list from C.

Note that for (4), a node does back off before beacon transmission to avoid possible hidden node problem due to multiple beacon transmissions. Also note that a device, say device A, leaves the group if there are two consecutive beacon transmissions without receiving an ACK. The reasoning for this operation is as follows. Suppose that during the update window of C, A removes B in the previous group TBTT. If device A does not receive an ACK, then it can be concluded that communication between device A and device C is lost. If C receives the beacon from B, then A will be removed from the group because B removes A in the previous group TBTT. If device C does not receive the beacon from device B, then it will remove device B, and during the next update window, it will wait for the beacon from device A. Again, device C will not hear anything because the communication between A and C is lost. Hence, device C will still remove device A and proceed. As a result, device A is highly probable to be removed, and let device A leave the group directly rather than leave latter. Note that for (5), device A accepts the group list from C if both B and C transmit a beacon. The reason is that device C will contain newer group list than device B.

The modifications of (4) and (5) essentially do not change the group behaviors in a static environment. The reason is that in a static environment, if a beacon does not receive the ACK from the down-chain, then the down-chain device must have stopped functioning. Hence, although an additional beacon is scheduled in the next group TBTT, there is still at most one beacon transmission in the next group TBTT. Further, the behavior of a device in the update window is the same. The only difference is that a device removes its up-chain due to a beacon transmission rather than the observation of no beacon transmission. Finally, when a device reschedules a beacon at the next group TBTT, it removes one device in the group list. As a result, the time for the scheduled update window and beacon transmission at next group TBTT remains the same as the original operation. In sum, under the same assumption of static environment, the group still functions correctly under the enhanced techniques.

For the third problem, theoretically, this protocol only requires the devices to maintain a low duty cycle, and the duty cycle will keep decreasing as the number of devices in a group increases. Actually, this is consistent with the simulation results. Specifically, the authors first simulate the protocol proposed in a static environment and compare this protocol with a basic protocol, where each device broadcasts a beacon every 100 ms and awakes 10 ms after the beacon transmission, that is, the duty cycle is 10%. In this setting, he randomly distributes

100 devices in an area of 70 m × 70 m. The time for devices to start functioning is a Poisson process with rate 1. The result shows that as the group size starts to increase, the power consumption also drops dramatically. When the group size is 100, the power consumption corresponds to that of a device with only 0.5% duty cycle, which is smaller than the 10% duty cycle of the basic protocol.

Besides energy efficiency, the simulation results also demonstrate that this protocol has fast discovery and scalability.

To show that this protocol is scalable, the authors compare the average total traffic of this protocol with that of the basic protocol, and the result shows that this proposed protocol has an improvement of 92% over the basic protocol.

To demonstrate that this protocol has fast discovery, a device is put in the center of the area, and the device starts to discover other devices. Simulation results show that under the proposed protocol, the new device only needs 0.09 s to receive only one beacon and discover all 100 devices. Next, the researcher adds the enhanced techniques and simulates the protocol in a mobile environment. The result shows that enhanced techniques indeed help the devices to maintain larger group size.

2.2.1.3 Combination of Wi-Fi and Bluetooth For peer discovery, Wi-Fi and Bluetooth have their own advantages and disadvantages. Wi-Fi takes advantage of its improved range over Bluetooth, but it does not address the problem of the higher energy cost of peer discovery. Bluetooth significantly reduces the energy costs. However, this reduction comes at the cost of a significant reduction in the discovery of neighbors due to the lower range of the low-power radios [12]. Instead of choosing between the radios and trading energy for range, CQuest [13] is proposed to coordinate the use of both radios and to achieve the energy-efficient neighbor discovery.

A naive approach to discovering near and distant neighbors would be to let each device in the network independently use both radios for discovery. However, this is inefficient due to node clustering and radio heterogeneity. Device mobility in real life often leads to the formation of temporary clusters, where the members of a cluster can reach each other through a low-power radio. Because of the range disparity between high-power and low-power radios, the devices

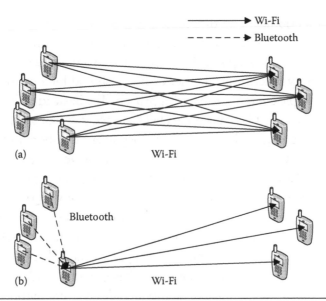

Figure 2.5   The impact of social clustering on device discovery: (a) direct discovery using a Wi-Fi only approach and (b) indirect discovery using CQuest.

within a low-power neighborhood will likely discover the same set of high-power neighbors (as shown in Figure 2.5a).

Based on this observation, instead of dismissing the use of the high-power radio, the CQuest neighbor discovery protocol embraces the use of the high-power radio during neighbor discovery without incurring the high overhead. To reach distant neighbors at an acceptable energy cost, CQuest lets devices equipped with multiple radios coordinate with nearby devices through the low-power radio and share the cost of discovery of distant neighbors through the high-power radio (as shown in Figure 2.5b). CQuest is designed around a simple distributed approach that enables devices to cooperate using only local knowledge about their neighborhood and incurring minimal overhead.

CQuest has three major components. First, it uses distributed scanner set selection (since discovery is performed by scanning the wireless channel). Only the selected devices (the set is called S_t at time t) can reach distant neighbors with high-power scanning using Wi-Fi. In a static network, S_t can be determined during network setup. However, in a network with mobile nodes, S_t needs to be updated periodically to maintain coverage. In addition, the composition of S_t needs to change periodically to distribute the load of scanning. Thus, time is divided into rounds of equal length and S_t is determined for each round. CQuest takes a purely local approach that enables each node to independently determine whether or not it should be a member of S_t based on simple contention-based approach. The basic idea for scanner selection in CQuest is that a node checks to see if any of its low-power neighbors is going to do a scan for high-power neighbors in the next round. If yes, the node does not scan for high-power neighbors in the next round. Otherwise, the node starts scanning in the next round.

In this case, there may be a contention collision; CQuest utilizes a contention phase at the end of each scanning round to solve the problem. The contention phase is divided into contention slots, where the number of slots is a configurable protocol parameter called *windowSize*. At the beginning of each contention phase, nodes randomly pick a slot between 0 and *windowSize* and broadcast a contention packet in that slot over the low-power radio. If a node receives a contention packet before its own selected slot, the node decides that is has *lost* that contention phase and one of its neighbors is in S_t. Otherwise, if a node does not get any contention packet before its own selected slot comes up, it broadcasts a contention packet itself and considers itself as the winner of that contention phase and includes itself in S_t.

Obviously, using this type of contention resolution may result in occasional collisions, and multiple low-power neighbors could end up scanning in the same round. To reduce control overhead and to maintain the completely distributed nature of the protocol, CQuest does not require a minimal dominating set and allows *redundant* scanning. Similarly, CQuest trades-off simplicity and reduced control overhead for the guarantee of coverage in the face of mobility. If a scanner node leaves a cluster, there may be a period of time without complete discovery coverage.

The final part of scanner selection is load balancing. If the same set of nodes always scans, their energy will drain quickly. Instead, the cost of high-power discovery needs to be shared across the low-power cluster. To achieve fairness, CQuest attempts to improve the chances of a node being a scanner in a round if it was not selected in the previous round(s). Although different techniques can be applied to achieve this objective, CQuest uses an exponentially decreasing contention window size. Initially, nodes start with a *windowSize* equal to *max-window-size.* Every time a node fails to win a contention phase, it decreases the *windowSize* by reducing it by half, until it reaches the *min-window-size.* On the other hand, if the node wins a contention phase, it resets the *windowSize* to *max-window-size.*

Second, after selection of the high-power scanner set through coordination over the low-power radio, the next step is to scan for neighbors using both of the radios. In each round, scanners use their HP radios to discover any node in their high-power neighborhood and nonscanners turn off their high-power radios. The actual protocol used for scanning and discovery is determined by the specific network environment. In a synchronous network, the discovery protocol can be as simple as using a beacon with a fixed period, and so a round could be defined as some predefined number of such periods. For asynchronous networks, existing asynchronous neighbor discovery protocols can be used.

Third, CQuest's neighbor maintenance lets devices exchange the results of high-power scanning over the low-power radio, so that nonscanner devices can learn about their potential high-power neighbors without scanning themselves. High-power discovery beacons include the IDs of all low-power neighbors. Similarly, low-power discovery beacons include the IDs of all high-power neighbors. Once a node gets added to the neighbor database, subsequent discoveries refresh that information. CQuest uses two thresholds to determine the staleness of an entry based on the last refresh time. If the time elapsed since the last refresh time is more than freshness threshold, the entry still remains in the database but is not included as part of neighbor information exchanged. On the other hand, if the time elapsed since the entry was last refreshed exceeds expiration threshold, the entry gets completely removed from the database.

The experiment results show that such cooperation dramatically reduces energy consumption—up to 50% over an uncoordinated high-power-based approach—without significantly giving up on the high-power radio contacts. In addition, this minimal reduction in contacts has little or no impact on opportunistic applications, and CQuest ultimately reduces average communication latency. Obviously, the integration of the high-power radio results in higher energy consumption than a low-power-only solution. However, the improvement in communication is dramatic. In certain test scenarios, the delivery ratio improved more than 100%.

2.2.2 Service Discovery without Network Assistance

This section will mainly focus on service discovery in Wi-Fi Direct, which is a popular technology of D2D in current.

As described in Chapter 1, in Wi-Fi Direct, P2P devices have first to discover each other, and then negotiate which device will act as P2P GO in the case of standard P2P group formation [14]. Wi-Fi Direct devices usually start by performing a traditional Wi-Fi scan (scan phase), by means of which they can discover the existent P2P groups and Wi-Fi networks. After this scan, a discovery algorithm is executed (find phase), which is aimed at discovering the services. Once the scan phase ends the P2P device proceeds to the find phase. The aim is to quickly ensure that two P2P devices in device discovery are located in the same channel to exchange device information and determine if a connection should be attempted. This is achieved by letting the P2P devices alter between the two states listen and search. Listen state is when the P2P device waits for a probe request on one fixed channel, and search state is when the P2P device sends probe requests to a fixed set of channels. If a suitable P2P device receives a probe request, it could respond with a probe response to proceed to the next phase. The probability for two devices to find each other depends on the number of channels and the time they spend in each of the two states, which are randomly generated within a boundary. This is optimized by choosing the set of channels recommended for Wi-Fi Direct, known as the three Social Channels 1, 6, and 11 in the 2.4 GHz band [15].

2.2.2.1 Supported Types of Service in Wi-Fi Direct The services make interoperability possible across Wi-Fi Direct capable devices, allowing different brand devices running different applications to interoperate as long as they support the same WFDS (Wi-Fi Direct services). The following four services are defined by Wi-Fi Alliance:

1. Send Service: The send service enables the transferring of bulk data from a send transmitter to a send receiver. The data plane that provides the transport path uses hypertext transfer protocol (HTTP) for managing the data transfer. The send transmitter utilizes the HTTP client functionality to send files to the send receiver, which utilizes as a HTTP server.

2. Play Service: The play service enables a digital living network alliance (DLNA) connection, offering the user a control interface to play encoded media through other P2P devices. DLNA is an organization responsible for defining guidelines for interoperable sharing of digital media between multimedia devices. The play transmitter utilizes the DLNA push controller functionality to stream encoded video, audio, and images to a play receiver, which utilizes the DLNA media render functionality.

3. Display Service: The display service leverages Wi-Fi Miracast certification with the addition of application service platform (ASP), to setup and manage multiple services operating at the same time. This makes WFDS display devices not only compatible with other WFDS display devices, but also with Miracast devices without ASP, support (i.e., Miracast devices that do not support WFDS). The display transmitter utilizes the Wi-Fi display source functionality and supports a user interface to display media contents on a display receiver. The display receiver utilizes the Wi-Fi display sink functionality to display the media content and has the ability to accept or reject an incoming connection request.

4. Print Service: The print service specification is based on the Internet printing protocol (IPP), a standard network protocol for remote printing. IPP is implemented using HTTP as transfer protocol. The print transmitter utilizes the IPP client functionality, which discovers and starts a print session with the print receiver. The print receiver utilizes the IPP server functionality, which receives data from the transmitter and performs print jobs.

2.2.2.2 Service Discovery Entity in Wi-Fi Direct Wi-Fi Alliance has defined the application service platform (ASP), which is a logical entity implementing necessary functions needed by services. The ASP is responsible of coordinating the advertisement and discovery of services. In addition, it is also responsible of ASP-sessions management, connection-topology management, and security.

The ASP and services communicate through an interface described in terms of method and event primitives. Methods and events are asynchronous, which means that one action does not necessarily follow the occurrence of another. Methods are actions initiated by a service, to send information to the ASP through method parameters. One example is AdvertiseService(), which triggers service advertisement by the ASP. Another example is SeekService(), which requests the ASP to search for WFDS services. A method call always returns a value back to the service immediately, which for instance indicates if the advertisement or search was successfully initiated. Events are one-way actions that provide information from the ASP to a service about information obtained over a network. One example is the SearchResult() Event, which calls the service whenever the ASP has found a service name that matches the search criteria.

2.2.2.3 Service Discovery in Wi-Fi Direct Compared with Service Discovery in Bluetooth The common smartphone stacks for Bluetooth do not support broadcast or multicast messaging. Therefore, in order to exchange information about opportunistic services (beyond Bluetooth service discovery protocol that advertises simple connectivity services such as the type of device), the Bluetooth devices need to first establish a mutual communication context in order to exchange information about the services, which is different from Wi-Fi Direct. This context is formed by establishing a socket to another device. If the device is not part of the opportunistic network, the resources used for scanning and socket-formation procedures are wasted. This might suggest blacklisting the devices to prevent future wasting of resources. However, from the client's standpoint it may be difficult to differentiate whether the device was not an opportunistic service provider, whether it was simply busy serving another client, whether it just had its service stopped for a moment, or whether the connection failed for another reason. In case of a successfully established socket communication, the process continues by first exchanging the service and security information with a custom protocol. Later, a convergence layer can be established to allow the communication in an overlay network [7].

2.2.2.4 Service Discovery in Wi-Fi Direct Compared with Service Discovery in LTE Direct For LTE Direct, an Expression(s) is used by peers in discovery of proximate services. Expression(s) are also used by

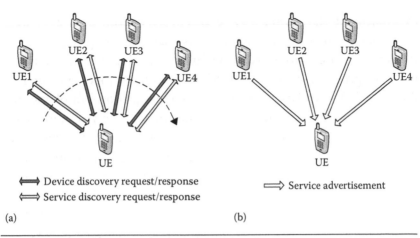

Figure 2.6   (a) Wi-Fi Direct discovery and (b) LTE Direct discovery.

proximate peers to establish direct communications. Expressions at the application/service layer are referred to as *expression names* that are mapped to 128 bits at the physical layer and referred to as *expression code* [16]. In addition, expressions are classified as public or private expressions based on the type of mapping. UEs transmit and receive expressions within the discovery resources.

As shown in Figure 2.6, Wi-Fi Direct discovery adopts two-step asynchronous message-based discovery: device exchanges device discovery request response messages followed by service discovery with every Wi-Fi Direct in range; while LTE Direct adopts broadcast-based discovery: devices wake up periodically and synchronously to discover all devices within range by broadcasting their services [17]. In addition, compared to LTE Direct, the discovery range of Wi-Fi Direct is rather shorter [18].

A salient feature of Wi-Fi Direct is the ability to support service discovery at the link layer. In this way, prior to the establishment of a P2P group, P2P devices can exchange queries to discover the set of available services and, based on this, decide whether to continue the group formation or not. In this case, all nodes that have a registered interest for a service, but may not yet be known to the provider, are able to discover the service without requiring the two-way communication establishment. Notice that this represents a significant shift from traditional Wi-Fi networks, where it is assumed that the only service clients are interested in is the Internet connectivity. In order to implement the above, service discovery queries generated by a higher layer protocol, for example, universal plug and play (UPnP) or Bonjour [19], are transported at the link layer using the generic advertisement protocol (GAS), an OSI link layer protocol specified by 802.11u. GAS is a layer two-query/response protocol implemented through the use of public action frames that allows two nonassociated 802.11 devices to exchange queries belonging to a higher layer protocol (e.g., a service discovery protocol). GAS is implemented by means of a generic container that provides fragmentation and reassembly, and allows the recipient device to identify the higher layer protocol being transported. GAS is used as a container for ANQP (access network query protocol) elements sent between clients and APs [20].

2.3 Peer and Service Discovery with Network Assistance

2.3.1 Peer Discovery

With network-assisted peer discovery can be faster and more efficient, since UEs can register with their services (will be described in subsection 2.3.2) in the network, and the network can complete location tracking of UEs instead of scanning and announcing UE periodically at the user side. The main task of network involves location tracking of UEs and alerting the corresponding users, once the registered UEs with D2D services for discovery are close enough to each other.

2.3.1.1 Direct Discovery with Network Assisted As described in Section 2.1, in order to find D2D peers, a UE could have two roles in the direct D2D discovery: (i) announcing (i.e., periodically transmitting beacons) and (ii) monitoring (i.e., periodically listening to beacons and possibly sending discovery response to an announcing UE). The D2D discovery and response messages contain D2D UE identity and D2D application layer identity. Regardless of its benefits such as permanent availability at any time as well as operation in and out of macro network coverage situations, the main challenge for direct D2D discovery is the energy consumption during the announcing and monitoring procedures. In particular, announcing UEs would consume a lot of energy for permanent periodic transmission of the discovery message.

Typically, the location of one of the ProSe UEs is known in advance, for example, the UE belongs to a shop, or the location, where the UEs would meet is known a priori. Hence, it is not necessary to constantly query the location of both UEs in order to analyze the potential of proximity. In case of the D2D functionality being turned on constantly regardless of the probability of success in proximity detection, it results in unnecessary UE battery drain for announcing and listening for other UEs in proximity. The current direct discovery concept has some drawbacks in battery drain or signaling overhead. To tackle such a problem, the concept of proximity-area (P-Area) [21] is introduced, as the area where two or more devices subscribed to the same ProSe service can meet with a high probability. The P-Area may also be described similar to the *fingerprint* (FP) mechanism for small or home eNBs (HeNBs) for more accurate location detection within a cell. A *fingerprint* maybe based on the unique combination of the available information of, for example, PLMN ID, location area/tracking area, and cell ids of the surrounding cells. But the key difference here is that, while small cells are stationary, D2D devices could be mobile as well, thereby requiring a dynamic location-tracking mechanism. It is assumed that this information is present at the ProSe function, within the 3GPP network. The ProSe function plays the role of location services client (SLP agent) to communicate with the SLP (the secure user plane location platform) and be aware of the UEs' locations to determine their proximity (in Reference [22], an entity SLP is specified by 3GPP whose responsibility is to keep track of the UE's locations).

To resolve such a discovery challenge, P-Area(s) is defined on a per ProSe application basis and according to various criteria including time, type of UE, group membership, and so on. Each P-Area can be mapped to a geographical region described by one or more cell IDs, or other location information. This geographic information can then be used by a UE autonomously or in a network-assisted manner, in order to compare whether it is entering or leaving its P-Area(s).

As shown in Figure 2.7, it is an example of how a ProSe UE seeking for a specific service could discover devices providing the desired service when it enters a P-Area. Once a ProSe UE detects that it enters a P-Area, it turns on its ProSe discovery and announces itself, while it is listening to the announcements of other UEs (Step 1). Although

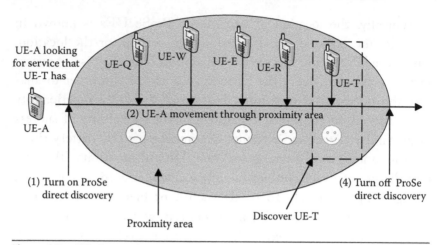

Figure 2.7   D2D discovery inside a P-Area.

the UE moves around the P-Area, it receives advertisements from other UEs in the P-Area (Step 2), and hence detects another device, named UE-T, which provides the desired service when UE-A gets close enough (Step 3). When the UE-A leaves the P-Area it turns off its ProSe discovery to save battery power (Step 4).

This process mainly involves two problems: the determination of the proximity area and the location tracking of UE.

A. Determination of the Proximity Area

Once a UE registers to the ProSe application server (AS) for a specific application, the ProSe AS in coordination with the ProSe function should determine the right P-Area(s) for the UE's ProSe service. For this purpose, the ProSe AS gathers location information about the users subscribed to the same ProSe application and takes several parameters into account.

First, the ProSe AS differentiates whether the ProSe application is, for example, a chat service, where the presence of the buddy list needs to be tracked, or whether the application is, for example, a shoe shop that announces its best offers of the day. In addition, the mobility of the users has to be known, in order to be able to predict potential proximity service opportunities. Specifically, the ProSe AS may try to determine specific location patterns over time based on empirical mobility information of the UE from the network. The ProSe AS could also estimate P-Areas for a specific time window, for example, when the subscriber is at work; determine closest UEs of the subscribed service and estimate/lookup P-Areas for each of them. The ProSe function then provides the P-Area(s) to the UE. For the calculation of a P-Area, the ProSe function considers the location of UEs subscribed to the same ProSe application. The ProSe AS/function updates the closest UEs of the subscribed services (excluding UEs with low potential of involvement) with the *own* P-Area(s) of the UE.

B. Location Tracking of User Equipment

Concerning location determination, the secure user plane location (SUPL) platform is a new entity that is specified in Reference 22, whose responsibility is to keep track of the UE's locations by sending updates of the locations of the UE (periodically, or in response to triggering events) to its corresponding ProSe function, given that the UE was already registered. Once UE-A (consumer) chooses UE-B (provider), ProSe function-A will be responsible for monitoring its own proximity with respect to UE-B. ProSe function-A will have to collect the location updates from ProSe function-B, provided by SLP-B. When the user of UE-A has requested to be informed when a provider UE is in proximity, UE-A will receive proximity alert messages when such a UE becomes near, so that UE-A can turn on direct discovery.

For UEs located in a fixed place or having low-mobility characteristics, P-Areas can easily be decided. However, for UEs moving with a high speed the complexity to provide P-Areas updates increases significantly. Handling such scenario should involve UEs location tracking to be performed regularly, or alternatively allowing such UEs to announce/listen permanently whether other UEs with the same subscriber services/applications are within proximity [23].

In the simulation analysis, the researcher evaluated the energy consumption for D2D discovery mechanisms, and the results show that by using the proposed proximity area-based D2D discovery approach, the overall power consumption of a UE can be reduced up to 78% compared to other cases where UEs use conventional D2D direct discovery mechanisms.

2.3.1.2 Network-Dependent Peer Discovery Schemes In the previous section, a proposal of P-Area(s) is described. It is noted that when the two UE get close they initiate a direct discovery, in which network is not involved to the process of peer discovery. In fact, network can continue to play a very good role. In this case, although the UEs are aware that they are in proximity according to the network, they do not know the actual situation of channel. Due to D2D UEs and cellular UEs may use overlapping time and frequency resources [24] when using sharing uplink channel, there are interference between them [25]. Message exchanges between the UEs of a D2D pair should satisfy an signal to interference plus noise ratio (SINR) threshold, which indicates the correct reception of a message. The network can get the path gains between UEs, and then determine whether they are able to carrying out D2D performance.

In order to take better use of network in the process of peer discovery, the centralized fully network-dependent algorithm is proposed by Thanos et al. [26], which focuses on an accurate description of the signaling message exchanges between the entities of the network with respect to the necessary information for identifying a new D2D pair. The exchange of discovery messages will provide the network with information about the measurement results of path gains between the entities of the cell. Knowledge of the path gains will allow the network to identify the proximity between two devices and to estimate the possible interference that will be created between the new D2D pair and existing users.

As shown in Figure 2.8, the fully centralized network-dependent algorithm is displayed. In the first step, UE1 informs the BS that it wants to communicate with UE2. In the following step, the BS requests UE2 to expect a discovery message from UE1. Then, it requests UE1 to send the discovery message to UE2. In the fourth step, UE1 sends the discovery message to UE2. In this step, if the message is not received by UE2, which means that γ_{D2D} (SINR threshold between the devices of a D2D pair) was not satisfied, and retransmissions occur. However, it is more efficient to define a maximum time interval during which UE1 may keep retransmitting. In the fifth step, after UE2 receives the discovery message from UE1, it reports the measured SINR value of the message to the BS. In the sixth step, the BS requests both UE1 and UE2 to listen for interference from existing users in the cell. This means that, if there are N existing pairs in the cell, they will have to listen for N time slots. In the seventh step, both UE1 and UE2 report the measured interference to the BS. In this step, both UE1 and UE2 transmit to the BS. In the final step of

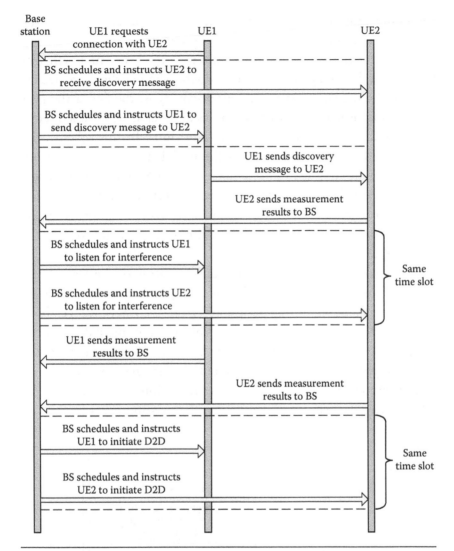

Figure 2.8 Fully centralized network-dependent peer discovery algorithm.

the algorithm, the BS requests the devices to communicate with each other using D2D mode. It is assumed that both messages to UE1 and UE2 are transmitted simultaneously by the BS.

After the successful transmission of all messages as defined by the algorithm, the two devices are considered to be a D2D pair. Throughout the process, the BS receives information about the path gains between itself and the devices. This information allows it to decide whether the link between the devices is favorable for D2D communication. However, if it is not, the method still requires the first three steps of the algorithm to be executed.

Given the above scheme that makes peer discovery more efficient than direct discovery, there may be a doubt whether the more dependent on the network-assistance peer discovery is, the better the peer discovery is? Reference 26 also proposes another scheme called semicentralized seminetwork-dependent algorithms. In this scheme, the role of the BS is less dominant, as the initial steps of the algorithm do not include message transmissions to the BS. The method is less dependent on the network for discovering D2D pairs. However, the role of the network is still important, as it receives information by the D2D pair and decides on whether the devices should work in the D2D mode or not.

As shown in Figure 2.9, a visual representation of the second discovery algorithm is displayed. The discovery process here is initiated by UE1, without requesting permission from the BS. Moreover, both devices listen for interference from other users immediately on the reception of the first message sent by UE1 to UE2. In the first step of the algorithm, UE1 broadcasts a discovery message, requesting to communicate with UE2. Upon the reception of this message by UE2, both UEs listen for interference by existing users in the cell

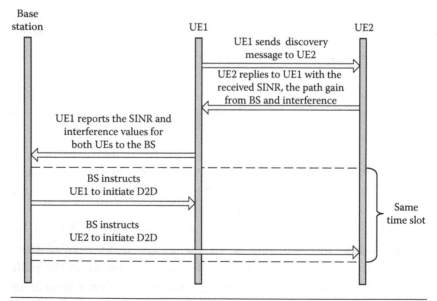

Figure 2.9 Semicentralized seminetwork-dependent algorithm.

and estimate their path gains to the BS using system signals like synchronization signals.

In the second step, UE2 reports the received SINR and the path gain to the BS, and the received interference by existing users to UE1. In the third step, UE1 reports to the BS the SINR and interference measurements for both itself and UE2. In the fourth step, the BS requests both UEs to initiate the D2D communication.

This inactive discovery algorithm requires the transmission of fewer messages compared to the active one, while succeeding in providing information about the path gains of the links between almost all entities in the cell. Moreover, in the case that the new pair proves not to be a D2D pair, only the first step is required to be executed. The disadvantage of the algorithm is that UE2 needs to constantly listen for transmissions from UE1.

In both the centralized fully network-dependent algorithm and the semicentralized seminetwork-dependent algorithms. One separate channel is assumed to be reserved for the purposes of discovery of D2D pairs. New D2D pairs share the same channel for discovery. As multiple potential D2D users may send discovery signals at the same time, users transmit in the beginning of each time slot with a certain transmission probability to avoid collision. The probability of a successful transmission of a message in one time slot for one user is the probability that only one user is transmitting, whereas all the other users are idle. Otherwise, a user should attempt retransmitting the discovery message. So, there is a limited time for retransmitting the discovery message. If this time is exceeded, the discovery process fails and the new pair cannot transmit in the D2D mode.

The simulation results show that the second discovery algorithm has much better performance than the first one. This means that the second algorithm needs less time to discover D2D pairs than the first algorithm, for example, it can be more than 50% faster. The reason lies in that smaller number of total message exchanges is needed in the second algorithm.

2.3.2 Service Discovery with Network Assistance

In the first family of service discovery approaches described in Section 2.2.2, the service discovery is carried out without network coordination, in which the UE starts to blindly decode discovery signals associated to a service immediately after being triggered by an application. Uncoordinated approaches with blind decoding require significant processing, and in general are undesirable from a power consumption perspective. However, service discovery can benefit from network coordination since network can provide essential information for it. In this section, how the network works in the process of service discovery is described.

2.3.2.1 App and Service Registration   First the cloudlets concept is brought to the D2D framework: cloudlets are *smaller clouds*, which are powerful computers that usually serve nearby users and excel at offloading content and tasks from mobile devices. By analogy, with proximity-based services in the D2D framework, a mobile device can also share its resources and provide its services to other devices in its vicinity. By this, the mobile device sharing its resources will play the role of a mobile cloudlet providing software as a service, data as a service, and network as a service. This can be beneficial economically where UEs can lend their extra resources to other UEs that be charged for this. Normally, cloud services can be defined as: Infrastructure as a Service (IaaS), Platform as a Service (PaaS), and Software as a Service (SaaS). Among all of these, researchers generally believe that only SaaS is applicable, and furthermore add Data as a Service (DaaS) and Network as a Service (NaaS), where in the latter the device may, for example, play the role of a hotspot (Wi-Fi Access Point) for other devices that do not have access to the Internet [23].

A. Application registration

The user equipped with a ProSe-enabled device and registered in the network as a ProSe subscriber, can download a ProSe application (offering services to other ProSe-enabled devices) from an application server through an operator application distributor or an application store. To activate the ProSe features on these applications, the user should also authenticate and authorize it through the ProSe function that caches a list of all the IDs of the applications allowed to use ProSe features along with their corresponding authorized range classes. The discovery range class can be short, medium, or maximum based on geographical distance or radio conditions. This range defines how far a UE holding this app can discover another radio signal or can be discoverable. This set of ranges is sent back to the UE by the ProSe function in the application registration acknowledgment message. The user has the freedom to choose one of these allowed ranges while requesting to communicate with a nearby device holding this app. Note that the authentication is done on a per-application basis.

Most designs propose to develop the applications on the application server to run in two modes: consumer mode (default mode) and provider mode. An app runs in consumer mode when the user is requesting services, and runs in provider mode when the user is willing to share his device's resources. Note that after downloading the app on his device, the user can choose which mode to work with via an app user interface. There should be a mechanism that determines how the user will inform the network about which mode he or she intends to use the application, and the required criteria that qualifies the device to run in such a mode. Depending on which of the three intended mobile cloudlet services (SaaS, DaaS, and NaaS) the device will offer, the criteria will be different. For example, in the case of SaaS, the device is expected to host software apps that are to be run under the supervision of the downloaded ProSe App (provider) as possible services to same-type consumer apps. In case of DaaS, the device is supposed to host the necessary data (e.g., song files) to be sent to consumer apps. Finally, for NaaS, the device must have, for example, an active Wi-Fi connection to the Internet (e.g., via subscription or privilege) and is able to configure itself as a hotspot for requesting devices running the same-type consumer app to connect through to the Internet. After developing the criteria for each class of cloudlet services, another related subject that is to be investigated is the mechanism according to which the network, and specifically the ProSe function, will verify the suitability and conformance of the device. When the user configures the downloaded ProSe app as a provider app, a protocol will be initiated to enable the ProSe function to test the suitability of the device and its readiness to offer cloudlet proximity services. In the case of Data as a Service (DaaS), for instance, the user, through an interface, could configure the local SQLite database to store the names and other metadata of available video and image files that are to be shared with nearby LTE-A users. Second, the consumer mode does not require an elaborate registration or configuration, as it simply can be accomplished [26].

B. Service registration

A service in order to be discovered first needs to register its own service across the network, so other devices may discover it. Part of this functionality is identifying the general D2D service type, which is implicitly provided through same-type ProSe-compliant applications (services). That is a device running a ProSe-compliant app can only discover devices running the same app but in provider mode. However, discovery has to be more specific. For example, a SaaS cloudlet could offer parking information services, whereas another SaaS cloudlet may offer Mexican food ordering services. It follows that a ProSe-compliant SaaS app interested in finding free parking spots in nearby parking garages should only communicate with SaaS provider apps running on nearby devices that offer parking information.

To realize the above capability, a device, on registration in provider mode, should provide to the network (application server) a set of keywords in order to help it (i.e., the network) find nearby cloudlets that offer the particular services. With this setup, a user wanting a particular proximal service will have to supply search keywords that are compared against the registered keywords to determine the appropriate cloudlets, and present them to the user. Obviously, multiple matches could occur, in which case, the user is free to select which one to connect to first. In more technical terms, a device sends a proximity request to the network through messages that ask the ProSe function for help in finding nearby targeted devices, or to alert it when other devices come around (specific methods will be described in detail in Section 2.3.2.2).

The proximity criteria are defined by the user when he chooses a range class for this app. That is, a user who has chosen *short*-range class (e.g., short class corresponds to 90 m) will not be informed about a UE 150 m away. In one scenario, when UE-A is interested in finding UE-B, it contacts the network in order to be alerted when this device comes to its vicinity. It is therefore assumed that UE-A knows the application user ID of UE-B. This assumption is removed by making UE-A search for an application ID along with identifying keywords for the desired service. By this, the network will alert it when a UE holding this application and matching the requested keywords is close, or becomes close. The alert will be in a form of a list containing the IDs of all the providers of interest, leaving it to the UE to choose one. The available services listed in the directory on the ProSe server are provided by service platforms that may be owned and managed by the operator or by third-party service providers. The allowed third-party services are those for which there is a commercial agreement between the operator and a third-party service provider (e.g., social networks and commercial services). It could happen, though no providers are in the range, in which case the network responds to UE-A's request by sending an empty list mentioning that it will alert it whenever a provider enters this range.

In addition, in order to start a service discovery it is necessary to implement listening and then to implement the update of ProSe service when new services are found or existing services are lost. This way the listening that has been implemented can receive updates on the ProSe service findings. The update function runs every time since finding and losing a service on the network is of high frequency. Therefore it is needed to keep track of all the services that have been currently found on the network and have not been lost yet [23].

2.3.2.2 Approach for Network-Assisted Service Discovery　In the network-assisted service discovery approaches, the ProSe function assists the monitoring UE by indicating if the service is available in the ProSe area. If the service is available, the network provides the corresponding radio frequency (RF) resource [27,28] used for discovering signal transmission. In this coordinated case, there are several ways for the ProSe function to provide the service information to the monitoring UE. In this section two service discovery strategies are described: the *PULL* and *PUSH* [29].

A. PULL service discovery

In PULL service discovery, after being triggered by a ProSe application, a monitoring UE (client) transmits a monitoring request to the ProSe function (server). The ProSe function responds with a unicast message indicating whether or not the service is available and if available provides the corresponding RF discovery resource. As shown in Figure 2.10, PULL service discovery is illustrated, and S1 means the service that UE2 wants.

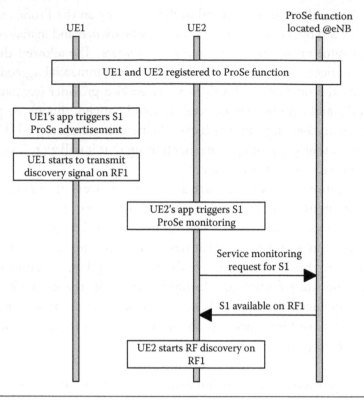

Figure 2.10 PULL service discovery.

B. PUSH service discovery

In PUSH service discovery, the ProSe function (server) (through one or several eNBs) periodically transmits a ProSe broadcast channel. This channel indicates the list of proximity services currently advertised in the ProSe area and the associated RF discovery resources. This channel will be referred to as the PUSH channel.

After being triggered by a ProSe application, the monitoring UE (client) starts to scan the PUSH channel to determine the service availability and the corresponding RF discovery resource in the ProSe area. In practice, a change notification may be needed to indicate when the list of proximity services changes. In this case, if the monitored service is not available in the first scan, the monitoring UE does not have to rescan the PUSH channel until the change notification is received. As shown in Figure 2.11, PUSH service discovery is illustrated.

In Reference 3, the author finally proposes that the network uses service statistics to select the best discovery strategy mentioned above

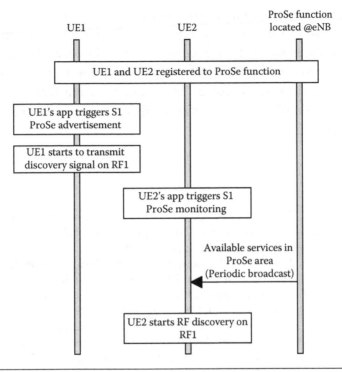

Figure 2.11 PUSH service discovery.

to be applied for each service (i.e., some services will be advertised with the PUSH approach, whereas the other will be advertised with the PULL mechanism). This method is called *combined PUSH/PULL* service discovery. In this method, an optimal threshold for the selection of PUSH or PULL discovery is defined. This threshold is based on the control overhead and the UE energy consumption.

2.4 Conclusion

This chapter describes the methods of peer and service discovery, which is one of the key underlying technologies of proximity services. In general, there are two types of peer and service discovery: without network assistance and discovery with network assisted. Both have their own advantages and disadvantages.

Without network assistance, energy consumption of devices is a very headache problem since discovery procedure has to be always on. Although many solutions are proposed, compared to discovery with network assisted, energy consumption is still considerable. With network assistance, the discovery is mainly processed at core network, so that it makes the burden of devices reduced, and thus makes discovery more energy-efficient. In addition, due to the assistance of network, discovery becomes faster and more efficient. However, in the scenario of no infrastructure, only discovery without network assistance can work.

References

1. Wang, Y., A.V. Vasilakos, Q. Jin, and J. Ma. Survey on mobile social networking in proximity (MSNP): Approaches, challenges and architecture. *Wireless Network*, 2014; 20(6): 1295–1311.
2. Wang, Y., J. Tang, Q. Jin, and J. Ma. BWMesh: A multi-hop connectivity framework on android for proximity service. In: *Proceeding of the IEEE 12th International Conference on Ubiquitous Intelligence and Computing (UIC)*, Beijing, China, August 10–14, 2015.
3. Fodor, G., S. Sorrentino, and S. Sultana. Network assisted device-to-device communications: Use cases, design approaches, and performance aspects. In: *Smart Device to Smart Device Communication*, Mumtaz S., and Rodriguez J., (Eds.), 2014, 135–163, Springer International Publishing, Switzerland.
4. Lei, L., Z. Zhong, C. Lin, and X. Shen. Operator controlled device-to-device communications in LTE-advanced networks. *IEEE Wireless Communications*, 2012; 19(3): 96–104.
5. Fodor, G., E. Dahlman, G. Mildh, S. Parkvall, N. Reider, G. Miklós, and Z. Turányi. Design aspects of network assisted device-to-device communications. *IEEE Communications Magazine*, 2012; 50(3): 170–177.
6. Gandotra, P. and R.K. Jha. Device-to-device communication in cellular networks: A survey. *Journal of Network and Computer Applications*, http://dx.doi.org/10.1016/j.jnca.2016.06.004.
7. Pitkänen, M., T. Kärkkäinen, and J. Ott. Mobility and service discovery in opportunistic networks. In: *Proceedings of the IEEE International Conference on Pervasive Computing and Communications Workshops (PERCOM Workshops)*, Lugano, Switzerland, March 19, 2012, pp. 204–210.
8. Kravets, R.H. Enabling social interactions off the grid. *IEEE Pervasive Computing*, 2012; 11(2): 8–11.
9. Huang P.K., E. Qi, M. Park, and A. Stephens. Energy efficient and scalable device-to-device discovery protocol with fast discovery. In: *Proceedings of the 10th Annual IEEE Communications Society Conference on Sensor, Mesh and Ad Hoc Communications and Networks (SECON)*, New Orleans, LA, June 24–27, 2013.
10. Shang, T. A comparison between neighbour discovery protocols in low duty-cycled wireless sensor networks. *International Journal of Computer Science and Mobile Computing*, 2015; 4(2): 265–271.

11. Bakht, M., M. Trower, and R. Kravets. Searchlight: Helping mobile devices find their neighbors. *In: Proceedings of the 3rd ACM SOSP Workshop on Networking, Systems, and Applications on Mobile Handhelds*, Cascais, Portugal, October 23–26, 2011.

12. Pyattaev, A., O. Galinina, K. Johnsson, A. Surak, R. Florea, S. Andreev, and Y. Koucheryavy. Network-assisted D2D over WiFi direct. In: *Smart Device to Smart Device Communication*, S. Mumtaz, and Rodriguez J., (Eds.), 2014, 165–218, Springer International Publishing, Switzerland.

13. Bakht, M., J. Carlson, A. Loeb, and R. Kravets. United we find: Enabling mobile devices to cooperate for efficient neighbor discovery. In: *Proceedings of the Twelfth ACM Workshop on Mobile Computing Systems & Applications*, San Diego, CA, February 28–29, 2012.

14. Camps-Mur, D., A. Garcia-Saavedra, and P. Serrano. Device-to-device communications with Wi-Fi Direct: Overview and experimentation. *IEEE Wireless Communications*, 2013; 20(3): 96–104.

15. Jimmy, T., S. Ye. Wi-Fi direct services, Lund University, 2014, available online: http://www.eit.lth.se/sprapport.php?uid=818

16. Tsolkas, D., N. Passas, L. Merakos, and A. Salkintzis. A device discovery scheme for proximity services in LTE networks. In: *Proceedings of the IEEE Symposium on Computers and Communication (ISCC)*, Madeira, Portugal, June 23–26, 2014.

17. Balraj, S. *LTE Direct Overview*. San Diego, CA: Qualcomm Research, 2012, available online: http://s3.amazonaws.com/sdieee/205-LTE+Direct+IEEE+VTC+San+Diego.pdf

18. Baccelli, F., N. Khude, R. Laroia, J. Li, T. Richardson, S. Shakkottai et al. On the design of device-to-device autonomous discovery. In: *Proceedings of the Fourth IEEE International Conference on Communication Systems and Networks (COMSNETS)*, Phuket, Thailand, January 3–7, 2012.

19. Edwards, W.K. Discovery systems in ubiquitous computing. *IEEE Pervasive Computing*, 2006; 5(2): 70–77.

20. Hughes Systique Coporation, Wi-Fi Direct™, White Paper, available online: http://hsc.com/Portals/0/Uploads/Articles/WFD_Technology_Whitepaper_v_1.7635035318321315728.pdf

21. Prasad, A., A. Kunz, G. Velev, K. Samdanis, and J. Song. Energy-efficient D2D discovery for proximity services in 3GPP LTE-advanced networks: ProSe discovery mechanisms. *IEEE Vehicular Technology Magazine*, 2014; 9(4): 40–50.

22. 3GPP TR 23.703, Study on architecture enhancements to support Proximity-based Services (ProSe) (Release 12), available online: http://www.3gpp.org/ftp/specs/archive/23_series/23.703/23703-c00.zip

23. Doumiati, S., H. Artail, and D.M. Gutierrez-Estevez. A framework for LTE-A proximity-based device-to-device service registration and discovery. *Procedia Computer Science*, 2014; 34: 87–94.

24. Yang, Z.J., J.C. Huang, C.T. Chou, H.Y. Hsieh, C.W. Hsu, P.C. Yeh, and C.C.A. Hsu. Peer discovery for device-to-device (D2D) communication in LTE-A networks. In: *Proceedings of IEEE Globecom Workshops*, Atlanta, GA, December 9–13, 2013.

25. Zhao, Y., B. Pelletier, P. Marinier, and D. Pani. D2D neighbor discovery interference management for LTE systems. In: *Proceedings of IEEE Globecom Workshops*, Atlanta, GA, December 9–13, 2013.

26. Thanos, A., S. Shalmashi, and G. Miao. Network-assisted discovery for device-to-device communications. In: *Proceedings of IEEE Globecom Workshops*, Atlanta, GA, December 9–13, 2013.

27. Simsek, M., A. Merwaday, N. Correal, and I. Guvenc. Device-to-device discovery based on 3GPP system level simulations. In: *Proceedings of IEEE Globecom Workshops*, Atlanta, GA, December 9–13, 2013.

28. You, L., X. Zhu, and G. Chen. Neighbor discovery in peer-to-peer wireless networks with multi-channel MPR capability. In: *Proceedings of IEEE International Conference on Communications (ICC)*, Ottawa, Canada, June 15, 2012.

29. Poitau, G., B. Pelletier, G. Pelletier, and D. Pani. A combined PUSH/PULL service discovery model for LTE direct. In: *Proceedings of the IEEE 80th Vehicular Technology Conference (VTC Fall)*, Vancouver, Canada, September 14–17, 2014.

3

MESSAGE FORWARDING
STRATEGIES

3.1 Introduction

Recently, mobile social networks (MSNs) have gained tremendous attention, which free the users from face-to-monitor life, while users still can share information and stay in touch with their friends on the go. However, most MSN applications regard mobile terminals just as entry points to existing social networks, in which continual Internet and centralized servers (e.g., for storing and processing of all application/context data) connectivity are prerequisites for mobile users to exploit MSN services, even though they are within proximity area (such as campus, event spot, and community), and can directly exchange data through various wireless technologies (e.g., Bluetooth, Wi-Fi Direct). Today, modern mobile phones have the capability to detect proximity of other users and offer means to communicate and share data ad hoc with the people in the proximity, which leads to the bloom of mobile social networking in proximity (MSNP). MSNP can be explicitly characterized as: a wireless peer-to-peer (P2P) network of spontaneously and opportunistically connected nodes utilizing both social relationships and geoproximity as the primary filter in determining who can be discoverable [1]. MSNPs are infrastructure-free and self-organized, in which there exists no centralized server (e.g., communication stations), and a pair of devices can directly send and receive messages (e.g., pictures, videos, advertisements, and software updates) when they move into each other's communication range. Thus, message forwarding/data routing in MSNPs relies on the movement of individuals and their encounter opportunities to relay data to the destination, that is, nodes mobility is seen as a resource to bridge disconnections, rather than a problem to be dealt with.

Most of the existing message forwarding and dissemination protocols employ so-called store-carry-forward (SCF) fashion to carry messages between the network nodes. If there is no connection available at a particular time, a mobile node can store and carry the data until it encounters other nodes. When the node has such a forwarding opportunity, all encountered nodes could be the candidates to relay the data. Thus, relaying selection and forwarding decision need to be made by the current node based on certain forwarding strategies.

Various studies that collect and analyze real mobility traces have been conducted in literature. These studies have shown that (1) the movement of humans in real life is not random, but exhibits repetition, to some extent; (2) the mobility of humans is influenced by their social relationships. Based on this observation and exploring the recent advances in social network analysis, exploiting users' social behaviors to forward messages in MSNP, have drawn tremendous interests.

In general, existing social information-based message forwarding strategies can be divided into location-based (e.g., geographical coordination) and social encounters-based (e.g., the strength of social ties). These two kinds of information indeed represent two different levels of human behaviors: the location information is concrete and corresponds to the physical property of human activity; whereas the social behavior information is logical and represents virtual human interactions.

Location-based message-forwarding strategies were widely studied in the past decades. Those approaches made forwarding decision according to the geographical information such as GPS coordinates and geographical distance, and guided the data to the destination gradually. The encounter-based message-forwarding strategy utilizes the fact that users in MSNPs are interactively connected by social encountering events. In this book, we further divide encounter-based strategies into social property-based strategy and community-based strategy.

Figure 3.1 pedagogically illustrates three message-forwarding patterns in MSNP, where node A wants to send a message to a destination D. The location-based scheme indicates the physical locations of the mobile devices. Based on the measurement of geographical distance, it tends to choose A→B→C→E→D as the shortest routing path. The medium layer indicates the social connections of nodes.

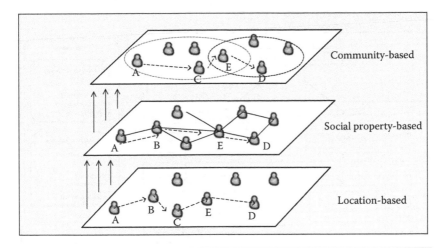

Figure 3.1   Example of various message-forwarding strategies in MSNP.

Based on the measurement of their social property (i.e., social centrality), the node A tends to forward data via the path A→B→E→D (since the value of social properties follow: A<B<E). In the community-based layer, A may tend to forward by the path A→C→E→D (users A, C, and E are in the same community, and users E and D are in another community).

The main goal of this chapter is to provide deep understanding on SCF-based message forwarding, which properly incorporate the social properties of individuals in MSNP, and bring new visions to MSNP research and applications. This chapter is organized as follows: The taxonomy of social behavior-based forwarding strategies is presented in Section 3.2. In Section 3.3, we describe and compare the typical message-forwarding schemes in those categories. Section 3.4 summarizes the open research issues and gives some potential solutions. Some SCF-based implementation prototypes are discussed in Section 3.5. Finally, we briefly conclude this paper.

3.2 Taxonomy of Social Behavior-Based Forwarding Strategies

Forwarding policy in MSNPs varies from epidemic replication of all the messages to every node, through multicopy and single-copy forwarding. Flooding-based protocols with unlimited replicas of messages cause high demand on network resources, such as storage and

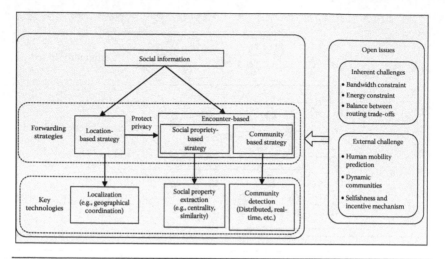

Figure 3.2   Taxonomy of social behaviors-based forwarding strategies in MSNPs.

bandwidth and cause congestion. However, multicopy protocols typi-cally aim to limit the number of replicas of the message in order to leverage a trade-off between resource usage and probability of mes-sage delivery. Single-copy strategies require routing algorithms to implement a next-best-hop heuristic that forward the messages to those nodes with a highest probability to deliver the message to its destination. Here, we focus on the heuristic single-copy/multicopy-forwarding policies [2].

Figure 3.2 illustrates the taxonomy of social behaviors-based for-warding strategies in MSNPs and open issues, which will be dis-cussed in detail as follows. Note that we intentionally divide those open issues into two categories: (1) internal challenges caused by the limitation of physical devices (including bandwidth and energy con-straints) and (2) external challenges caused by human's social behav-iors including mobility prediction, dynamic communities, selfishness and incentives, and so on [3].

3.2.1 Location-Based Strategies

Location-based message-forwarding strategies were widely studied in the past decades. They use the geographical information (geo-graphical coordination) and mobility patterns to make forward-ing decision. Geographic information provides accurate position of nodes in the network. Mobility patterns (trajectories, visiting histories) provide many characteristic of user movements. If this information could be used properly, it can improve message-forwarding performance in MSNPs. To describe location social information and mobility of nodes, the nodes' movements are always considered as discrete time-varying events. Each event suggests the location of a node with time label. Typically, an event can be described by four elements: node ID, location ID, start time, and time duration [4].

Nodes in the network travel from one location to another. Generally speaking, two nodes with similar mobility pattern and being close in geographical locations (they share many common visited places and their visited places are close) are more likely to meet each other in the future. As shown in Figure 3.3, suppose node A and B are possible relays to deliver data to node D. Three squares represent the visiting areas of three nodes respectively. The circles inside of squares are locations they visit. Overall, the average distance between visiting locations of node A (circles in the left square) and visiting locations of node D (circles in the middle square) is longer than the distance between node B's visiting locations (circle in the right square) and node D's visiting locations. Besides, node A shares 3 locations with node D as shown in gray circle, whereas node B shares 5 locations with node D as shown in black circles. Based on the similarity of

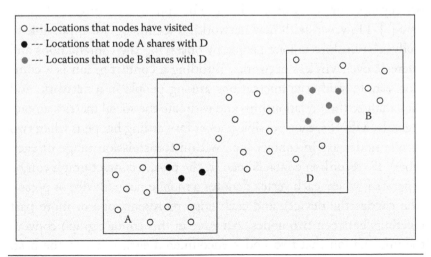

Figure 3.3   Illustration of locations visited by individuals.

mobility pattern and geographical distance, node B will most likely be selected for data delivery to node D.

However, location information is obtained from the detailed movements of users. Given such information, a user's visited positions and movement trajectory are revealed, which are normally considered as user privacy and sensitive. It has drawn so much attention on location privacy protection. Therefore, location information needs to be applied very carefully.

3.2.2 Encounter-Based Strategies

Encounter-based SCF strategies forward messages relying on the social information derived from encounter-based social graph [5,6]. This chapter explicitly divides the encounter-based strategies into social property-based and community structure-based strategies.

3.2.2.1 Social Property-Based Strategies   The intuitive way to study the social relations among people and extract their social properties, is building a social graph. A social graph is a global mapping of everybody and how they are related. With a social graph, a variety of social properties (e.g., centrality and similarity) can be calculated or estimated, and these properties can be used to forward data in MSNP.

A social graph is an intuitive source for many social metrics such as friendship. Unfortunately, it is not always available (due to either privacy or security reasons) or hard to be obtained via disclosed social data [7]. However, with new networking technology, it is possible to study relationships among people by observing their interactions and interests over wireless networks. Building a contact graph is a common way to study the interactions among people in a network, and thus analyze their relationships and estimate the social metrics among them. In MSNPs, each possible packet forwarding happens when two mobile nodes are in contact (i.e., within transmission range of each other). By recording contacts seen in the past, a contact graph can be generated, where each vertex denotes a mobile node (device or person who carries the device), and each edge represents one or more past meetings between two nodes. An edge in this contact graph conveys the information that two nodes encountered each other in the past. Thus, the existence of an edge intends to have predictive capacity for future contacts. A contact graph can be constructed separately for each single time slot in the past, or it can be constructed to record the encounters in a specific period of time by assigning a set of parameters to each edge to record the time, the frequency, and the duration of these encounters. From the observation that people with close relationships, such as friends, family members, and so on, tend to meet more often, more regular and with longer duration, we can extract nodes' relationships from the recorded contact graph, estimate their social metrics, and use such information to choose relays with higher probabilities of successful forwarding.

In graph theory and network analysis, centrality is a quantitative measure of the topological importance of a vertex within the graph. There are several ways to define centrality in a graph: degree centrality, betweenness centrality, and closeness centrality [8]. Degree centrality is the simplest centrality measure, which is defined as the number of links (i.e., direct contacts) incident on a given node. A node with a high-degree centrality is a popular node with a large number of possible contacts, and thus it is a good candidate of a message forwarder for others (i.e., a hub for information exchange among its neighborhood). Betweenness centrality measures the number of shortest paths passing via certain given node. Nodes that occur on many shortest paths between other nodes have higher betweenness than those that do not. A node with high betweenness centrality can control or facilitate many connections between other nodes, thus it is ideal for a bridge node during message exchange. The closeness centrality of a node is defined as the inverse of its average shortest distance to all other nodes in the graph. If a node is near to the centre of the graph, it has higher closeness centrality and is good for quickly spreading messages over the network.

Similarity is a measurement of the degree of separation. It can be measured by the number of common neighbors between individuals in social networks. Sociologists have known that there is a higher probability of two people being acquainted if they have one or more other acquaintances in common. In a network, the probability of two nodes being connected by a link is higher when they have a common neighbor. When the neighbors of nodes are unlikely to be in contact with each other, diffusion can be expected to take longer than when the similarity is high (with more common neighbors).

Friendship is another concept in sociology, which describes close personal relationships. In delay-tolerant networks (DTNs), friendship can be defined between a pair of nodes. On the one hand, to be considered as friends of each other, two nodes need to have long-lasting and regular contacts. On the other hand, friends usually share more common interests as in real world. In sociology, it has been shown that individuals often befriend others who have similar interests, perform similar actions, and frequently meet with each other. This observation is called homophily phenomenon. Therefore, the friendship in DTNs can be roughly determined by using either contact history between two nodes or common interests/contents claimed by two nodes.

Selfishness has also been well studied in sociology and economics and has recently been considered in the design of computer networks [10]. In MSNs, selfishness can describe the selfish behaviors of nodes controlled by rational entities. Selfish nodes can behave selfishly at individual level and aim to only maximize their own utilities without considering system-wide criteria. They can also behave selfishly in a social sense and are willing to forward packets for nodes with whom they have social ties but not the others. A selfish node may drop others' messages and excessively replicate its own messages to increase its own delivery rate while significantly degrading other users' performance or even cause starvation.

3.2.2.2 Community-Based Strategies Social structure is also appropriately applied to enhance the performance of message forwarding in MSNPs. Sociologists have studied the interactions between people in communities at many spatial and temporal scales. It has been shown that a member of a given community is more likely to interact with another member of the same community than with a randomly chosen member of the population. Therefore, communities naturally reflect social relationship among people.

Communities were typically assumed to be densely connected internally but sparsely connected to the rest of the network. In other words, a community should be considered as a densely connected subset in which the probability of an edge between two randomly picked vertices is higher than average. Moreover, a community should also be well connected to the remaining network, that is, the number of edges connecting a community to the rest of the graph should be significant.

The design of the community-based strategy is motivated by the observation that the mobility of people concentrates on a local area and the communication occurs in the form of communities. To apply such characteristics for elevating message-forwarding performance in MSNPs, many community-based protocols are proposed. By dividing MSNP into multiple communities (regarding user locations and interaction routines), community-based forwarding strategies may adopt different routing strategies to handle the intracommunity and intercommunity data delivery due to the fact that the connections within a community are rich, whereas the connections between different communities are relatively weak.

The basic community detection can happen on static networks and the usual approach is to collect data during a long period and then to aggregate all these data to create a large static network. By doing this, much information is lost since real data are always evolving: people launch new relationships, update/delete old relationships, and so on. Thus, recent researches have been investigating the case of communities in evolving networks. Through the time, characteristics of mobile nodes such as their locations, social relationship, interests, and movement patterns can be changed dynamically, which make community discovery a challenging issue in MSNPs. Therefore, community detection algorithms in MSNPs according to their mechanisms can be categorized into temporally independent and temporal community tracking and evolution analysis algorithms.

3.2.2.2.1 Temporally Independent Community Detection Many traditional community detection methods are borrowed or inspired from graph-clustering algorithms. Partitioning the nodes in a network into a predetermined number of disjoint communities is one of the traditional methods for identifying communities. However, since the community structure of real-world networks is not usually known, making assumptions about the number of communities or the size of the communities is not realistic. Moreover, many real-world networks have a hierarchical structure, where meaningful communities at different scales can exist and such community structures cannot be captured by partitioning algorithms. Therefore, another group of community-detection algorithms have been introduced that can identify hierarchical communities. Hierarchical clustering techniques can be divided into agglomerative and divisive methods. Agglomerative algorithms use a bottom-up approach where clusters are iteratively merged. Divisive algorithms use a top-down approach where the clusters are iteratively split. Overall, using hierarchical algorithms allow us to choose the suitable level of hierarchy and study the communities at that level of hierarchy. In many real-world networks, nodes can naturally belong to multiple communities, therefore the communities can overlap. In social networks, an individual can belong to a community of family members, to a community of friends, and to a community of colleagues. Traditional community-detection algorithms fail to uncover the community overlaps. Not being able to identify community overlaps in networks with naturally overlapping communities' means missing valuable information about the structure of the network. Therefore, overlapping community-detection algorithms have gained a lot of attention.

The majority of existing community-detection algorithms implicitly assumes that the entire structure of the network is known and is available. They are called global algorithms, since they require a global knowledge of the whole network in order to uncover all the communities in that network. As such knowledge might not be available for large networks; local algorithms are gaining more popularity. Local algorithms typically start from a number of given seed nodes and expand them into possibly overlapping communities by examining only a small part of the network. As it is possible to find local communities from each seed independently, they are very suitable for being parallelized and therefore can scale well. The local communities identified from each seed can be aggregated in order to uncover the global community structure of the network. However, if the local community-detection algorithm is naively started from each node in a network, it can lead to many redundant communities, and therefore it is computationally expensive. Therefore, it is important to identify a number of good seeds that are well distributed over the network by using a seeding algorithm before running the local community detection. On the other hand, if the seeding algorithm does not select enough seeds, the communities might only cover a subset of the nodes in a network and therefore, the problem of selecting a reasonable number of seeds that are well distributed over the network is challenging.

3.2.2.2.2 Temporal Community Tracking and Evolution Analysis

3.2.2.2.2.1 Using Static Algorithms on Several Snapshots An evolving network is often defined as a sequence of static networks, each of them representing the state of the network at different timestamps. As each snapshot is a static graph, the first approach to compute communities on an evolving network is to use a classical algorithm on each snapshot more or less independently. Specifically, many attempts have been made to use static algorithms on dynamic networks: for each timestep, a partition of the nodes is computed and the main issue is then to characterize the evolution of the communities: what happened to a community between two timesteps. Indeed, a community may merge with another one, split, or disappear. Therefore, identifying some common parts between two partitions is equivalent to solving a matching problem [9].

The first intuitive idea to perform the matching between partitions is to use set theory and rules or methods to decide whether sets of different partitions are similar or not. For example, if two communities of successive snapshots share many nodes, they are related. The main problem is that given two partitions, one can find many different valid matchings between these partitions. Therefore the matching problem can be rephrased as the maximization of a quality function that tells whether or not a matching is good. Another way to use a community-detection algorithm to perform the matching is to build a temporal network representing the relations between communities at each timestep. The communities are then computed on this network and it therefore gives communities that span across several snapshots. Thus, such algorithms are decomposed in two phases: first, each snapshot is decomposed into static communities and then these static communities are joined in temporal communities that span over several timesteps. The static communities for each snapshot are sometimes called community instances to distinguish them from community.

Those works have two main problems: the instability of the algorithms and the difficulty of choosing many rules and parameters. As approaches using static algorithms are limited, many different algorithms have proposed to use directly the temporal information during the community detection and not afterward.

3.2.2.2.2.2 Using Temporal Information Directly to Find Better Communities Since classical community-detection schemes are based on static quality functions, modularity for instance, a solution of community detection in evolving networks using temporal information should modify such quality functions to integrate evolution. Usually, the traditional quality function is split into two terms: a part for the quality of the current snapshot and a part to ensure stability (e.g., a term evaluating the distance of the new partition with the precedent). This new quality function allows to obtain a series of more interesting clustering and to reduce the number of artifacts caused by the optimization algorithm, which can be used to extend some existing algorithms like k-means and hierarchical agglomerative algorithms for community detection in evolving networks.

The temporal information can also be coded in the graph itself. Given several snapshots of an evolving network, they may be placed side-by-side in one temporal graph. So, nodes that exist at several timestamps appear several times in this graph. Each snapshot is a slice of the dynamic network. Then temporal links between nodes in different slices can be added, typically between a node at time t and the same node at time $t + 1$. The result is one network with two kinds of links: the links that really exist at some moment in the evolution of the network and the links between slices. The communities on this graph thus contain nodes of various timestamps and are consequently communities over time. The matching problem is automatically solved since communities span over timestamps.

3.3 Typical Message-Forwarding Schemes for Proximity Service

3.3.1 Location-Based Message-Forwarding Schemes

Loc [4] takes the combination of the similarity of mobility pattern and geographical distance to the destination node as the utility to construct the message-forwarding path. The similarity matrix of mobility pattern S_{ij} measures the extent that two different nodes i and j stay at various (same or difference) locations. The distance matrix of m locations D_{ij} measures the physical separation between users i and j, when they are located in those locations.

Those two matrixes are formulated as follows, as shown in Equations 3.1 and 3.2:

$$S_{ij} = \begin{bmatrix} s_{1_i 1_j} & \cdots & s_{1_i m_j} \\ \vdots & \ddots & \vdots \\ s_{m_i 1_j} & \cdots & s_{m_i m_j} \end{bmatrix} \qquad (3.1)$$

where $s_{x_i y_j} = c_{x_i} * c_{y_j}$, c_{k_i} is the time proportion that node i at location k. Therefore, $s_{x_i y_j}$ represents the probability that node i stays at location x AND node j stays at location y, and the elements on the diagonal of matrix suggest the time proportion of common places that the two nodes have visited. It reveals the similarity of their mobility patterns.

$$D_{ij} = \begin{bmatrix} d_{11} & \cdots & d_{1m} \\ \vdots & \ddots & \vdots \\ d_{m1} & \cdots & d_{mm} \end{bmatrix} \qquad (3.2)$$

where $d_{xy} = (1/1 + \|s_x - s_y\|)$ and s_x is the GPS coordinates of location x.
The overall utility is combined as Equation 3.3:

$$H_{ij} = S_{ij} \circ D_{ij} = \begin{bmatrix} s_{1_i 1_j} d_{11} & \cdots & s_{1_i m_j} d_{1m} \\ \vdots & \ddots & \vdots \\ s_{m_i 1_j} d_{m1} & \cdots & s_{m_i m_j} d_{mm} \end{bmatrix} \qquad (3.3)$$

The matrix H_{ij} presents both similarity of mobility pattern and the geographical distance of two nodes i and j. Loc scheme uses the average of the sum of all elements in the matrix (3.3) as the geographical metric between node i and j. The larger of the average, the closer geographical relation two nodes have, and therefore the more chance they will encounter. The message forwarding relying on geographical distance mainly initiates their relay selection by greedy forwarding.

That is, when two nodes (i.e., i and j) encounter, for messages carried by $i(j)$, they decide whether to take $j(i)$ as the next relay by comparing their utilities to the destination, which is the average of the sum of elements in $H_{id}(H_{jd})$. If the average value of $i(j)$ is smaller than that of $j(i)$, the message will be forwarded from $i(j)$ to $j(i)$.

The Loc method can represent several location-based strategies for the message forwarding in MSNPs. On the one hand, it will use the similarity of mobility pattern to choose relays in the case that two nodes have visited the same set of locations or the geographical distances among locations are unknown. On the other hand, it can select relays by the geographical distance in the case that two nodes have no common visited places or the movement of nodes is rare. The weak point lies in that collecting location information needs dedicated equipment, privacy preserving is a serious problem to be concerned.

3.3.2 Encounter-Based Forwarding Schemes

3.3.2.1 Social Property-Based Schemes
Soc [4] incorporates social similarity and social centrality to design MSNP message-forwarding scheme, in which social similarity indicates the trustiness and cohesive of social links. Social centrality is the quantification of relative importance of nodes in the social network.

Specifically, the metric of similarity is defined as follows:

$$S_{i,j}(\tau) = 1 + \left| F_i(\tau) \cap F_j(\tau) \right| \tag{3.4}$$

where $F_i(\tau)$ is the set of friends of user i at time τ. Intuitively, if a node i has higher $S_{i,d}(\tau)$ value, it shares more common friends with the destination d, thus more likely to transmit the message successfully.

The centrality is defined as follows:

$$C_i(\tau) = \frac{\displaystyle\sum_{k=1}^{N} d_{ik}(\tau)}{N} \tag{3.5}$$

where $d_{ik}(\tau) = 1$ if a direct link exists between users i and k at time τ, and N is the number of nodes in the network.

The comprehensive utility is defined as follows:

$$Y_{i,d}(T) = S_{i,d}(T) * \left(\frac{C_i(\tau)}{T} \right) = \int_{t=0}^{T} S_{i,d}(\tau) \cdot \frac{C_i(\tau)}{T - \tau} \tag{3.6}$$

The convolution operation provides a time-decaying description of all prior values of social similarity and social centrality. The utility is updated each time by accumulation of social similarity when a new encounter occurs. The decay function suggests that the most recent encounters typically have the more influence, and the impacts of previous encounters decrease as elapsed time and social centrality.

The Soc method interprets both social similarity and social centrality. It can reflect the impact of social similarity on selecting relays as long as the network are evenly distributed, where most of the nodes have equal social centrality. It can also express the impact of social centrality on forwarding decision-making when two nodes have similar number of common friends with destination nodes.

However, the weak point of Soc lies in that: usually, global MSNP network information is required to accurately estimate individual's social centrality, which is extremely difficult to obtain, thus, instead of node's global social centrality, Soc roughly uses individual's local degree centrality to approximate his/her global social centrality.

3.3.2.2 Community-Based Forwarding Scheme The identification of social communities in MSNs can be used to improve the delivery of information by selecting appropriate forwarders instead of performing naive oblivious flooding. First, mobile nodes are grouped into the communities by the certain community-detection algorithm. The phase following the detection of existing communities is how to efficiently exploit the detected community structure. To forward data between the communities (intercommunity forwarding), network overlays constructed using hubs or brokers can be used. In this phase, if the relay nodes are out of destination community, the intercommunity-forwarding strategy relays data to the destination community. Otherwise, the intracommunity-forwarding strategy is used to pass data to the centrality node until the data reaches the destination. However, these approaches suffer with the overhead of community formation.

3.3.2.2.1 Social and Mobile Aware Routing Strategy A social and mobile aware routing strategy for DTN called SMART is proposed by Zhu et al. [11], in which a DTN is divided into a number of communities using an adaptive community-partitioning algorithms. Two data routing processes are introduced: intracommunity communication and intercommunity communication. For intracommunity communication, a utility function convoluting social similarity and social centrality with a decay factor is used to choose relay nodes. For intercommunity communication, the nodes moving frequently across communities are chosen as relays to carry the data to destination efficiently. It is shown that such message-forwarding strategy significantly alleviates the blind spot and dead end problems. It adapts to the community structure by enhancing performance for intercommunity communication. The detailed introduction of SMART is as follows.

3.3.2.2.1.1 Distributed Community Partitioning Community is defined as a social unit that shares a common value. It is a tight and cohesive social entity. Intuitively, communities are formed based on locations or interests. People in the same geographic location or sharing the same interest are likely to be in the same community. In the context of DTNs, SMART uses geographic locations to study community structure and investigate the relation between encounters and geographic distances for the discovery of communities. It is likely that there is correlation between encountering and geographic location. The closer the two nodes are, the more often they meet each other. The number of encounters rapidly decreases when the distance increases, which implies that, when the distance becomes longer, the number of encounters becomes smaller. Inspired by the earlier observation, a dynamic and distributed community-partitioning algorithm was proposed. The basic idea is adaptively grouping nodes into communities starting from a random partition (i.e., m communities) of the network. The detailed community-partitioning process is described as follows:

The community construction process is divided into two stages: the bootstrap stage and the evolution stage. In the bootstrap stage, m nodes are randomly selected, and each node is assigned with a unique community ID. Node without community affiliation will choose one community through encounters until every node in the network is assigned with a community ID. After this stage, the network has m communities. Then, in the evolution phase, each node counts the affiliation parameters (APs), which indicate the number of encounters with nodes in different communities. Then it adjusts the community affiliation according to updated APs. The f vector is used to represent the APs of any individual n_i, that is, $E_i = \left\{ ap_{1_i}, ap_{2_i}, ap_{3_i}, \ldots, ap_{m_i} \right\}$, where ap_{j_i} is the AP that n_i connects to community C_j, denoting the number of encounters between n_i and C_j. When n_i encounters a node in community C_j, it updates its AP value accordingly, and adaptively changes its community affiliation to the community with the maximal AP value in the vector. The algorithm is called m-partition.

The m-partition algorithm dynamically runs as each encounter occurs in the network in a distributed fashion. Therefore, the community structure may change from time to time and is maintained dynamically. We show that the communication cost of m-partition for maintaining community members is low in DTNs. Given two communities A and B, there are m nodes in A and n nodes in B. Suppose a node n_i in community A needs to switch its community from A to B. It first obtains the new community list from the encountered node in community B. If the traffic overhead for transmitting one node ID is assumed as 1, the communication overhead for obtaining community members will be n. It then floods its ID to its new community B to make other nodes in community B aware of n_i. The communication cost will also be n. Furthermore, when a node in community A meets a node in community B, it checks the community member list to see whether any node changes their community identity. In this case, n_i changes from A to B. Suppose there are k encounters between community A and B. The cost for transmitting node ID of n_i is k. The node in community A floods this information to make the remaining nodes (m-k nodes) in A exclude the membership of n_i from their local community, which needs m-k transmissions of ID of n_i. Overall, the communication overhead for maintaining community members caused by one community switch action is $O(m + 2n)$.

3.3.2.2.1.2 Intracommunity Communications   Social similarity indicates the trustiness and cohesiveness of social links. Social centrality is the quantification of relative importance of nodes in the social network. SMART incorporates social similarity and social centrality to design message-forwarding scheme. Specifically, the metric of similarity is defined as Equation 3.4, where $F_i(\tau)$ is the set of friends of user i at time τ. Intuitively, if a node i has higher $S_{i,d}(\tau)$ value, it shares more common friends with the destination d, thus more likely to transmit the message successfully.

The centrality is defined as Equation 3.5 where $d_{ik}(\tau) = 1$ if a direct link exists between users i and k at time τ, and N is the number of nodes in the network.

The encounter effect between the two nodes is therefore denoted by the social similarity. To model the decaying effect, we introduce a decay function with respect to social centrality and time as follows:

$$D_i(t - \tau) = \frac{C_i(\tau)}{t - \tau} \tag{3.7}$$

The comprehensive utility is defined as follows:

$$Y_{i,d}(T) = S_{i,d}(T) \otimes D_i(T) = \int_{t=0}^{T} S_{i,d}(\tau) \cdot D_i(t - \tau) \tag{3.8}$$

However, the encounter only occurs in several time units. Therefore, the accumulative effects of encounters are represented by a discrete convolution as

$$U_{i,d}(T) = \sum_{\tau=0}^{T} X(\tau)_{i,d} \cdot S_{i,d}(\tau) \cdot D_i(T - \tau) \tag{3.9}$$

where $X(\tau)_{i,d} = 1$ when an encounter occurs at time τ or when $\tau = 0$ (to initialize the utility value); otherwise, $X(\tau)_{i,d} = 0$. The utility function describes that when each encounter occurs, it yields an encounter effect represented by social similarity. Each effect occurs at different time decays as a decay function is composed of social centrality and time, indicating that the encounter effects of a node with higher social status decay slower than a node with poor connection to the network and that a recent encounter effect decays slower than an older encounter effect.

3.3.2.2.1.3 Intercommunity Communications   If a destination node n_d does not belong to the community of source node n_s, it is needed to choose some relay nodes to forward the message among communities. The idea is to use *fringe nodes* to bridge the communication of intercommunities. A fringe node is a node that is capable of remotely contacting other communities. It is measured by the number of links that it connects to other communities. Nodes with higher links outside the local community are chosen as fringe nodes. Each fringe node can be represented by its ID and the remote contact table to indicate its links to other communities.

To enable efficient intercommunity communications, a utility function extended from intracommunity utility is proposed to forward data from the fringe node to the destination community. Namely, the utility function from node-to-node is extended to node-to-community for intercommunity communications. To construct the utility between a fringe node f to the destination community C', SMART considered the social relation between f and C' as similarity between the node f and a set of nodes in C'. However, knowing the friends of all nodes in C' would suffer too much overhead in DTNs. Therefore, an approximate estimation is provided, which only counts the friends of nodes who have ever encountered with f. The similarity is defined as

$$S_{f,C'}(\tau) = 1 + \left| F_f(\tau) \cap F_{C'}(\tau) \right| \qquad (3.10)$$

where $F_{C'}(\tau)$ indicates the friends of a set of nodes in C' that ever encountered with f until time τ.

To formulate the social status of node f, SMART extended the concept of centrality from the local community to the entire network, so-called community centrality, denoted by $C_\Gamma(\tau)$, which is defined as the proportion of the number of communities that is connected with $[M_c(\tau)]$ to the total number of communities $[M(\tau)]$ at time τ. It is defined as

$$C_\Gamma(\tau) = \frac{M_c(\tau)}{M(\tau)} \qquad (3.11)$$

The decay function in the node-to-community utility becomes $D_\Gamma(t - \tau) = [(C_\Gamma(\tau)/t) - \tau]$. The overall utility function from the node f to community C' is thus defined as

$$U_{f,C'}(T) = \sum_{\tau=0}^{T} X(\tau)_{f,C'} \cdot S_{f,C'}(T) \cdot D_\Gamma(t - \tau) \qquad (3.12)$$

where $X(\tau)_{f,C'} = 1$ when an encounter occurs between node f and community C' at time τ or when $\tau = 0$ [to initialize $U_{f,C'}(T)$]. According to the utility function, the fringe node finds the next relay by choosing a node with a higher utility value with the destination community. The procedure continues until the data reach the destination community.

3.3.2.2.2 Homing Spread Homing spread (HS) [12] is a zero knowledge multicopy-routing algorithm that exploits the fact that nodes usually have a common interest and visit some locations, called community homes, frequently, whereas the other locations are visited less frequently. It is assumed that each home supports a virtual throw-box, a mechanism that can store a message at a local storage device, or at another node currently at the same home. A message holder is either a mobile node or a home that has message copies. HS consists of three phases: homing, spreading, and fetching. In the homing phase, the source sends copies quickly to homes. On reaching the first home, the message holder (that includes the source) dumps all copies to the home. When roaming occurs (i.e., a message holder meets another node at another location), copies are equally split between the two nodes and both become message holders. In the spreading phase, homes with multiple copies spread them to other homes and mobile nodes. The home gives one copy to each node located at the same home, subject to the availability of the copies. However, the last copy is kept at the home through a virtual throwbox (a mechanism that can store a message at a local storage device or at another node currently at the same home). Each new message holder with one copy starts its homing phase. Then, in the third phase, the destination fetches the message when it meets any message holder for the first time, which can be either a home or a mobile node.

3.3.2.2.3 Community-Aware Opportunistic Routing Community aware opportunistic routing (CAOR) [13] is a single-copy routing algorithm based on community homes, which turn the routing in mobile nodes into a routing in community homes. This method is then extended to the case of the virtual throwbox by letting the members of a community with high centralities acting as the home of this community. In this method, mobile users with a common interest autonomously form a community, in which the frequently visited location is their common home. Each community has a star topology where its home is the center. The whole network is composed of some overlapped star-topology communities. CAOR first turns the routing between lots of nodes to the routing between a few community homes. Then, a reverse Dijkstra algorithm by maintaining an optimal relay set for each home is adopted to determine the optimal relays and compute the minimum expected delivery delay. Each home only forward its message to the node in its optimal relay set, and ignores the other relays.

3.3.2.2.4 Social Group-Based Routing   Social group-based routing (SGBR) [14] is proposed to develop a routing protocol that spreads a small number of packet copies to reduce network overhead, while guiding the packet copies using only local information to reach the destination. To achieve that goal, the authors exploit the social grouping characteristic of DTN nodes. Specifically, nodes frequently meeting each other are considered to belong to the same social group where they are expected to again meet each other frequently. They are also expected to have around the same social relation with other nodes. In that sense, each node may consider itself a representative of the group to distribute its packets to other groups. Therefore, a node that has a packet destined to other node outside its group tends to forward the packet copies to other groups. Furthermore, SGBR uses an exclusive social metric that sprays messages by excluding nodes that are not expected to add a significant value to the node carrying the message, which reduces the need to collect network-wide information, while improving the performance metrics.

*3.3.3 Comparisons of Social-Based Forwarding Strategies
      in Mobile Social Networking in Proximity*

Table 3.1 summarizes and compares various social-based forwarding strategies in MSNP.

The following three implications can be inferred from those schemes:

- It has been found that the encounter-based strategies and the location-based strategies have no significant difference in forwarding data performance [3]. This indicates that concrete location information is not always necessary to be the key consideration for routing design. As collecting location information needs dedicated equipment and arguably violates user privacy, the encounter-based forwarding strategy is safe and effective for MSNPs.

Table 3.1 Comparison of Various Social-Based Forwarding Strategies in MSNP

STRATEGIES \ METRICS		PROTOCOLS	KEY IDEAS	ADVANTAGE/DISADVANTAGE
Location-based strategies		Loc [4]	Forward data to node with more similar mobility pattern and closer distance to destination	Needs dedicated equipments for gathering user's positions; violates user's privacy
Encounter-based strategies	Social property-based	Soc [4]	Forward data to node with higher encounter frequency or social similar with destination	Do not use location information, preserve privacy; Estimating the global social properties in distributed and real-time way is challenging
	Community-based	SMART [11], HS [12], COAR [13], GBSR [14]	Forward data according community structure	Utilizing the inherent community structure to improve delivery probability; Distributed community detection during message forwarding is challenging

- Second, encounter-based forwarding strategies significantly rely on estimating the social properties (e.g., social centrality, social similarity, and community detection) in real time and distributed way, when forwarding schemes are working. However, due to the lack of information about whole network and time-varying topology, it is extremely challenging to accurately estimate those social properties.
- The structure-based strategies are motivated by the observation that the mobility of people concentrates on a local area and the communication occurs in the form of communities. To apply such characteristics can elevate message-forwarding performance in MSNPs.

3.4 Open Issues and Challenges

Social information-based MSNP message-forwarding strategies have lately received much attention in research and application fields. However, there are still several issues left without being deeply investigated. Here we summarize some open research issues and their preliminary solutions, which bring new visions into the horizon to MSNP research and application.

3.4.1 Internal Challenges

3.4.1.1 Bandwidth Constraint This bandwidth factor determines the number of messages that can be transmitted at each encounter opportunity. For instance, if the traffic load increases due to a large number of users or a larger size of messages being transmitted, the unsuccessful transmissions due to insufficient encounter duration should be taken into account. Therefore, estimation of the number of messages that can be successfully transmitted is useful to reduce the number of aborted messages due to insufficient encounter duration. In addition, to transmit messages according to a corresponding priority is beneficial to utilize the limited bandwidth.

3.4.1.2 Energy Constraint A mobile device often has limited energy and cannot be connected to the power supplier easily. Energy-efficient forwarding schemes are necessary for reducing energy cost of data communication in MSNPs, which can be considered from the following three aspects: energy-efficient underlying communications technology, adaptive peer discovery, and utilizing users' special context [15,16].

Mobile devices always have several wireless interfaces for direct D2D communications, for example, Bluetooth, Wi-Fi Direct, and so on. If permitted (i.e., the volume of transmitted message is small), instead of high-energy wireless interface, low-energy interface (Bluetooth) can be used. Furthermore, an enhanced power management for Wi-Fi Direct that recognizes the properties of applications, and dynamically adjusts the duty cycle of P2P devices can be used for transmitting messages [17].

For an individual node, performing neighbor discovery in MSNP can be too expensive with a high-power, long-range radio (e.g., Wi-Fi). On the other hand, relying only on a low-power, short-range radio for detecting neighbors (e.g., Bluetooth) results in significantly fewer available contacts. Therefore, for adaptive peer discovery in proximity that is a prerequisite for message forwarding, it is feasible to leverage the clustering of nodes as well as the radio heterogeneity of mobile devices for more efficient long-range neighbor discovery. Furthermore, if some central infrastructure exists, the infrastructure-assisted mode can make peer and service discoveries and pairing procedures faster, more efficient in terms of energy consumption and more users friendly.

Context is any information that can be used to characterize the situation of an entity, including special location, users' mobility, wireless signal strength, identities of nearby people, and so on. The user's context can be used to achieve high energy savings. For example, user-context information provides clues to estimate users' service and resource demands in near future by analyzing past usage history, user preferences, and resource-demanding patterns.

3.4.1.3 Balance of Various Trade-Offs There are several forwarding trade-offs that should be considered in a protocol design [18]: packet delivery versus delivery cost, node cooperation versus resource limitation, and information quality versus delay, have conflicting requirements and goals. It is critical to determine the balance to satisfy both points of view. For example, maximizing delivery ratio requires increasing the number of packet copies spread throughout the network.

Another trade-off is the compromise regarding the amount of information collected to guide the packets to their destinations. Collecting information from the network helps in selecting the relaying nodes to the destination, but requires time to collect the information that increases the packet delays. On the other hand, collecting little or no information decreases the probability of reaching the destination unless a large number of copies were spread. Consequently, new metrics and measures should be introduced for studying the relationships between the routing conflicting trades-offs and establishing the balance between them.

3.4.2 External Challenges

3.4.2.1 Human Mobility Prediction The prediction of a user's future steps can streamline data-forwarding decisions and improve the performance of the routing algorithms significantly. Some of the existing routing protocols focus on the prediction of whether two nodes would have contact, without considering the spatial and temporal properties of the contact. Exploiting temporal and spatial information of human mobility, in addition to contact prediction, allow better resource usage and a higher delivery ratio, as well as avoiding useless transmission when the probability of reaching the destination is very low. As an example, predict and relay (PER) [19] considers the time of the contact and determines the probability distribution of future contact times and choose a proper next-hop in order to improve the end-to-end delivery probability. Despite the above-mentioned solutions, further investigations in this area could be carried out in order to improve the efficiency and effectiveness of the data replication protocols, based on predicting the further walks of mobile users. For example, data forwarding based on prediction models that take into account human rhythms on a weekly basis could provide extra benefit.

3.4.2.2 Dynamic Communities   Most of the temporally independent community detection approaches are not suitable for highly dynamic MSNPs. An incremental community mining approach is proposed by Mitra et al. [20], which considers both current and historic information into the objective of mining processes. Nevertheless, new algorithms should be developed to detect the evolution of communities in highly dynamic MSNPs. One potential solution is the identification of critical events and transitions for the evolving social communities.

Several other challenges related to the temporal tracking of communities are highlighted in the literature. One major challenge is in the validation of communities, both with and without ground truth information. Another major challenge is the selection of the number of communities at each time step. A poor choice for the number of the communities may create the appearance of communities merging or splitting when there is no actual change occurring.

3.4.2.3 Selfishness and Incentive Mechanisms   User selfishness is a very challenging issue for message forwarding MSNPs. To tackle with this problem, many techniques have been used to detect selfish nodes or study the impact of selfishness in MSNPs. Numerous incentive schemes have been proposed to encourage selfish nodes to cooperate in data delivery. In general, incentives could be classified into three categories: extrinsic (economic), intrinsic (e.g., entertainment- or game-based), and internalized extrinsic (e.g., reputation-based) incentives [21].

Economic incentives are the real money or any other commodity that the users consider valuable. They are probably the most straightforward way to motivate participants; the idea of taking entertaining and engaging elements from computer games and using them to incentivize participation in other contexts is increasingly studied in a variety of fields; social psychological factor is another widely harnessed nonmonetary incentive mechanism to promote increased contributions to online systems. For message forwarding in MSNP, individuals should know that their work can be easily evaluated by others, and the unique value of each individual's contribution should be displayed in some proper way.

3.5 Several Typical Prototypes

3.5.1 Haggle

Haggle [22] is a content-sharing system for mobile devices, allowing users to opportunistically share content without the support of infrastructure. Mobile devices share content and interests over direct Wi-Fi or Bluetooth links, and may store-carry-forward content on behalf of others based on interests, bridging otherwise disconnected devices. Unlike traditional Internet-based content-sharing systems, Haggle faces disconnections, unpredictable mobility, and time-limited contacts, which pose unique challenges to the system's design and implementation.

Although similar content-sharing systems typically value every content item the same, Haggle uses a ranked search to judiciously decide which content to exchange, and in which order. The search matches a device's locally stored content against the interests of other users that the device has collected, prioritizing relevant content when contacts are time limited and resources scarce. Thus, search enables dissemination of content in order of how strongly users desire it, offering delay and resource savings by exchanging the content that matters. An optional content delegation mechanism allows Haggle to altruistically disseminate a limited amount of items based on the interests of third-party nodes, increasing the benefit of the network as a whole, and protecting against networks that are partitioned along interests. Ranked searches, combined with delegation, allow Haggle to balance the short-term benefit of exchanging a content item between two nodes against the long-term benefit to the network as a whole.

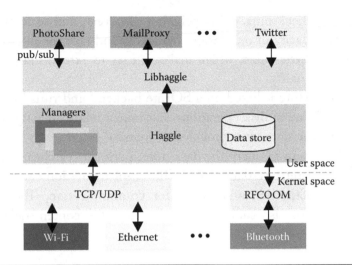

Figure 3.4   Implementation overview of Haggle system.

Haggle has been implemented Haggle on various smartphone plat-forms including Android, iOS, and Windows Mobile, and it can also run on Windows, Linux, or Mac OS X computers. The implementa-tion comprises around 20,000 lines of C and C++ code, excluding applications. The initial choice to develop the implementation in a native language has shown crucial for cross-platform development and for access to lower layer APIs. Higher level languages require vir-tual machines and libraries that are incompatible between platforms and they seldom expose complete APIs, for example, Bluetooth and Wi-Fi, which are required for an event-driven design.

Figure 3.4 gives a logical overview of our implementation. Applications link against a library, called libhaggle, which imple-ments an asynchronous pub/sub API. This library communicates with Haggle using local interprocess communication or intraprocess com-munication (IPC). The API makes application development straight-forward and has enabled a wide variety of applications.

Haggle itself runs as a user space process with a main thread in which a kernel and a set of managers run. The kernel implements a central event queue, while managers divide responsibility in areas such as security, node management, content dispatching, and integ-rity. Managers create and consume events and may run tasks in separate threads when they need to do work that require extended processing. This may include sending and receiving data objects, computing checksums, doing neighbor discovery, and so forth. Due to the modularity of our design, managers and task modules can be added with little effort, which makes it easy to extend Haggle with extra functionality.

The data store is based on a SQLite backend and runs in a separate thread due to disk operations that take a relatively long time to complete. The data store provides an internal interface to managers that allow them to query the data store, perform searches, and add, remove, and retrieve data objects. The implementation can leverage Bluetooth, Ethernet, and Wi-Fi for communication. The implementation can also support multiple pluggable content-delegation algorithms.

3.5.2 DMS: DTN-Based Mobile Social Network Application

DMS is a resource-aware DTN-based mobile social network application to get a better understanding of the altruistic nature of human behavior [23]. It is used to study how limited resources such as battery can affect the user's behavior toward the mobile application with varying levels of scrutiny in lab and daily life environment.

DMS is capable of microblogging and messaging by utilizing mainly close range communication via Bluetooth. DMS is built with the functionality to monitor and allocate battery power as per the user's desire. Battery is still a scarce resource in mobile world and serves as a limiting factor on usability of a mobile phone. DMS was used to evaluate the altruistic notion of human behavior when it comes to sharing battery under varying levels of scrutiny.

DMS supports data transmission in unicast, multicast, and broadcast mode. Short messages and contact information are sent in unicast mode via direct forwarding strategy. If a message has a high priority, and a direct Bluetooth connection is not present, DMS will use GSM (SMS) network, provided the user is ok with it. Friendship searching information and group messages are sent by multicast. Broadcast mode is configured to work on Bluetooth only, and the epidemic routing strategy is used to send the message to the whole network. With epidemic routing each node stores messages to be sent in its buffer, carries them, and sends the message to the nodes when it encounters. DMS consists of three parts, that is, the graphical user interface (GUI), the network interface

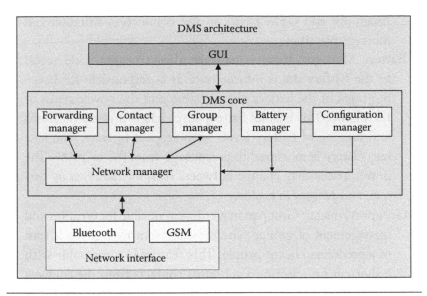

Figure 3.5 DMS Architecture.

(Bluetooth, GSM), and the DMS core. Illustrated as Figure 3.5, DMS core controls six managers that provide the main functionality to DMS, including contact, forwarding, and network managers, group and configuration managers.

Contact Manager: Contact manager stores contact information of the user (e.g., name, phone number) in the database. This is helpful in selecting the recipient's contact for creating new message, identifying nearby devices, and retrieving the sender of a message. It also records the Bluetooth MAC address for communication.

Forwarding Manager: Forwarding manager is used for forwarding the messages. It keeps a record of the message based on their unique ID and MAC addresses to avoid retransmission. It is also responsible to periodically check if friends are in range to forward the message. Once a message reaches its time to live (TTL), it is not forwarded any more.

Network Manager: Network manager controls and coordinates the network interface to send and receive data via Bluetooth and GSM. It scans for the nearby Bluetooth devices to exchange messages and friend requests to build-up social relationships. It is also responsible for switching between

Bluetooth and GSM based on the connection situation and message priorities.

Battery Manager: Battery manager allows system wide access to the battery status information. It is responsible for keeping track of the battery charge value and the power status of the phone. This enables battery level-based user notifications and logging of battery levels with respect to time. The battery charge is measured in percentage from 0% to 100%. The power status may change between *charging, powered by battery*, and *plugged in to charger*.

Groups Manager: Group manager is responsible for creation and management of groups. The message sent to a group is sent to a predefined list of people. This feature is comparable with a shortcut for selecting a subset of contacts from the contacts list to send a message to, with the exception that the list is managed by the group's owner.

Configuration Manager: Configuration manager manages the user preferences regarding different settings for DMS usage. This also includes the battery usage and resource sharing preferences. The user has to explicitly specify until what battery level percentage the subject would like to share the device as a relay and until what battery level percentage the subject would like to receive messages from other devices. This plays a vital role in the battery-monitoring process because the preference set by the user affect behavior of the handset, that is, whether or not to continue scanning for devices and whether or not to turn off receiving. First-time users have to set the preference before starting the application and after that the preferences can be edited anytime.

The experiment starts in a lab environment where the subjects are introduced to a paper-based survey/questionnaire to fill out their resource (battery in our case) sharing preferences under supervision. Next, the subjects are introduced to the mobile application with a demo account. Finally, the subjects create a real account and participate in a week long real experiment. The mobile phone application is used without supervision and the subjects may change their preference any time. After the experiment, researchers compared the results of the paper-based questionnaire with real-life experiment to see how the subjects behave with and without scrutiny, that is, whether the subjects act more or less altruistically in the real environment without supervision.

3.6 Conclusion

Proximity service is a newly emerging wireless networking paradigm, and its research is still in an early stage. Message-forwarding strategies have lately received much attention in the wireless communication network community. In this paper, we classify the message-forwarding strategies in MSNP into location-based strategies and encounter-based strategies. In addition, the encounter-based strategies are further divided into social properties-based strategies and community-based strategies. After that, we analyze and discuss the existing social aware-forwarding protocols and give a comparison of these three strategies. In addition, we also outline the open issues and their potential solution directions.

References

1. Wang, Y., A.V. Vasilakos, Q. Jin, and J. Ma. Survey on mobile social networking in proximity (MSNP): Approaches, challenges and architecture. *Wireless Network*, 2014; 20(6): 1295–1311.
2. Sobin, C.C., V. Raychoudhury, G. Marfia, and A. Singla. A survey of routing and data dissemination in delay tolerant networks. *Journal of Network and Computer Applications*, 2016; 67: 128–146.
3. Wang, Y., J. Chen, Q. Jin, and J. Ma. Message forwarding strategies in device-to-device based mobile social networking in proximity (MSNP), In: *Proceedings of the 2016 IEEE Cyber Science and Technology Congress (CyberSciTech)*, Auckland, New Zealand, August 8–12, 2016.
4. Zhu, K., W. Li, and X. Fu. Rethinking routing information in mobile social networks: Location-based or social-based? *Computer Communications*, 2014; 42: 24–37.
5. Vastardis, N. and K. Yang. Mobile social networks: Architectures, social properties and key research challenges. *IEEE Communications Surveys & Tutorials*, 2013; 15(3): 1355–1371.
6. Zhu, Y., B. Xu, X. Shi, and Y. Wang. A survey of social-based routing in delay tolerant networks: Positive and negative social effects. *IEEE Communications Surveys & Tutorials*, 2013; 15(1): 387–401.
7. Asgari, F., V. Gauthier, and M. Becker. A survey on human mobility and its applications. *arXiv Preprint* arXiv:1307.0814, 2013.

8. Wei, K., X. Liang, and K. Xu. A survey of social-aware routing protocols in delay tolerant networks: Applications, taxonomy and design-related issues. *IEEE Communications Surveys & Tutorials*, 2014; 16(1): 556–578.
9. Aynaud, T., E. Fleury, J.L. Guillaume, and Q. Wang. Communities in evolving networks: Definitions, detection and analysis techniques. In *Dynamics On and Of Complex Networks*, Vol. 2, Ganguly N., Mukherjee, A., Mitra, B., Peruani, F., and Choudhury, M. (Eds.). Springer, New York, pp. 159–200, 2013.
10. Mei, A. and J. Stefa. Give2get: Forwarding in social mobile wireless networks of selfish individuals, In: *Proceedings of the IEEE 30th International Conference on Distributed Computing Systems (ICDCS)*, Genova, Italy, June 21–25, 2010.
11. Zhu, K., W. Li, and X. Fu. SMART: A social-and mobile-aware routing strategy for disruption-tolerant networks. *IEEE Transactions on Vehicular Technology*, 2014; 63(7): 3423–3434.
12. Wu, J., M. Xiao, and L. Huang. Homing spread: Community home-based multi-copy routing in mobile social networks. In: *Proceedings of INFOCOM*, Turin, Italy, April 14–19, 2013.
13. Xiao, J.M., J. Wu, and L. Huang. Community-aware opportunistic routing in mobile social networks. *IEEE Transactions on Computers*, 2014; 63(7): 1682–1695.
14. Abdelkader, T., K. Naik, A. Nayak, N. Goel, and V. Srivastava. SGBR: A routing protocol for delay tolerant networks using social grouping. *IEEE Transactions on Parallel and Distributed Systems*, 2013; 24(12): 2472–2481.
15. Hu, X., T.H.S. Chu, V. Leung, E.C.H. Ngai, P. Kruchten, and H.C.B. Chan. A survey on mobile social networks: Applications, platforms, system architectures, and future research directions. *IEEE Communications Surveys and Tutorials*, 2015; 17(3): 1557–1581.
16. Vazifehdan, J., R.V. Prasad, and I. Niemegeers. Energy-efficient reliable routing considering residual energy in wireless ad hoc networks. *IEEE Transactions on Mobile Computing*, 2014; 13(2): 434–447.
17. Lim, K.-W., W.-S. Jung, H. Kim, J. Han, and Y.-B. Ko. Enhanced power management for Wi-Fi direct. In: *Proceedings of IEEE Wireless Communications and Networking Conference*, Shanghai, China, April 7–10, 2013.
18. Cao, Y. and Z. Sun. Routing in delay/disruption tolerant networks: A taxonomy, survey and challenges. *IEEE Communications Surveys & Tutorials*, 2013; 15(2): 654–677.
19. Yuan, Q., I. Cardei, and J. Wu. An efficient prediction-based routing in disruption-tolerant networks. *IEEE Transactions on Parallel and Distributed Systems*, 2012; 23(1): 19–31.
20. Mitra, B., L. Tabourier, and C. Roth. Intrinsically dynamic network communities. *Computer Networks*, 2012; 56(3): 1041–1053.
21. Wang, Y., X. Jia, Q. Jin, and J. Ma. QuaCentive: A quality-aware incentive mechanism in mobile crowdsourced sensing (MCS). *Journal of Supercomputing*, 2016; 72: 2924–2941.

22. Nordström, E., C. Rohner, and P. Gunningberg. Haggle: Opportunistic mobile content sharing using search. *Computer Communications*, 2014; 48: 121–132.
23. Hameed, S., A. Wolf, K. Zhu, and X. Fu. Evaluation of human altruism using a DTN-based mobile social network application. In: *Proceedings of the 5th workshop on Digital Social Networks (DSN)*, Braunschweig, Germany, September 16–21, 2012.

4

PROFILE MATCHING AND RECOMMENDATION SYSTEMS IN PROXIMITY SERVICE

4.1 Introduction

Recently, the explosive growth of mobile-connected and location-aware devices enables various proximity services (ProSe). In general, there exist two paradigms enabling the proximity awareness: over-the-top (OTT) and device-to-device (D2D) [1]. In the OTT mode, a centralized server located in the cloud receives periodical location updates from users' mobile devices (e.g., using GPS). The server then determines proximity based on location updates and interests. However, the constant location updates not only result in significant battery impact because of GPS power consumption and the periodical establishment of cellular connections, but also causes serious privacy problem.

With smartphones being equipped with various sensors and multiple wireless interfaces (e.g., Bluetooth, Wi-Fi, and LTE), users can sense, discover, and make new social interactions directly with physical-proximate mobile users, not only through the third-party centralized server, so-called D2D paradigm. Different from OTT mode, D2D schemes forego centralized processing in identifying relevancy matches, instead autonomously determining relevance at the device level by transmitting and monitoring for relevant attributes. This approach offers crucial privacy benefits. In addition, by keeping discovery on the device rather than in the cloud, it allows for user-level controls over what are shared.

Instead of replacing OTT-based ProSe, decentralized D2D mode is effectively complementary to OTT mode under the scenario of direct wireless communications range. If users are too far remote with each other, OTT is the only choice. Furthermore, using OTT, ProSe users can link to others' SNS (e.g., Facebook, Twitter, and Google+) to acquire others' profiles over the Internet and to enable common profile exchange. With this feature, proximity services are capable of performing common interest matchmaking and content recommendation for users.

Irrespective of OTT or D2D paradigm, with the rapid increase in huge and miscellaneous data, the first step toward effective proximity service is for mobile users to choose whom to interact with. As an example, Alice wants to have a small talk with nearby passengers at the airport. Since she can simultaneously interact with only one or a few persons, it is crucial for her to select those who can lead to the most meaningful social interactions. In brief, it has become more and more important to effectively search the desired information, and show appropriate contents on the relatively small screen of mobile terminals. Therefore, recommendation systems (RS) has developed rapidly in recent years to solve the issue. RS are widely used in various fields to provide personalized services, such as travel route recommendation, location and friend recommendation, and so on. Moreover, it is imperative for RS to take into account and leverage the users' behaviors and contexts in social networks to target at accurate prediction and recommendation. Among others, profile matching is one of the most important underlying functionalities for both D2D- and OTT-based proximity services.

This chapter focuses on the relevant frameworks, algorithms, and applications of profile matching and RS in D2D- and OTT-based proximity services. The contents are divided into two major parts: the first part investigates the profile matching in mobile social networking in proximity (MSNP); the second summarizes the latest works on RS in OTT-based mobile social networks.

The remainder of this chapter is organized as follows: Section 4.2 presents the overview of profile matching in D2D-based proximity service (especially in MSNP). Section 4.3 discusses the prevalent methods about RS and the emerging location RS for location-based social networks (LBSNs). Finally, we briefly conclude this chapter.

4.2 Profile-Matching Schemes Device-to-Device-Based Proximity Services

In D2D-based ProSe, users can communicate directly with each other and enjoy some services, such as making some comments on other users' status, sharing photos and videos, playing games, and so on. This kind of applications can provide users more opportunities to discover and make new social interactions within some public places such as airports, bars, or other social spots. In particular, when smart devices communicate with others via P2P communications for socialization purposes, lubricating social interaction is one of the most important MSNP applications. When people meet new friends or attend social events, they may be eager to find out some common topics or backgrounds for initiating conversations with others. For example, a group of strangers may want to find common hobbies, friends, or countries they visited before to chat, and a group of students may want to know the common courses they have taken to discuss. Observing the rapid growth of smart mobile devices, it is possible to greatly improve such social interactions or social experiences.

Basically, a profile characterizing the individual's attributes is stored in the associated mobile devices. For example, a profile may look like interest = {NBA, swimming}, course-taken = {linear algebra, operating system, C language}, country-visited = {China, USA, Slovak, Austria}. Here, *interest, course-taken,* and *country-visited* are called attribute profiles, and *NBA, swimming,* and so on are called attribute items. Common profile matching (CPM) refers to the need of finding some common attributes of a group of users in a local region.

Advances in sensing and tracking technology enable proximity services applications but they also create significant privacy risks. With a third-party server, by submitting location-based service queries to the server, users can enjoy the convenience provided by proximity service. However, since the third-party server which may not be trusted has all the information about users such as where they are at which time, what kind of queries they submit, what they are doing, and so on, he may track users in various ways or release their personal data to other (commercial or even malicious) parties. Thus, we need to pay more attention to user's privacy. Especially, a major challenge for profile matching is to ensure the privacy of personal profiles, which often contain highly sensitive information related to gender, interests, political tendency, health conditions, and so on. This challenge necessitates privacy-preserving profile matching, in which two users compare personal profiles without disclosing them to each other.

4.2.1 Bloom Filter-Based Profile Matching

In this chapter, we introduce several communication-efficient solutions to the issue of CPM based on basic and iterative Bloom filters. Since Bloom filter is a space-efficient data structure for comparing the items owned by two parties without publishing their entire attribute profiles. In addition, it can provide some privacy against eavesdroppers due to its property of one-way hashing. It is difficult to reconstruct the filter information without exhaustively searching the input space.

4.2.1.1 Basic Concepts about Bloom Filters Bloom filter [2] is a time- and space-efficient probabilistic data structure for testing whether an element is a member of a set. It is composed of a vector of m bits, each of which is initially set to "0." When adding a new element to the Bloom filter, the elements over k independent hash functions are computed to generate k hash values as the indices to the vector. The corresponding k entries are set to "1." To insert a set of n elements, this procedure is repeated n times until all the elements are encoded in the Bloom filter. During the procedure, if a bit is already set, we leave it as "1." To query an element against a given Bloom filter, the k hashing indices are computed: The element is a member of the set only if all the k corresponding bits are "1" in the vector. Figure 4.1 illustrates an example of inserting an *interest* attribute profile to a Bloom filter and two query examples.

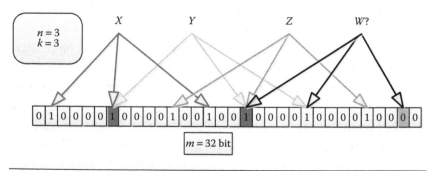

Figure 4.1 Overview of a Bloom filter.

As shown in Figure 4.1, the Bloom filter consists of a bit string of length 32. Three elements have been inserted, namely x, y, and z. Each of the elements have been hashed using $k = 3$ hash functions to bit positions in the bit string. The corresponding bits have been set to 1. Now, when an element not in the set, w, is looked up, it will be hashed using the same 3 hash functions into bit positions. In this case, one of the positions is zero and hence the Bloom filter reports correctly that the element is not in the set. It may happen that all the bit positions of an element report the corresponding bits have been set. When this occurs, the Bloom filter will erroneously report that the element is a member of the set.

As a probabilistic data structure, Bloom filters are subject to false positives, that is, they may mistakenly confirm the membership of a given element in lookup. The false-positive rate f is defined by the probability that all the corresponding k bits for any given element are "1" in the Bloom filter, although it is not really a member of the represented set.

Assuming that a hash function selects each position in a Bloom filter with equal probability, then the quantitative measurement of the false-positive rate is defined by the following formula:

$$f = \left(1 - \left(1 - \frac{1}{m}\right)^{k \cdot n}\right)^k \approx \left(1 - e^{-(k \cdot n/m)}\right)^k \qquad (4.1)$$

Obviously, to guarantee a given false-positive rate f, the size of a Bloom filter m should grow linearly with the size of a dataset n.

4.2.1.2 Existing Bloom Filter-Based Profile-Matching Schemes Two bloom filter-based profile-matching schemes are discussed in this subsection. The first is a distributed mobile communication system aiming to facilitate more effective social networking among strangers in physical proximity. Without requiring the Internet access, this system directly exchanges user information between two phones and performs matching locally through Bluetooth, and is implemented using Java ME. Hence, it can be deployed on most commercial off-the-shelf (COTS) mobile phones shipped with Bluetooth and Java. The second focuses on mobile D2D social networks (i.e., MSNP), in which CPM is one of the most important underlying components, especially for the scenario that a group of smartphone users meet in a small region (such as a ball room) and these users are interested in identifying the common attributes among them from their personal profiles efficiently via short-range (such as D2D) communications. For example, a group of strangers may want to find common hobbies, friends, or countries they visited before, and a group of students may want to know the common courses they have ever taken.

4.2.1.2.1 Common Profile Matching for One-to-One Scenario E-SmallTalker, a distributed mobile communications system, is proposed by Champion et al. [3], to facilitate social networking in physical proximity, which can automatically discover and suggest topics such as common interests for more significant conversations. The system is built on Bluetooth service discovery protocol (SDP) to exchange potential topics by customizing service attributes to publish nonservice-related information without establishing a connection. Specifically, a novel iterative Bloom filter protocol is designed, which encodes topics to fit in SDP attributes and achieves a low false-positive rate. In each round, a Bloom filter with some false-positive rate is published and a subset of common data items is computed; this smaller subset is then encoded in another new Bloom filter with much lower false-positive rate; eventually, the commonalities are reported with the desired rate f.

The detailed protocol is introduced as follows:

For simplicity, it assumes that there are only two parties: A and B. Let the interest datasets of A and B be $SetU_A$ and $SetU_B$, respectively. It is not difficult to extend the protocol specification to allow for multiple parties.

Step 1: Initially (round 0), both devices encode their own datasets, $SetU_A^0 = SetU_A$ and $SetU_B^0 = SetU_B$, in two static Bloom filters, BF_A^0 and BF_B^0, respectively, and publish them using a special static attribute ID in the SDP service record.

Step 2: In the $(r+1)$th round, user A first retrieves BF_B^r via Bluetooth SDP and then checks the membership of each data item in $SetU_A^r$ against BF_B^r to obtain a matching set $SetU_A^{r+1} \in SetU_A^r$. Next, A encodes $SetU_A^{r+1}$ into a new dynamic Bloom filter BF_A^{r+1}. Finally, A publishes BF_A^{r+1} and the current step number r using a new dynamic attribute ID calculated from B's Bluetooth ID (e.g., the last 4 bytes of a hash of B's ID), making it specific to B. Symmetrically, B takes the same action.

Step 3: In the following $(r+2)$th round, A first retrieves BF_B^r that is specially generated for A. Step 2 is repeated similarly to generate a new matching set $SetU_A^{r+2}$ and a new Bloom filter BF_A^{r+2} especially for B published with the same attribute ID as in Step 2. This process is repeated until the new matching set is empty or the same as that of the last round or the desired false-positive rate is reached. A dynamic attribute is removed from the SDP service record when a predefined lifetime is reached after the end of the above process.

In each round, an old hash function is replaced with a new independent hash function, which can iteratively eliminate the case in which two items have the same set of hash values. Thus, for any two honest parties A and B, the resulting $SetU_A^r$ in round r ($r \geq 1$) between A and B converges to $SetU_A \cap SetU_B$ as $r \to \infty$.

Between two strangers A and B, it is reasonable to assume that the intersection of their datasets is a proper subset of either original set, and the intersection is much smaller. As the dataset size n decreases, so does the Bloom filter size m when the false-positive rate f and number of hash functions k are fixed. In each round, when a new Bloom filter is constructed, we can either dynamically decrease the filter size m according to the current n and resulting f, or decrease f dramatically by keeping the filter size m.

4.2.1.2.2 Common Profile Matching for Many-to-Many Scenario For social interactions among multiparty users, that is, many-to-many social interactions, Reference 4 considers the CPM issue in a MSNP, where users are willing to cooperate to find some common attributes among them without privacy violation. Specifically, three versions of the CPM problem, namely all-common, β-common as well as top-γ-popular, are formulated to present solutions based on the iterative Bloom filters.

In any user group U, an attribute profile is associated with each individual. An attribute profile could include the hobbies, the countries visited, the courses taken, or the celebrities followed by the individual, and so on. Solving the problem of CPM for multiple user groups and attribute profiles can be done by repeating this for each user group and attribute profile. Let $U = \{u_1, u_2, \cdots, u_q\}$ and the attribute profile of user $u_1 \in U$ be P_i, $i = 1 \cdots q$. Each element in P_i is called an attribute item. Let $n_i = |P_i|$. The universal set of the attribute items is denoted by P that is assumed to be known by all users.

Three versions of the CPM problem are described as follows:

- All-common: The goal is to find the intersection of all users' attribute profiles, that is, $C_{all} = \cap_{i=1 \cdots q} P_i$.
- β-common: The goal is to find the set of all attribute items such that each attribute item is in at least β users' attribute profiles. For any $a \in P$, a membership function $\eta(a, P_i)$ is defined, such that $\eta(a, P_i) = 1$ if $a \in P_i$ and $\eta(a, P_i) = 0$ otherwise. The β-common set is defined as $C_{\beta\text{-com}} = \{a | \Sigma_{i=1 \cdots q} \eta(a, P_i) \geq \beta\}$.
- Top-γ-popular: The goal is to find the set of top-γ hottest attribute items that are shared by all users' profiles. The top-γ set is defined as

$$C_{\text{top-}\gamma} = \left\{ a \,\middle|\, \text{the value} \sum_{i=1 \cdots q} \eta(a, P_i) \text{ of } a \text{ is ranked top-}\gamma \right\}.$$

Definition 1. Given any attribute profile P_i, define $\text{BF}(P_i)$ to be the m-bit array obtained by inserting each attribute item in P_i into the Bloom filter. Given any set $S \subseteq P$, define $Qry(S, \text{BF}(P_i))$ to be the set containing each element $a \in S$ that returns a positive answer when querying the existence of a Bloom filter $\text{BF}(P_i)$, that is, $Qry(S, \text{BF}(P_i)) = \{a | (a \in S) \wedge (a \in \text{BF}(P_i))\}$. (Here, $a \in \text{BF}(P_i)$ means that for each hashed value of a, the corresponding bit in $\text{BF}(P_i)$ is 1.)

As shown in Figure 4.2, the iterative solution is detailed as follows. These steps are executed by each user u_i, $i = 1, \ldots, q$, concurrently:

1. User u_i computes its Bloom filter array $\text{BF}(P_i)$ and broadcasts $\text{BF}(P_i)$ to all other users.
2. User u_i collects the Bloom filter arrays $\text{BF}(P_j)$ of all other users $u_j, j \neq i$.

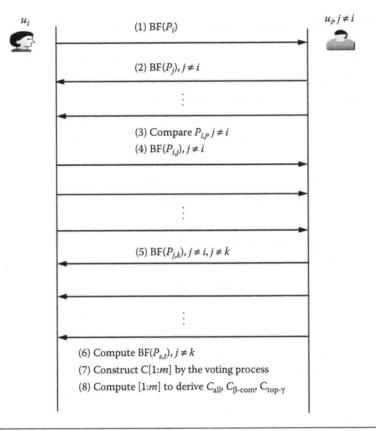

Figure 4.2 Workflow of two-iteration IBF solution. (From Tarkoma, S., C.E. Rothenberg, and E. Lagerspetz. *IEEE Comm. Surv. Tutor.*, 14, 131–155, 2012.)

3. For each $\mathrm{BF}(P_j)$ collected from u_j, u_i computes the set $P_{i,j} = Qry(P_i, \mathrm{BF}(P_j))$, which is an estimation of $P_i \cap P_j$. (It contains the set of attribute items that appear in P_i as well as in the Bloom filter $\mathrm{BF}(P_j)$).
4. In step 3, u_i already has the set $P_{i,j}$ for all $j \neq i$. Then, u_i computes $(q-1)$ Bloom filter arrays $\mathrm{BF}(P_j)$, $j \neq i$, and broadcasts these arrays to all other users.
5. User u_i collects $(q-1)$ Bloom filter arrays $\mathrm{BF}(P_{j,k})$ from each u_j, $j \neq i$, $k \neq j$. (There are totally $(q-1)^2$ arrays received.)
6. Combining the $(q-1)$ arrays in step 3 and the $(q-1)^2$ arrays in step 5, u_i now has $q \cdot (q-1)$ arrays, namely $\mathrm{BF}(P_{j,k})$ for all $s \neq t$. Then u_i computes the vector $\mathrm{BF}(P_{s,t}) = \mathrm{BF}(P_{s,t}) \wedge \mathrm{BF}(P_{t,s})$,

where the notation $\wedge$ is the bit-wise logical AND operator, obviously $\widetilde{BF}(P_{s,t}) = \widetilde{BF}(P_{t,s})$.

7. Next, u_i uses a voting process to construct an integer array $C[1:m]$. The value of each $C[d]$ is initially 0, $1 \le d \le m$. Intuitively, in the voting process, each user can cast one (and only one) vote to $C[d]$ if any hashed value of any attribute item i_{sd}. To calculate $C[d]$, Reference 4 adopted a tentative set T_d, which is set to $\emptyset$ initially. Then the d-th bit of each g $BF(P_{j,k})$ is checked. There are three cases:

a. If $BF(P_{j,k})[d] = 0$, do nothing.
b. If $BF(P_{j,k})[d] = 1$ and only one of u_j and u_k is in T_d, increase $C[d]$ by 1 and include one of the u_j and u_k, which is not in T_d.
c. If $BF(P_{j,k})[d] = 1$ and none of u_j and u_k is in T_d, increase $C[d]$ by 2 and include both u_j and u_k into T_d.

Note that the above process should be repeated for each $\widetilde{BF}(P_{j,k})$. This process avoids querying each $\widetilde{BF}(P_{j,k})$ by the universal set P to reduce computation cost. Compared to the basic Bloom filter solution, this step may slightly increase the false positive rate of the final results. To summarize, this voting process helps to quickly estimate the number of votes that $C[d]$ receives. Using array $C[1:m]$, u_i derives its answers to the three CPM problems as follows:

- All-common: Converts $C[1:m]$ to an array $\widetilde{C}[1:m]$ such that $\widetilde{C}[d] = 1$ if $C[d] = q$ (i.e., q votes) and $\widetilde{C}[d] = 0$ otherwise. The answer of C_{all} is $Qry(P, \widetilde{C})$.
- β-common: Converts $C[1:m]$ to an array $\widetilde{C}[1:m]$ such that $\widetilde{C}[d] = 1$ if $C[d] \ge \beta$ (i.e., β votes) and $\widetilde{C}[d] = 0$ otherwise. The answer of $C_{\beta\text{-com}}$ is $Qry(P, \widetilde{C})$.
- Top-γ-popular: The answer of $C_{\text{top-}\gamma}$ can be obtained by repeating the β-common problem by setting $\beta = q$ and gradually collecting the attribute items by decreasing the value of β by 1 each time, until γ attribute items are collected or β is equal to 1.

4.2.2 Privacy-Preserving Profile Matching

As described in the preceding subsection, Bloom filters can provide some privacy against passive attacks (e.g., eavesdroppers) as Bloom filters use one-way hashing to encode topics (contacts, interests, etc.).

This makes it very difficult to reconstruct the original topics in a filter without exhaustively searching the topic space. However, it fails to provide privacy against active attackers.

People have growing privacy concerns for disclosing personal profiles to arbitrary persons in physical proximity before deciding to interact with them. Although similar privacy concerns also exist in online social networking, preserving users' profile privacy is more urgent in proximity services, as attackers can directly associate obtained personal profiles with real persons nearby and then launch more targeted attacks.

Most of the existing proximity services employ third-party centralized servers, which are always trusted and acting as matching centers to serve users (keeps all users' profiles and computes the similarity between users when needed). Specifically, after each user sending his or her profile attributes information to the server, the server replies users with the matching result to indicate the potential *friends*. The servers need to know the users' profile attributes information to perform the matching process, so it is thus much dangerous when the servers are compromised. Social serendipity provided mobile users more opportunities to make social interactions with potential friends nearby.

To avoid the third-party servers, over recent years, many cryptographic tools-based solutions have been proposed for direct interactions among peers. There are two mainstreams of approaches to solving the privacy-preserving profile-based friend-matching problem, as shown in Figure 4.3 [5].

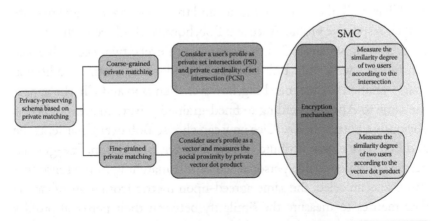

Figure 4.3 Two mainstreams of approaches of privacy-preserving schema based on private matching.

The first category called coarse-grained private matching assuming that each user's personal profile comprises multiple attributes chosen from a public set of attributes such as various interests, friends, or disease symptoms, it could provide well-designed protocols to privately match users' profiles based on private set intersection (PSI) and private cardinality of set intersection (PCSI) enabling two users to find the intersection or intersection cardinality of their profiles without disclosing additional information to either party; the second category called fine-grained private matching considers a user's profile as a vector and measures the social proximity by private vector dot product. Both categories rely on secure multiparty computation (SMC). The goal of secure multiparty computation is to evaluate a function (or algorithm) that takes an input value of each participating party. At the end of the protocol, each participant should know the result. However, none of the participants should know more about the other participants' input values than what can be derived from the result and their own input value.

In fine-grained private matching, every attribute value is an integer in [0, maxvalue] and indicates the level of interest from no interest (0) to extremely high interest (maxvalue). Every personal profile is then defined as a set of attribute values, each corresponding to a unique attribute in the public attribute set. Fine-grained personal profiles have significant advantages over traditional coarse-grained ones comprising only interested attributes from a public attribute set. First, fine-grained personal profiles enable finer differentiation among the users having different levels of interest in the same attribute. For example, Alice, Bob, and Charlie all like watching movies and thus have *movie* as an attribute of their respective profile. Alice and Bob, however, both go to the cinema twice a week, whereas Charlie does so once every two weeks. If Alice can interact with only Bob or Charlie, under the existing traditional coarse-grained private matching, however, then Bob and Charlie appear the same to Alice. According to fined-grained private matching, Bob is obviously a better choice. Alice now can choose Bob over Charlie, as she and Bob have closer attribute values for *movie*. In addition, fine-grained personal profiles enable personalized profile matching in the sense that two users can select the same agreed-upon metric from a set of candidate metrics to measure the similarity between their personal profiles or even different metrics according to their individual needs. Therefore, this section mainly emphasizes fined-grained private matching.

Reference 6 designed a suite of novel fine-grained private matching protocols. The protocols enable two users to perform profile matching without disclosing any information about their profiles beyond the comparison result. In contrast to existing coarse-grained private matching schemes for proximity services, the protocols allow finer differentiation between proximity services users and can support a wide range of matching metrics at different privacy levels.

Consider Alice with profile $u = (u_1, u_2, \cdots, u_d)$ and Bob with profile $v = (v_1, v_2, \cdots, v_d)$ as two exemplary users of the same proximity services application from here on. Assume that Alice wants to find someone to chat with, for example, when waiting for the flight to depart. As the first step (neighbor discovery), she broadcasts a chatting request via the MSNP application on her smartphone to discover proximate users of the same proximity services application. Suppose that she receives multiple responses including one from Bob who may also simultaneously respond to other persons. Due to time constraints or other reasons, both Alice and Bob can only interact with one stranger whose profile best matches hers or his. The next step (profile matching) is thus for Alice (or Bob) to compare her profile with those of others who responded to her (or whom she responded to).

In particular, let F denote a set of candidate matching metrics defined by the proximity services application developer, where each $f \in F$ is a function over two personal profiles that measures their similarity. The proposed private-matching protocols allow Alice and Bob to either negotiate one common metric from F or choose different metrics according to their individual needs. The authors focus on the latter, more general case henceforth, in which private matching can be viewed as two independent protocol executions, with each user initiating the protocol once according to her/his chosen metric. Assume that Alice chooses a matching metric $f \in F$ and runs the privacy matching protocol with Bob to compute $f(u, v)$. According to the amount of information disclosed during the protocol execution, the following two privacy levels can be defined from Alice's viewpoint, which can also be equivalently defined from Bob's viewpoint for her chosen matching metric:

- Definition 1. Level-I privacy: When the protocol ends, Alice only learns $f(u, v)$ and Bob only learns f.

- Definition 2. Level-II privacy: When the protocol ends, Alice only learns $f(u,v)$ and Bob learns nothing.

For all two privacy levels, neither Alice nor Bob learns the other's personal profile. The authors introduced a suite of private-matching protocols satisfying one of the three privacy levels. Besides privacy guarantees, other design objectives include small communication and computation overhead, which can translate into the total energy consumption and matching time, and thus are crucial for resource-constrained mobile devices and the usability of proximity services.

The proposed protocols rely on the Paillier cryptosystem, in which it is assumed that every proximity services user has a unique Paillier public/private key pair, generated via a function module of the proximity service application. How the keys generated and used for encryption and decryption are briefed as follows to help illustrate and understand the protocols:

Key generation: An entity chooses two primes p and q and compute $N = p \cdot q$ and $\lambda = \text{lcm}\,(p-1, q-1)$. It then selects a random $g \in Z_{N^2}^*$ such that $\gcd = (L \cdot (g^{\lambda} \bmod N^2), N) = 1$, where $L(x) = (x-1)/N$. The entity's Paillier public and private keys are (N, g) and λ respectively.

Encryption: Let $m \in Z_N$ be a plaintext and $r \in Z_N$ be a random number. The ciphertext is given by $E(m \bmod N, r \bmod N) = g^m r^N \bmod N^2$, where $E(\cdot)$ denotes the Paillier encryption operation on two integers modulo N.

Decryption: Given a cipher text $c \in Z_{N^2}$ the corresponding plaintext can be derived as $D(C) = (L(c^{\lambda} \bmod N^2)/L(g^{\lambda} \bmod N^2)) \bmod N$, where $D(\cdot)$ denotes the Paillier decryption operation.

The Paillier's cryptosystem has two very useful properties:

- Homomorphic: For any $m_1, m_2, r_1, r_2 \in Z_N$, the following equation holds:

$$E(m_1, r_1) \cdot E(m_2, r_2) = E(m_1 + m_2, r_1 r_2) \bmod N^2,$$

$$E^{m_2}(m_1, r_1) = E^{m_2}(m_1 m_2, r_1^{m_2}) \bmod N^2$$

- Self-blinding: $E(m_1, r_1) \cdot r_1^N \bmod N^2 = E(m_1, r_1 r_2)$, which implies that any ciphertext can be changed to another without knowing the plaintext.

4.2.2.1 Protocol 1 for Level-I Privacy Protocol 1 is designed for the l_1 distance as the matching metric. The l_1 distance (also called the Manhattan distance) is computed by summing the absolute value of the element-wise subtraction of two profiles and is a special case of the more general l_α distance defined as

$$l_\alpha(\hat{u},\hat{v}) = \left(\sum_{i=1}^{d}|v_i - u_i|^\alpha\right)^{1/\alpha},$$

where $\alpha \geq 1$. When $\alpha = 1$, the following equation holds:

$$l_1(\hat{u},\hat{v}) = \sum_{i=1}^{d}|v_i - u_i| = \sum_{i=1}^{(\gamma-1)d}|\hat{u}_i - \hat{v}_i| = \sum_{i=1}^{(\gamma-1)d}\left|\hat{u}_i - \hat{v}_i\right|^2 = l_2^2(\hat{u},\hat{v})$$

Moreover:

$$l_2^2(\hat{u},\hat{v}) = \sum_{i=1}^{(\gamma-1)d}\left|\hat{u}_i - \hat{v}_i\right|^2 = \sum_{i=1}^{(\gamma-1)d}\hat{u}_i^2 - 2\sum_{i=1}^{(\gamma-1)d}\hat{u}_i\,\hat{v}_i + \sum_{i=1}^{(\lambda-1)d}\hat{v}_i^2$$

Protocol 1 Details:

1. Alice does the following in sequence:
   a. Construct a vector $\hat{u} = h(u) = (h(u_1),\cdots,h(u_d))s = (\hat{u}_1,\cdots, \hat{u}_{(\gamma-1)d})$, $j \in [1,((\gamma-1)d)]$, where $\hat{u}_j$ is equal to one for every $j \in \tau_u = \{j\,|\,(i-1)(\gamma-1) < j \leq (i-1)(\gamma-1)+u_i, 1 \leq i \leq d\}$, and zero otherwise.
   b. Compute $E(\hat{u}_j, r_j)$ for every $j \in [1,((\gamma-1)d)]$ using her public key (N,g).
   c. Send $\{E(\hat{u}_j, r_j)\}_{j=1}^{(\gamma-1)d}$ and her public key (N,g) to Bob.
2. Bob does the following after receiving Alice's message:
   a. Construct a vector $\hat{v} = h(v) = (h(v_1),\cdots,h(v_d)) = (\hat{v}_1,\cdots, \hat{v}_{(\gamma-1)d})$, $j \in [1,((\gamma-1)d)]$, where $\hat{v}_j$ is equal to one for every $j \in \tau_v = \{j\,|\,(i-1)(\gamma-1) < j \leq (i-1)(\lambda-1)+v_i, 1 \leq i \leq d\}$ and zero otherwise.
   b. Compute $E(\hat{u},\hat{v},s) = E(\sum_{j\in\tau_v}\hat{u}_j, \prod_{j\in\tau_v}r_j) = \prod_{j\in\tau_v}E(\hat{u}_j, r_j)$ mod N^2, $s = \prod_{j\in\tau_v}r_j$
   c. Compute $E((N-2)\hat{u}\cdot\hat{v},s) = E^{N-2}(\hat{u}\cdot\hat{v}, \prod_{j\in\tau_v}r_j)$mod N^2, where $s = (\prod_{j\in\tau_v}r_j)^{N-2}$ mod N.
   d. Compute $E(\sum_{j=1}^{(\gamma-1)d}\hat{v}_j^2, r)$ with a random $r \in Z_N$.
   e. Compute $E(\sum_{j=1}^{(\gamma-1)d}\hat{v}_j^2 - 2\hat{u}\cdot\hat{v}, rs) = E(\sum_{j=1}^{(\gamma-1)d}\hat{v}_j^2, r)\cdot E((N-2)\hat{u}\cdot\hat{v},s)$ mod N^2.
   Finally, Bob returns $E(\sum_{j=1}^{(\gamma-1)d}\hat{v}_j^2 - 2\hat{u}\cdot\hat{v}, rs)$ to Alice.

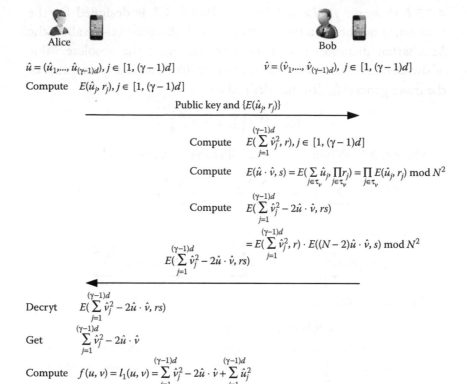

Figure 4.4 The detailed operations of Protocol 1. (From Zhang, R., Zhang, Y., Sun, J. et al. Fine-grained private matching for proximity-based mobile social networking. In *Proceedings of the IEEE INFOCOM*, Orlando, FL, March 25–30, 2012.)

3. Alice decrypts $E(\sum_{j=1}^{(\gamma-1)d} \hat{v}_j^2 - 2\hat{u}\cdot\hat{v}, rs)$, gets $\sum_{j=1}^{(\gamma-1)d} \hat{v}_j^2 - 2\hat{u}\cdot\hat{v}$, and then computes $l_1(u,v) = \sum_{j=1}^{(\gamma-1)d} \hat{v}_j^2 - 2\hat{u}\cdot\hat{v} + \sum_{j=1}^{(\gamma-1)d} \hat{u}_j^2$

The detailed operations of Protocol 1 are as shown as Figure 4.4.

4.2.2.2 Protocol 2 for Level-II Privacy Definition: A function $f(u,v)$ is additively separable if it can be written as $f(u,v) = \sum_{i=1}^{d} f_i(u_i, v_i)$.

In protocol 1, it is needed to compute $l_1(u,v) = \sum_{j=1}^{(\gamma-1)d} \hat{v}_j^2 - 2\hat{u}\cdot\hat{v} + \sum_{j=1}^{(\gamma-1)d} \hat{u}_j^2$. Specifically, Alice needs to compute $\sum_{j=1}^{(\gamma-1)d} \hat{u}_j^2$ and Bob needs to compute $\sum_{j=1}^{(\gamma-1)d} \hat{v}_j^2 - 2\hat{u}\cdot\hat{v}$.

However, now, assume that $f(u,v)$ is additively separable and $f(u,v) = \sum_{i=1}^{d} f_i(u_i, v_i) = \sum_{j\in[0,\gamma d]} \tilde{u}_j = \sum_{j=1}^{\gamma d} \tilde{u}_j \tilde{v}_j = \tilde{u}\cdot\tilde{v}$.

So, Protocol 2 only needs to compute $\tilde{u}\cdot\tilde{v}$, which can reduce the complexity of computing.

Protocol 2 Details:

1. Alice does the following in sequence:
   a. Construct a vector $\tilde{u} = h(u) = (h(u_1), \cdots, h(u_d)) = (\tilde{u}_1, \cdots, \tilde{u}_{\gamma d})$, where $\tilde{u}_j = f_i(u_i, k), i = [(j-1)/\gamma] + 1,$ and $k = (j-1) \bmod \gamma$, for all $j \in [1, \gamma d]$.
   b. Compute $E(\tilde{u}_j, r_j)$ for every $j \in [1, d]$ using her public key (N, g). Send $\{E(\tilde{u}_j, r_j)\}_{j=1}^{(\gamma-1)d}$ and her public key to Bob.
2. Bob does the following after receiving Alice's message:
   a. Construct a vector $\tilde{v} = h(v) = (h(v_1), \cdots, h(v_d)) = (\tilde{u}_1, \cdots, \tilde{v}_{\gamma d})$, $j \in [1, \gamma d]$, where $\tilde{v}_j$ is equal to one for every $j \in \tau_v = \{j \mid j = (i-1)\gamma + v_i + 1, 1 \leq i \leq d\}$, and zero otherwise. Compute $E(\tilde{u}, \tilde{v}, \prod_{j \in \tau_v} r_j) = \prod_{j \in \tau_v} E(\tilde{u}_j, r_j) \bmod N^2$. (10)
   b. Compute $E(\tilde{u} \cdot \tilde{v}, r_B \prod_{j \in \tau_v} r_j) = \prod_{j \in \tau_v} E(\tilde{u} \cdot \tilde{v}, \prod_{j \in \tau_v} r_j) r_B^N \bmod N^2$. (11)
   c. Finally, Bob returns $E(\tilde{u} \cdot \tilde{v}, r_B \prod_{j \in \tau_v} r_j)$ to Alice.
3. Alice decrypts $E(\tilde{u} \cdot \tilde{v}, r_B \prod_{j \in \tau_v} r_j)$ and finally gets $\tilde{u} \cdot \tilde{v}$, then compute $f(u, v) = \tilde{u} \cdot \tilde{v}$.

The detailed operations of Protocol 2 are shown as Figure 4.5.

4.2.3 Light-Weighted Privacy-Preserving Profile Matching

Although the above fine-grained approaches for proximity service could effectively enforce privacy-preserving profile matching among nearby users without the support of the trusted third party, they have the following disadvantage: Always rely on heavy public-key cryptosystem and homomorphic encryption. Usually, multiple rounds of interactions are required to perform the public key exchange and private matching between each pair of parties, which incurs high communication and computation costs to resource-limited mobile terminals in MSNP. Based on nonhomomorphic encryption-based privacy-preserving scalar product computation, an efficient weight-based private matching (EWPM) protocol was proposed to employ confusion matrix transformation algorithm instead of computation-consuming homomorphic cryptographic system, to achieve the privacy preserving goal with a higher efficiency [7]. The main weakpoint in EWPM is that the inferred matching value does not have strict semantic meaning, and can only roughly represent the profile similarity among

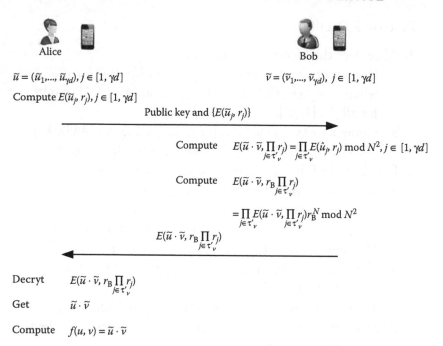

Figure 4.5 The detailed operations of Protocol 2. (From Zhang, R., Zhang, Y., Sun, J. et al. Fine-grained private matching for proximity-based mobile social networking. In *Proceedings of the IEEE INFOCOM*, Orlando, FL, March 25–30, 2012.)

users. Therefore a lightweighted fine-grained privacy-preserving profile-matching mechanism for D2D based MSNP, Light-weighted fIne-grained Privacy-Preserving Profile matching mechanism (LIP3) was proposed by Wang et al. [8], which, in comparison with the existing CMT schemes (e.g., EWPM), can provide strict and accurate profile matching value-cosine similarity result among individuals.

4.2.3.1 System Architecture of LIP3   Commonly, each user's interest profile is defined from a public attribute set consisting of n attributes. The number of n may range from several tens to several hundreds. Each attribute is associated with a user-specific integer value $i \in [1,l]$ (called as the weight of an attribute) indicating the corresponding user's association with this attribute. The higher the value of this attribute is, the more interest the user has in the attribute. Usually, letting l equal to 10 may be sufficient to differentiate user's interest level.

Suppose two users Alice' and Bob's interest sets are characterized as the following profile vectors: $\vec{u}_A = (u_{A1}, u_{A2}, \cdots, u_{An})$ and

$\vec{u}_B = (u_{B1}, u_{B2}, \cdots, u_{Bn})$, respectively. Each individual can modify her/his profile later on when needed.

The most widely applied similarity metric to infer the matching value between individuals, say Alice and Bob, is cosine similarity:

$$\text{similarity}(A, B) = \frac{\vec{u}_A \cdot \vec{u}_B}{\|\vec{u}_A\| \cdot \|\vec{u}_B\|} = \frac{\sum_{i=1}^{l} u_{Ai} \cdot u_{Bi}}{\sqrt{\sum_{i=1}^{l}(u_{Ai})^2} \cdot \sqrt{\sum_{i=1}^{l}(u_{Bi})^2}} \quad (4.2)$$

A ProSe session usually involves two users and consists of three phases. First, two users need to discover each other in the neighbor-discovery phase. Second, they need to compare their personal profiles in the matching phase. Third, two matching users enter the interaction phase for real information exchange.

Figure 4.6 illustrates the system architecture of the proposed privacy-preserving profile-matching scheme LIP3, which is composed of two mobile users with specific interest profiles, and several components that facilitate the similarity calculation in LIP3 scheme.

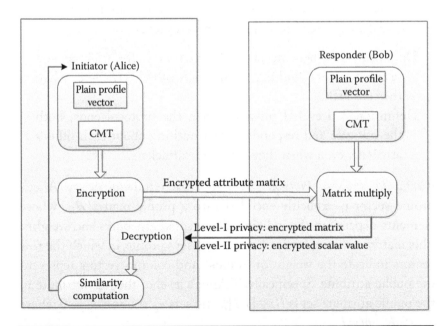

Figure 4.6 System architecture LIP3 scheme. (From Wang, Y., Cheng, X., Jin, Q., and Ma, J. LIP3: Privacy-preserving profile-matching mechanism for mobile social networks in proximity. In *Proceedings of the 5th International Symposium on Trust, Security and Privacy for Emerging Applications (TSP)*, Zhangjiajie, China, November 18–20, 2015.)

The plain profile vectors in the initiator (say Alice) and responder (say Bob) are firstly transformed into corresponding attribute matrices through CMT, which can completely describe users' profiles. Then the initiator encrypts her attribute matrix and sends it to responder. The responder Bob will calculate the multiplication between the received encrypted matrix and her attribute matrix. The obtained matrix (in Level-I privacy) or the scalar value (in Level-II privacy) will be sent to initiator who then decrypts and obtains the cosine similarity between the initiator and responder. Note that, in our proposal, the module of responder's profile vector should be explicitly sent to initiators.

There exist attacks from outside adversaries, such as eavesdropping the wireless communication channel or modifying, replying, and injecting the captured messages. We assume that the users in our protocol are honest-but-curious (HBC), which means they will comply with the protocol but they are curious about other users and try to learn more information than allowed. Furthermore, some users may be inside attackers who monitor the matching process and obtain the intermediate results without complying the agreements. They try to infer users' profiles through these observations. Based on the adversary models, we define the following privacy level:

Definition 1. When the protocol ends, both the initiator and responder learn nothing about each other's attribute, when they are HBC.

Definition 2. Level-II privacy: when the protocol ends, both the initiator and responder learn nothing about each other's attribute, even when they are inside attackers.

4.2.3.2 The Proposed Protocols In LIP3, each individual, say Alice's profile vector is explicitly encoded into a profile matrix $A_{l\times n}$ whose elements depend on the individual's personal attributes and weights. This matrix can completely describe an user's profile, in which the row vectors indicate the weight of interest and column vectors represent the public attribute. Specifically, if Alice's level of the j-th attribute in the public attribute set is i ($i \in [1,l]$), she sets $a_{ij} = 1$ and $a_{mj} = 0$ where $a_{mj} \in A_{l\times n}, m \neq i$.

4.2.3.2.1 Protocol Satisfying the Level-I Privacy LIP3 explicitly defines a weight matrix $W_{l\times l} = (w_{ij})_{l\times l}$ through which the accurate cosine similarity can be inferred, without revealing individuals' private profiles. Specifically, the element w_{ij} is given as the equation

$$(w_{ij})_{l \times l} = i \cdot j \qquad (4.3)$$

In LIP3, the initiator (Alice) and responder (Bob) respectively hold the attribute matrices $A_{l \times n}$ and $B_{l \times n}$, which are transformed from both the users' plain profile vectors. p and q are two large primes. $C_{l \times n}$ and $R_{l \times n}$ are two matrixes used for hiding personal information. The vector $\vec{k}$ is the secret key kept by initiator to decrypt the original results.

The detailed procedure of LIP3 is given as follows:

1. The initiator initializes her personal profile according to Algorithm 1, which can be run offline, and broadcasts her friend discovery request to others. When Algorithm 1 ends, the initiator keeps $\bar{k} = [k_1, k_2, \cdots, k_l]$ and q secretly and sends $A_{l \times n}^*$ to the responder.

Algorithm 1

Input: Initiator's attribute matrix $A_{l \times n}$
Output: Encrypted matrix $A_{l \times n}^*$
Choose two large primes p and q, where $|p| = 256$ and $q > (n + 1) \cdot l^2 \cdot p$;
Randomly generate two matrixes $C_{l \times n}$ and $R_{l \times n}$, $\forall c_{ij} \in C_{l \times n}$, $\forall r_{ij} \in R_{l \times n}$, $\sum_{i=1}^{l} \left(\sum_{j=1}^{n} c_{ij} \right) < (p - l \cdot n)$, $|r_{ij} \cdot q| \approx 1024$;

$\forall a_{ij} \in A_{l \times n}, \forall a_{ij}^* \in A_{l \times n}^*, k_i \in \vec{k}$,

The following operations are done:
FOR (i=1; $i \le l$; i++) **DO**
$k_i = 0$;
FOR (j=1; $j \le n$; j++) **DO**
IF $a_{ij} = 1$ **THEN** $a_{ij}^* = p + c_{ij} + r_{ij} \cdot q$;
ELSE $a_{ij}^* = c_{ij} + r_{ij} \cdot q$;
ENDIF
$k_i = k_i + r_{ij} \cdot q - c_{ij}$;
ENDFOR
ENDFOR

2. On receiving $A^*_{l \times n}$, the responder computes $D_{l \times l} = (d_{ij})_{l \times l}$ according to Algorithm 2 and sends $D_{l \times l}$ to the initiator.

Algorithm 2

Input: $A^*_{l \times n}$, $B_{l \times n}$
Output: $D_{l \times l} = (t_{ij})_{l \times l}$

The following operations are done
FOR (i =1; i ≤ l; i++) **DO**
    FOR (j=1; j ≤ l; j++) **DO**
        $d_{ij} = 0$;
        FOR (m=1; m ≤ n; m++) **DO**
            IF ($b_{im} = 1$) **THEN** $d_{ij} = d_{ij} + p \cdot a^*_{im}$;
      ELSE $d_{ij} = d_{ij} + a^*_{im}$;
      ENDIF
        ENDFOR
    ENDFOR
ENDFOR

---

3. Initiator operates the following steps: $T_{l \times l} = (t_{ij})_{l \times l} = (d_{ij} + k_i)$ mod q. It is shown that the above equation implies that $T_{l \times l} = A_{l \times n} \times B^T_{l \times n}$. Moreover, let $T^*_{l \times l} = (t^*_{ij})_{l \times l}$ and $t^*_{ij} = (t_{ij} - (t_{ij} \bmod p^2))/p^2$.

4. The initiator considers the corresponding weights and computes

$$H_{l \times l} = W_{l \times l} .^* T^*_{l \times l} = \begin{pmatrix} w_{11} \cdot t^*_{11} & \cdots & w_{1l} \cdot t^*_{1l} \\ \vdots & \cdots & \vdots \\ w_{l1} \cdot t^*_{l1} & \cdots & w_{ll} \cdot t^*_{ll} \end{pmatrix}$$

in which the operator .* denotes multiplying the corresponding elements of two matrices $W_{l \times l}$ and $T^*_{l \times l}$ to obtain the matrix $H_{l \times l}$.

5. The initiator calculates the matching value $\tau = \sum_{i=1}^{l} \sum_{j=1}^{l} h_{ij}$, which equals the value $\vec{u}_A \cdot \vec{u}_B$, and then the cosine similarity between two interacting individuals can be obtained.

4.2.3.2.2 Protocol Satisfying Level-II Privacy Note that the above procedures can only satisfy the privacy level I. In order to resist the malicious users to achieve the Level-II privacy, instead of directly sending the matrix $D_{l\times l}$ to initiator, the responder can send the scalar value $\sigma = \sum_{i=1}^{l}\sum_{j=1}^{l} t_{ij}$ to initiator, in which $(t_{ij})_{l\times l} = D_{l\times l} \cdot^{*} W_{l\times l}$. Then, on receiving the message σ, the initiator decrypts the matching value τ via the following operators:

$$\tau_1 = \left(\sigma + l\left(\sum_{i=1}^{l} k_i\right)\right) \bmod q; \quad \tau = \vec{u}_A \cdot \vec{u}_B = \frac{\tau_1 - (\tau_1 \bmod p^2)}{p^2}$$

Then the cosine similarity between Alice and Bob can be obtained:
similarity$(A,B) = \left(\tau / \|\vec{u}_A\| \cdot \|\vec{u}_B\|\right)$

The following simple pedagogical example verifies the property of LIP3. Assuming three users Alice, Bob, and Charles are within the communication range. The number of attributes n, is 3, and the maximal attribute value l, is 2. Suppose Alice is the initiator, with profile $\vec{u}_A = (1,1,2)$, translate to matrix is

$$A_{2\times3} = \begin{pmatrix} 1 & 1 & 0 \\ 0 & 0 & 1 \end{pmatrix}$$

Bob and Charles are the responders, and the profiles of Bob and Charles are $\vec{u}_B = (1,1,1)$, matrix

$$B_{2\times3} = \begin{pmatrix} 1 & 1 & 1 \\ 0 & 0 & 0 \end{pmatrix}$$

$\vec{u}_C = (1,2,1)$ matrix

$$C_{2\times3} = \begin{pmatrix} 1 & 0 & 1 \\ 0 & 1 & 0 \end{pmatrix}$$

respectively. Since the calculation process between Alice and Bob is similar to that of Alice and Charles, we just describe the process between Alice and Bob in detail, and give the matching value between Alice and Charles directly. Similarly as [7], we can get

$$T^{*}_{2\times2} = \begin{pmatrix} 2 & 0 \\ 1 & 0 \end{pmatrix}$$

which numerically equals the result as $A_{2\times3} \times B^{T}_{2\times3}$.

Then, according to Equation 4.3, we obtain

$$W_{2\times2} = \begin{pmatrix} 1 & 2 \\ 2 & 4 \end{pmatrix}$$

then

$$H_{l\times l} = W_{2\times2} \cdot^* T_{2\times2}^* = \begin{pmatrix} 2 & 0 \\ 2 & 0 \end{pmatrix}; \tau_{AB} = \sum_{i=1}^{l}\sum_{j=1}^{l} h_{ij} = 4$$

Note that, interestingly, the term τ equals the value of $\vec{u}_A \cdot \vec{u}_B$. Therefore, the similarity value between Alice and Bob is

$$\text{similarity(A,B)} = \left(\tau_{AB} / \|\vec{u}_A\| \cdot \|\vec{u}_B\|\right) == \left(4/\sqrt{3} \times \sqrt{6}\right) = 0.943$$

Similarly, we can get the value $\tau_{AC} = 5$, and the similarity value between Alice and Charles is

$$\text{similarity (A,C)} = \left(\tau_{AC} / \|\vec{u}_A\| \cdot \|\vec{u}_B\|\right) == \left(5/\sqrt{6} \times \sqrt{6}\right) = 0.833$$

Obviously, for initiator Alice, Bob is the better matching person than Charles. However, using the protocol EWPM proposed in Reference 8, we can only obtain $S_{AB} = 3$ (the matching value between Alice and Bob), and $S_{AC} = 3$ (the matching value between Alice and Charles). Those values neither have strict semantic meaning nor distinguish whether Alice matches more with Bob or Charles. Thus, LIP3 obviously advantages over EWPM in terms of matching accuracy (measured with profile similarity). Furthermore, LIP3 scheme only brings additional computation of the modules of the initiator's and responder's profile vectors, and additional transmission of a scalar value, which are all constant operations, independent of the parameters used in LIP3, for example, the number of attributes n, and the maximal attribute value l. Those trivial additional overhead can be totally negligible, thus our computation and communication overhead is same as EWPM.

4.3 Over-the-Top-Based Recommendation Systems

In general, recommender systems (RSs) are techniques providing suggestions for items to a user. The suggestions provided are aimed at supporting their users in various decision-making processes, such as what items to buy, what music to listen, or what news to read, and so on. Recommender systems have proven to be valuable means for online users to cope with the information overload and have become one of the most powerful and popular tools in electronic commerce. Correspondingly, various methods for recommendation generation have been proposed and during the last decade, many of them have also been successfully deployed in commercial environments.

RSs [9,10] collect information on the preferences of its users for a set of items (e.g., movies, songs, books, jokes, gadgets, applications, websites, travel destinations, and e-learning material). The information can be acquired explicitly (typically by collecting users' ratings) or implicitly (typically by monitoring users' behavior, such as songs heard, applications downloaded, websites visited, and books read). RS may exploit the demographic features of users (e.g., age, nationality, and gender). Social information, like followers, followed, twits, and posts, is commonly used. Recently, there is a growing tend toward the use of information from the Internet of things (e.g., GPS locations, RFID, and real-time health signals).

RS must be able to predict the utility of items, or at least compare the utility of some items, and then decide what items to recommend based on this comparison. Consider, for instance, a simple, nonpersonalized, recommendation algorithm that recommends just the most popular songs. The rationale for using this approach is that in absence of more precise information about the user's preferences, a popular song, that is, something that is liked (high utility) by many users, will also be probably liked by a generic user, at least more than another randomly selected song. Hence the utility of these popular songs is predicted to be reasonably high for this generic user.

With the popularity of social networking applications, users are no longer mere consumers of information, but the *producer of information* [11]. Through their interactions in the social networking, they will upload their photos and personal information, check-in point-of-interests (POIs), write blogs, and communicate with online friends. All activities would contribute rich data to RS such that we can extract the preferences of users, their circles of friends, and even their areas that they often walk around in their daily life.

Utilizing the rich information created by social networking, personalized RS can be effectively designed and evaluated. Actually, there are many algorithms such as collaborative filtering (CF), content-based and hybrid approaches, and many models established to evaluate users' preference for items. These algorithms and models have been continuously improved, so that they can effectively analyze users' ratings to construct the neighbors that have great similarities.

In addition to the improvements of the algorithms and models, content-aware RS is an interesting research area. Specifically, an extensive overview was presented by Verbert et al. [12], which focused on incorporating contextual information in the recommendation process such as time, physical conditions, activity, and social relations, and so on. Among existing contextual dimensions, time information can be considered as one of the most useful ones. It facilitates tracking the evolution of user preferences. For example, it can identify periodicity in user habits and interests. Dataset from LBSN can bridge the gap between the physical and digital worlds and enables a deeper understanding of users' preference and behavior. Especially, with the rise of Foursquare, Gowalla, and other LBSNs, location-RS systems have been a hot topic and been well studied.

Several typical recommendation schemes and location RS are described as follows.

4.3.1 Existing Recommendation Techniques

4.3.1.1 Content-Based Filtering Algorithms   Content-based filtering algorithms [13] make recommendations based on user's choices made in the past. The objects recommended are similar to these items that users have bought and selected. They are mainly used in a Web-based e-commerce RS. For instance, in a movie recommendation application, content-based recommendation system will rely on information such as genre, actors, director, producer, and so on, and match this against the learned preferences of the user in order to select a set of promising movie recommendations.

Content-based filtering algorithms usually utilize user profiles to describe users' tastes, preferences, and so on, which can be elicited from users explicitly, for example, through questionnaires or implicitly by learning from their transactional behavior over time. Usually, users' profiles can be characterized by some keywords, and each item can also be represented by several keywords. Then the similarity between items recommended and items that users' profiles, can be calculated.

In general, measuring the *importance* of keyword in one item can directly affect the performance of RS. The *importance* of keyword k_i in some item s_j is determined with some weighting w_{ij} that can be defined in several different ways. One of the best-known measures for specifying keyword weights is the term frequency/inverse document frequency (TF-IDF) measure. Therefore, ContentBasedProfile of user c and *Content* of item s can be represented as vectors $w_c = \{w_{c1}, \ldots, w_{ck}\}$ and $w_s = (w_{1j}, \ldots, w_{kj})$ of keyword weights, in which each weight w_{ci} $(i = 1, \ldots k)$ denotes the importance of keyword k_i to user c and can be computed from individually rated content; each $(w_{1j}, \ldots, w_{kj})$ denotes the importance of keyword k_i in item s_j.

In content-based systems, the utility function $u(c, s)$ that measures the usefulness of item s to use u is usually defined as $u(c, s) = \text{score}(\text{ContentBasedProfile}, \text{Content})$.

Moreover, the utility function $u(c, s)$ is usually represented by some scoring heuristic defined in terms of vectors w_u and w_s, such as the cosine similarity measure denoted as Equation 4.2.

Content-based filtering algorithm has some advantages. Specifically, content-based recommenders exploit mainly ratings provided by the active user, thus it can be a good response to the situation of data sparse and new items. However, content-based filtering algorithm also has some of the limits that cannot be ignored. First, it cannot extract keywords automatically from multimedia data, for example, graphical images, audio streams, and video streams. Besides, if two different items are represented by the same set of features, they are indistinguishable. Since text-based documents are usually represented by their most important keywords, content-based filtering algorithm cannot distinguish between a well-written article and a badly written one, if they happen to use the same terms. What is worse, a new user, having very few ratings, would not be able to get accurate recommendations.

4.3.1.2 Collaborative Filtering Recommendation Algorithm The definition of collaborative filtering [14,15] can be expressed as: Supposedly, we have a finite set $U = \{u_1, u_2, \ldots, u_N\}$ is the set of N users, $P = \{p_1, p_2, \ldots, p_M\}$ is the set of M items. Each item $p_x \in P$ can be paper, news, merchandise, movie, service, or any informational types that the users need. Relationship between the users set U and the items set P are represented by an evaluative matrix $R = \{r_{ix}\}$, $i = 1,...,N$, $x = 1,...,M$. Each value $r_{ix} \in \{\varnothing,1,2,...,V\}$ represents the evaluation of the user $u_i \in U$ with the item $p_x \in P$. The value r_{ix} can be collected directly by inquiring user's opinion or indirectly by user's feedback. The value $r_{ix} = \varnothing$ can be interpreted as that the user u_i has never given evaluation or known the item p_x yet. Tasking of CF is predicting evaluation of the current user $u_i \in U$ with the new item $p_x \in P$, therefore generating recommendation for the user u_i with items that are appreciated highly.

Unlike content-based approaches, which use the content of items previously rated by a user u, CF approaches rely on the ratings of u as well as those of other users in the system. The key idea is that the rating of u for a new item i is likely to be similar to that of another user v, if u and v have rated other items in a similar way. Likewise, u is likely to rate two items i and j in a similar fashion, if other users have given similar ratings to these two items. The two methods are called user-based CF and item-based CF. The similarity measure between user u and v, $\text{sim}(u,v)$, is essentially a distance measure and is used as a weight, that is, the more similar users u and v are, the more weight rating r_{vx} will carry in the prediction of r_{ux}. The two most commonly used similarity measures are correlation and cosine-based. For instance, in the correlation-based approach, the Pearson correlation coefficient is used to measure the similarity:

$$\text{sim}(u,v) = \frac{\sum_{p \in P_{uv}} \left(r_{u,p} - \overline{r_u}\right) \cdot \left(r_{v,p} - \overline{r_v}\right)}{\sqrt{\sum_{p \in P_{uv}} \left(r_{u,p} - \overline{r_u}\right)^2 \cdot \sum_{p \in P_{uv}} \left(r_{v,p} - \overline{r_v}\right)^2}} \quad (4.4)$$

where P_{uv} are the set of all items corated by both users u and v.

In Figure 4.7, item network contains all items in social networks and then user network contains all users. The links between users and items represent the ratings of items that users have given. Then, the similarities between users are obtained according to the Pearson correlation coefficient.

In general, CF-based RS consists of three processes: (1) candidate selection, (2) similarity inference, and (3) recommendation-score predication. Taking location-based recommendation as an example, these three processes are briefly introduced.

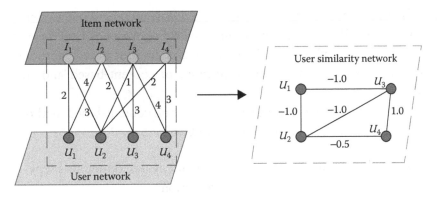

Figure 4.7 Graph representation of the process of obtaining user similarity.

Candidate selection: This step is to select a subset of candidate nodes to reduce the computational overhead. The traditional CF-based recommendation algorithms limit to use the most similar users (or locations, activities, etc.) as the candidates. CF-based recommender systems in LBSNs can also use geographic bounds and associations to constrain the candidate selection process.

Similarity inference: Similarities between users (or locations, activities, etc.) are inferred from users' ratings in traditional RS. The CF models can be divided into two subgroups: (1) user-based models that use similarity measures between each pair of users and (2) item-based models that use similarity measures between each pair of items (media content, activities, etc.).

Recommendation Score Predication: Finally, CF systems predict a recommendation score for each object (locations, social media, etc.) in the candidate set. These scores are calculated from ratings given by the set of users (U) and the similarity measures between individual users.

Collaborative approaches overcome several limitations of content-based ones. For instance, items for which the content is not available or difficult to obtain can still be recommended to users through the feedback of other users. Furthermore, collaborative recommendations are based on the quality of items as evaluated by peers, instead of relying on content that may be a bad indicator of quality. However, it also creates new limits, such as data sparsity, cold start, and so on.

In view of the above advantages and disadvantages about content-based and collaborative approaches, hybrid approaches have been a hot topic by combining collaborative and content-based methods, which helps to avoid certain limitations of content-based and collaborative systems.

Different ways to combine collaborative and content-based methods into a hybrid RS can be classified as follows:

- Implementing collaborative and content-based methods separately and combining their predictions.
- Incorporating some content-based characteristics into a collaborative approach.
- Incorporating some collaborative characteristics into a content-based approach.
- Constructing a general unifying model that incorporates both content-based and collaborative characteristics.

Hybrid RS have produced many excellent research results. Reference 16 integrated content-based filtering with CF using coclustering model, which solved the data sparsity problem by grouping similar objects in the same cluster. This paper first conducted coclustering with augmented matrices (CCAM) algorithm using rating information, item features with categorical attributes and user profiles with categorical attributes to generate the proper coclustering model. Then, it estimated an unknown rating of a given user-item pair from the coclustering model using a similar formula. Finally, items can be recommended according to the prediction scores. Reference 17 proposed a novel hybrid approach for effective Web services recommendation, which exploited a three-way aspect model systematically combining classic CF and content-based recommendation, and simultaneously considering the similarities of users' ratings and semantic Web-service content.

4.3.1.3 Improvement to Recommendation Systems

4.3.1.3.1 *Community Detection* A community is basically defined as a group of network nodes, within which the links connecting the nodes are dense but between which they are sparse. The links can be represented as the similarity of user preferences, social relations,

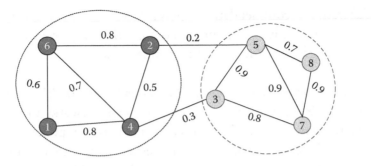

Figure 4.8 The basic idea of community detection.

and so on. Communities can be detected according to the similarities between users in some aspects such as the preferences of music, food, and so on. Based on the similarities, users can be divided into different communities where users share common preferences in some aspects. Naturally, community-based RS can significantly reduce the complexity of recommendation algorithm and improve efficiency. Moreover, due to the similarities of users in the same community in some kind of behavior, recommendations based on community can also improve performance in precision.

The methods of community detection that are used to RS include methods based on clustering-like *k*-means, and modularity-based algorithms similar to Louvain algorithm [18].

Figure 4.8 describes the basic idea of community detection where dots represent different users and the links between dots represent the similarities between users in some aspects. According to the definition of community, communities are formed where the links connecting the nodes are dense but between which they are sparse.

Reference 19 proposed collaborative RS based on clustering, which provided an efficient implementation of clustering by adapting a specifically tailored clustering-skipping inverted index structure. This paper mined the links between users in items that users had rated and then used clustering-skipping inverted structure to classify users. This method is a good compromise that yields high accuracy and reasonable scalability figures. Reference 20 proposed an improved CF recommendation algorithm based on community detection. In this book, a novel discrete particle swarm optimization algorithm is applied to find communities according to the similarities between users. Then collaborative recommendation approach was used to recommend items in communities. The proposed algorithm improved the precision, coverage, and efficiency of recommendation.

4.3.1.3.2 Developing Unifying Recommendation Model by Mining Information in Social Networks   Users in social networks produce lot of information, which can be used to predict user preferences and recommend items that users are interested in. Contexts and social network information have been proven to be valuable information for building accurate recommender system. Contextual information (e.g., time, mood, and weather) has been recognized as an important factor that influences the accuracy of recommendations. Besides, the fast growth of online social networks has brought a trend of so-called social recommendation that relies on the opinions of the target user's friends who are assumed to share similar interests. What is more, social recommendation can help to mitigate the issues of data sparsity (i.e., a user's preference for unrated item can be inferred from his/her friends).

In Reference 21, a unified ranking framework fusing social contextual information and common social relations was proposed to apply for the scenarios where users did not provide explicit ratings or users' feedback was implicit in some real social networks. This book extended the user latent features by the implicit interest deduced from social context, and then integrated the common social relations to future improved recommendation quality. A context-aware recommender system was proposed by Li et al. [22] to elaborately incorporate and process social information to improve quality of recommendations, in which various contexts, and random decision trees algorithms were adopted to form user-item ratings with similar contexts, thus imposing higher impact on users. Then, social relationships were introduced to infer a user's preference for an item by learning his/her friends' tastes.

4.3.2 Location Recommendation Systems in Location-Based Social Networks

With the rapid development of mobile devices and wireless network, a number of location-based social networking services have merged in recent years. LBSNs provide rich information to RS; among other

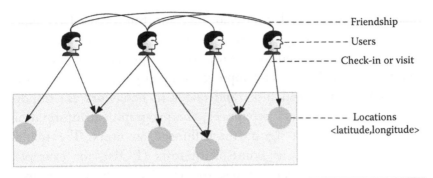

Figure 4.9   The graph representation of information in LBSNs.

things, the most important information is location, which bridges the gap between the physical world and the digital online social networking services, giving rise to new opportunities and challenges in recommendation systems. With the emergence of LBSNs such as Foursquare and Gowalla, and so on, it is prevalent to recommend some specific locations for users, which not only help users explore new places but also makes LBSNs more attractive to users. What is more, LBSNs provide more kinds of information for recommendation system.

As we can see from Figure 4.9, social links, check-in behaviors, and the geographic coordinates of locations that have been visited by users are included in LBSNs that can be used to predict users' preferences and habits. With these kinds of information, new RS are proposed to improve the performance in precision and scalability.

Over the few years, a great deal of location recommendation algorithms has been proposed. Among these researches, tradition recommendation algorithms, such as user-based CF are common approaches to location recommendation in LBSNs. Some studies have considered the social link and the interactions between users in check-in activities in CF algorithm to improve the accuracy in the location recommendation. Reference 23 proposed algorithms that created recommendations based on four factors: (a) past user behavior (visited places), (b) the location of each venue, (c) the social relationships among the users, and (d) the similarity between users. The location-friendship bookmark-coloring algorithm (LFBCA) first merged social interaction links and check-in behaviors to form the new similarity edges.

Then, the improved bookmark-coloring algorithm was used to predict the scores of unvisited places. This paper exploited the network structure, user behaviors, and social links, so as to produce recommendations for every active user and improve the quality of recommendation.

Although the social links and check-in behaviors are important to mine the users' preferences, the geographical information of users also plays a significant influence on users. The spatial locations are known as point-of-interests (POIs), for example, restaurants, stores, and museums, and are distinct from other non-spatial items, such as books, music, and movies in conventional RS, because physical interactions are required for users to visit or check-in locations. In order to utilize the geographical influence of users, many studies have been produced. A personalized and efficient geographical location recommendation framework called GeoSoCo was proposed by Zhang et al. [24], to take full advantage of the geographical influence on location recommendations. GeoSoCo exploited geographical correlations, social correlations, and categorical correlations among users and POIs, and personalized the geographical influence to accurately predict the probability of a user visiting a new location. Specifically, the geographical influence is modeled as a personalized distance distribution for each user based on a nonparametric method kernel density estimation (KDE) to predict the probability of users to new locations. Then social correlations were used to form social check-in frequency distribution to predict the probability of users to new locations. Similarly, categorical correlations were used to predict the probability of users to new locations. Finally, the above three probabilities are multiplied to obtain the final prediction of the probability of users to new locations.

LBSNs have unique network structures that contain different relations between the individuals, such as friendships, common interests, and check-in locations. Users' location histories contain a rich set of information reflecting their preferences, once the patterns and correlations in the histories have been analyzed. Due to the unique properties of location, RS in LBSNs exist new opportunities and challenges.

4.4 Conclusion

This chapter mainly introduces profile-matching mechanisms in D2D-based proximity services and RS in OTT-based proximity service. Those functionalities are becoming more and more important for proximity services. For profile matching in D2D paradigm, we present some methods mainly using Bloom filter to conduct efficient matching between one–one and many–many users. Moreover, we also summarize several typical fine-grained privacy-preserving profile-matching schemes including heavy-weighted homomorphic encryption, that is, Paillier cryptosystem, and light-weighted Confusion Matrix Transformation-based scheme.

In the OTT-based RS, we discuss some popular algorithms in RS such as content-based filtering algorithms, CF algorithms, and hybrid recommendation algorithms. Moreover, some techniques for improving RS performance are summarized, including the approaches of community detection and developing unifying recommendation model by mining and integrating multiple-modal information in social networks. In the end, the popular RS based on LBSNs are introduced, which bring new research and application topics to RS.

References

1. Wang, Y., A.V. Vasilakos, Q. Jin, and J. Ma. Survey on mobile social networking in proximity (MSNP): Approaches, challenges and architecture. *Wireless Network*, 2014; 20(6): 1295–1311.
2. Tarkoma, S., C.E. Rothenberg, and E. Lagerspetz. Theory and practice of bloom filters for distributed systems. *IEEE Communications Surveys & Tutorials*, 2012; 14(1): 131–155.
3. Champion, A.C., Z. Yang, B. Zhang, J. Dai, and D. Xuan. E-SmallTalker: A distributed mobile system for social networking in physical proximity. *IEEE Transactions on Parallel and Distributed Systems*, 2013; 24(8): 1535–1545.
4. Chen, Y.A., W.H. Lin, and Y.C. Tseng. On common profile matching among multiparty users in mobile P2P social networks. In: *Proceedings of IEEE Wireless Communications and Networking Conference (WCNC)*, Istanbul, Turkey, April 6–9, 2014.

5. Wang, Y. and J. Xu. Overview on privacy-preserving profile-matching mechanisms in mobile social networks in proximity (MSNP). In: *Proceedings of the 9th Asia Joint Conference on Information Security (AsiaJCIS)*, Wuhan, China, September 3–5, 2014.

6. Zhang, R., Y. Zhang, J. Sun. et al. Fine-grained private matching for proximity-based mobile social networking. In: *Proceedings of the IEEE INFOCOM*, Orlando, FL, March 25–30, 2012.

7. Zhu, X., J. Liu, S. Jiang, Z. Chen, and H. Li. Efficient weight-based private matching for proximity-based mobile social networks. In: *Proceedings of IEEE International Conference on Communications*, Sydney, Australia, June 10–14, 2014.

8. Wang, Y., X. Cheng, Q. Jin, and J. Ma. LIP3: Privacy-preserving profile matching mechanism for mobile social networks in proximity. In: *Proceedings of the 5th International Symposium on Trust, Security and Privacy for Emerging Applications (TSP)*, Zhangjiajie, China, November 18–20, 2015.

9. Bobadilla, J., F. Ortega, A. Hernando, and A. Gutierrez. Recommendation systems survey. *Knowledge-Based Systems*, 2013; 46: 109–132.

10. Bao, J., Y. Zheng, D. Wilkie, and M.F. Mokbel. Recommendations in location-based social networks: A survey. *Geoinformatica*, 2015; 19(3): 525–565.

11. Zhou, X., Y. Xu, Y. Li, A. Josang, and C. Cox. The state-of-the-art in personalized recommender systems for social networking. *Artificial Intelligence Review*, 2012; 37(2): 119–132.

12. Verbert, K., N. Manouselis, X. Ochoa, M. Wolpers, and H. Drachsler. Content-aware recommender systems for learning: A survey and future challenges. *IEEE Transactions on Learning Technologies*, 2012; 5(4): 318–335.

13. Castellanos, A., J. Cigarrán, and A. García-Serrano. Content-based recommendation: Experimentation and evaluation in a case study. In: *Proceedings of the 15th Conference of the Spanish Association for Artificial Intelligence (CAEPIA)*, Madrid, Spain, September 17–20, 2013.

14. Yang, X., Y. Guo, Y. Liu, and H. Steck. A survey of collaborative filtering based social recommender systems. *Computer Communications*, 2014; 41: 1–10.

15. Lien, D.T. and N.D. Phuong. Collaborative filtering with a graph-based similarity measure. In: *Proceedings of the International Conference on Computing, Management and Telecommunications*, Da Nang, Vietnam, April 27–29, 2014.

16. Wu, M., C. Chang, and R. Liu. Integrating content-based fitering with collaborative filtering using co-clustering with augmented matrices. *Expert Systems with Applications*, 2014; 41: 2754–2761.

17. Yao, L., Q.Z. Sheng, A. Segev, and J. Yu. Recommending web services via combining collaborative filtering with content-based Features. In: *Proceedings of the IEEE 20th International Conference on Web Services (ICWS)*, Santa Clara, CA, June 27– July 2, 2013.

18. Miche, P. and C. Michel. Survey on social community detection. In *Social Media Retrieval*, Ramzan, N., van Zwol, R., Lee, J.-S., Clüver, K., and Hua, X.-S. (Eds.). Springer, London, pp. 65–85, 2013.

19. Altingovde, I.S., O.N. Subakan, and O. Ulusoy. Cluster searching strategies for collaborative recommendation systems. *Information Processing & Management*, 2013; 49: 688–697.

20. Cao, C., Q. Ni, and Y. Zhai. An improved collaborative filtering recommendation algorithm based on community detection in social networks. In: *Proceedings of the 2015 Annual Conference on Genetic and Evolutionary Computation (GECCO'15)*, Madrid, Spain, July 11–15, 2015.

21. Guo, L., J. Ma, Z. Chen, and H. Zhong. Learning to recommend with social contextual information from implicit feedback. *Soft Computing*, 2015; 19: 1351–1362.

22. Li, X. and K. Aberer. SoCo: A social network aided context-aware recommender system. In: *Proceedings of the International World Wide Web Conference (WWW)*, Rio de Janeiro, Brazil, May 13–17, 2013.

23. Wang, H., M. Terrovitis, and N. Mamoulis. Location recommendation in location-based social networks using user check-in data. In: *Proceedings of the 21st ACM SIGSPATIAL International Conference on Advances in Geographic Information Systems (SIGSPATIAL'13)*, Orlando, FL, November 5–8, 2013.

24. Zhang, J.D. and C.Y. Chow. GeoSoCa: Exploiting geographical, social and categorical correlations for point-of-interest recommendations. In: *Proceedings of the 38th International ACM SIGIR Conference on Research and Development in Information Retrieval (SIGIR 2015)*, Santiago, Chile, August 9–13, 2015.

5

PROXIMITY-SERVICE APPLICATION DEVELOPMENT FRAMEWORK

5.1 Introduction

Recently, proximity services (ProSe) has become a promising mobile industry that is capable of creating new mobile service opportunities and offload traffic. The purpose of ProSe applications is to discover instances of the applications running on devices within proximity of each other, and ultimately exchange application-related contents. With ProSe, people's virtual interactions become more location-centric and tied to their current physical neighborhood (typically <150 ft). This is different from the rather static *friend lists* in typical online social interactions [1].

Existing technologies used to serve the proximity awareness can be broadly divided into over-the-top (OTT) and device-to-device (D2D) (peer-to-peer [P2P]) solutions. In the OTT model, a server located in the cloud receives periodic location updates from user mobile devices (using GPS). The server then determines proximity based on location updates and interests. The constant location updates not only result in significant battery impact because of GPS power consumption and the periodic establishment of cellular connections, but also causes serious privacy problem. Moreover, OTT approaches may incur undesired network overheads and latency for discovery and communication [2].

Different from OTT, D2D schemes forego centralized processing in identifying relevancy matches, instead autonomously determining relevance at the device level by transmitting and monitoring for relevant attributes. This approach offers crucial privacy benefits. In addition, by keeping discovery on the device rather than in the cloud, it allows for user-level controls over what are shared. Typical D2D-enabled communications technologies in unlicensed spectrum include Bluetooth (including Bluetooth classic and Bluetooth low energy [BLE]) and Wi-Fi Direct.

Although D2D communications have much great potential in some particular scenario, it should not be seen as a replacement to OTT. In some scenarios, for example, though there are no proximity devices to communicate directly, it is imperative to support OTT, such that users can find others. Actually, both OTT and D2D technologies have their own advantages and disadvantages, and are complementary with each other, enabling proximity service more efficient and robust.

The purely local, ephemeral, and decentralized characteristics of proximity service can facilitate users to share experiences in real time and in a specific place. It gives users additional spatial and temporal semantics, that is, a sense of *here-and-now*, on which proximity applications can be easily built. Such applications are more fun, and greatly increase user engagement [3]. The arising of ProSe applications represents a recent and enormous sociotechnological trend that has been generating innovative business models for the mobile networks. It is estimated that proximity-based service market has grown to U.S.$1.9 billion in revenues by 2016. This trend not only provides new opportunities for application vendors but also has the potential to disrupt the current social networking market [4].

On account of the popularity of proximity service, there has been several proximity solutions or applications presented. However, the glaring absence of the practical proximity solutions on the market is still alarming, probably, because there is lack of comprehensive understanding of special requirements in designing ProSe applications. Moreover, despite knowing all the requirements, building an application from scratch considering both communication and application requirement is not a trivial task. In order to help developers to implement communication primitives more easily, a developing framework is necessary for the developers in general.

The rest of this chapter is structured as follows: In Section 5.2, we introduce some basic concepts of ProSe and provide two different viewpoints of development of ProSe applications from two viewpoints: the ProSe designer's view and the framework builder's view; the ProSe designer's view focuses on requirements of ProSe system and the development framework builder's view aim to provide a convenient framework to help developing ProSe application. Section 5.3 summarizes the requirements that should be taken into account while developing ProSe applications and compares some present works. In Section 5.4, the challenges and their corresponding solutions to building ProSe development framework are provided. Some prominent development frameworks are exploited in Section 5.5. Finally, we briefly conclude this chapter.

5.2 Overview of Proximity Service (Basic Concept and Development View Point)

Recently, with rich connectivity and features, smartphones have greatly extended our presence, which are not only simply used as a tool accessing to the Internet, but also can enable many contextual properties in personal mobile device. For instance, the rich collection of sensors in the users' mobile phones can enable novel classes of applications that sense the users' environment and provide a unique view on the surrounding world [5].

As described above, existing technologies to provide proximity service can be broadly divided into OTT and D2D. In this subsection, differences of OTT and D2D will be discussed, which leads to the inference that a hybrid architecture of OTT and D2D will be the trend of future. Next, some typical ProSe scenarios will be presented, which are mainly classified into two categories: public safety communication and discovery mode.

5.2.1 Over-the-Top and Device-to-Device Development Pattern

The purpose of ProSe applications is to discover instances of the applications running on devices within proximity of each other, and ultimately exchange application-related information [2]. Currently in cellular networks, proximity service has been realized in a limited scope and functionalities through OTT mode, for example, Foursquare, Facebook, and so on, which enable proximity service via centralized servers, by collecting the subscribers' location information supplied by GPS. However, GPS is unreliable in indoor environment;

better accuracy is also needed to determine if two users very close to each other, for example, in the same room. Moreover, the OTT approach with centralized servers incurs undesired network overheads and latency for discovery and communication and maintaining a constant connection to the online service consumes energy and reduces the battery life of the device [2,6].

OTT approach fails to meet the ever-increasing demand of proximity-based social/commercial services and applications, whereas direct communication between mobile devices can solve these problems, which is also called D2D communication. With D2D capability, devices can detect and identify other devices in close proximity using wireless broadcast transmissions and by scanning of other devices. New technologies like Wi-Fi Direct are systematizing the interfaces and increasing the performance of media transfers between user devices. Modern phones can also monitor their environment using various sensors to gather context and recognize the activities of the associated user [6]. Apart from the general social/commercial usage, D2D communication is further expected to address public safety communities [7]. Table 5.1 illustrates the differences between these two communication models.

From a technical perspective, exploiting the natural proximity of communicating devices may provide multiple performance benefits. First, D2D user equipment may enjoy high data rates and low end-to-end delay due to the short-range direct communication. Second, it is more resource-efficient for proximal devices to communicate directly with each other than routing through a base station and possibly the core network. In particular, compared to normal downlink/uplink cellular communication, direct communication saves energy

Table 5.1 Comparisons between OTT and D2D

MODEL DIFFERENCES	OTT	D2D
Architecture	C/S	Decentralized
Latency	High	Low
Privacy	Unsafe	Safe
Location accuracy	GPS is inaccurate if when users are indoor or very close to each other	Accurate in short range
Stability	Stable	Unstable

and improves radio resource utilization. Third, switching from an infrastructure path to a direct path offloads cellular traffic, alleviates congestion, and thus benefits other non-D2D devices as well. Other benefits may be envisioned such as range extension via D2D relaying.

From an economic perspective, proximity service could create huge new business opportunities. For example, many social networking applications rely on the ability to discover users that are in proximity, but the device discovery processes (e.g., Facebook Places) typically work in a nonautonomous manner, that is, users first register their location information in a central server on launching the application; the central server then distributes the registered location information to other users using the application. It would be appealing to service providers if device discovery can work autonomously without manual location registration. Other examples include e-commerce, whereby private information need only be shared locally between two partners and large file transfers (e.g., just-taken video clips shared among other nearby friends) [8].

Moreover, there are many scenarios when we lack adequate access to mobile telecommunications because of encountering extraordinary events that can disable, disrupt, or overwhelm mobile telecommunications infrastructure. This covers a variety of situations including: (a) war or terror attack, where infrastructure may be purposefully targeted; (b) adverse weather that may disrupt the logistical supply chain to mobile communications infrastructure, for example, energy supply, or indeed affect the infrastructure itself, or both; (c) disaster, such as earthquake, flood, bush-fire, or fire-storm that may damage, inundate, or isolate mobile communications assets via a variety of means, including disabling the back-haul that connects the assets to the rest of the network, reducing the effective range of the cellular signal, for example, due to heat and smoke from fire-storms; and (d) civil or other emergency where a surge in demand results that overwhelms the capacity of the infrastructure to provide service [9]. In these scenarios, D2D communication is the only approach to realize proximity communication.

D2D have unique promising features in many specific scenarios of proximity services. However, D2D cannot be seen as the replacements of OTT. Using OTT, ProSe users can link to others' SNS (e.g., Facebook, Twitter, and Google+) to acquire others' profiles over the Internet, and enable common profile exchange. With this feature,

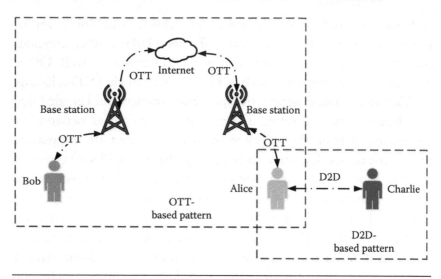

Figure 5.1   A hybrid architecture of future proximity services.

proximity services are capable of performing common interest match-making and content recommendation for users.

OTT and D2D have their own superiority and specifically suitable scenarios (e.g., if there exist no infrastructure, D2D is the only way to communicate; whereas if users are too far remote with each other, OTT is the only choice). Actually, future proximity service will likely to have a hybrid architecture integrating the traditional Internet (OTT) and opportunistic networks (D2D). As shown in Figure 5.1, Alice is too far away from Bob, he may need to get connected with Bob through OTT, while he can connect directly with others, such as Charlie, in D2D mode.

The future mobile devices can communicate with each other with or without infrastructure. In addition to data such as news, weather forecasts, traffic alerts, and social media that can be retrieved from the Internet, user-generated data such as messages, photos, and microblogs can also be collected and shared between mobile devices [10].

5.2.2 Typical Scenarios of Proximity Services

In general, most proximity service applications can be classified as two categories of scenarios: Public safety communication and social discovery (Figure 5.2). Here we list some typical applications of each category:

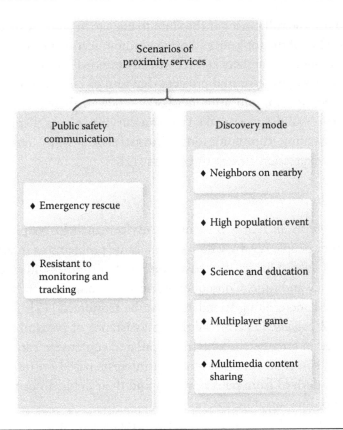

Figure 5.2   Scenarios of proximity services.

5.2.2.1 Public Safety Communication

- *Emergency rescue*: In a war, terror attack or disaster, it is important in such situations to support people to communicate to seek help and assist emergency management coordination processes. As proximity service can work both within and without the Internet, they are attractive for supporting interactions and collaborations between people in these emergency situations [10].

- *Resistant to monitoring and tracking*: The past few years witnessed some popular uprising against repressive regimes such as the 2011 Arab Spring. At the same time, the actions of these regimes demonstrated the fragility of the infrastructure that connects people to these services, as well as their willingness to use the full power of the state to engage in large-scale censorship of the Internet and other communication networks. In response, researchers and technically minded activists around the world have started projects that aim to build censorship-resistant communication networks. They all share a common goal—building networks that can survive serious disruption to existing communications infrastructure while ensuring free expression among their users [11].

5.2.2.2 Discovery Mode

- *Neighbors on nearby*: Proximity service can help people find other people nearby who have the same problems or interests in a fast, reliable, and easy way, with little expenditure of money and time. Consequently, new relationships can be initiated and existing friendships can be reinforced [12].
- *High population event*: With the assistance of proximity service, an attendee in a highly populated conference can easily find someone who has common interests based on information derived from public profiles and their public information on online social networks [13].
- *Science and education*: Scientific groups can use proximity service to share information and knowledge anytime and anywhere. Thus, proximity service can help the groups to expand both their knowledge base and their flexibility of organization in ways that would not be possible within a self-contained hierarchical organization. Proximity service can also support educators by extending discussions with and among students beyond the classrooms [10].
- *Multiplayer game*: Even without an Internet connection, mobile users still can find players for a multiplayer game in proximity, like in a train or room [14].
- *Multimedia content sharing*: Users can share songs, movies, photos, and other multimedia content in their own storage with people around who have similar tastes and interests.

In addition, more specific usage scenarios could be envisioned by both developers and users in the future.

5.2.3 Two Different Viewpoints for ProSe Development

It is believed that ProSe market would have enormous potential in the future. Recently, the research and application of proximity service has attracted the attention of academia and industrial fields. However, current solutions or applications cannot solve all problems in it. This is because there have not been a global analysis of the requirements in building a proximity service application, whereas these requirements are essential for a proximity service system designer to grasp. Furthermore, even after understanding how to design a proximity application, it is still a hard task to easily develop it, for issues such as address assignment and peer/service discovery do not have standard precedents in proximity service. Thus, it is important to build a reusable framework to help developers to realize these functions or features of ProSe with minimal knowledge of the underlying technologies and connectivity.

Figure 5.3 illustrates the workflow of designing and developing a proximity service application from two viewpoints: the ProSe designer's view and the framework builder's view. The ProSe service designers (or App developers) can develop such applications directly or using a fully functional development framework, which is provided by ProSe framework builder. In direct development, every aspect of

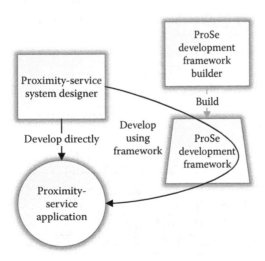

Figure 5.3 Workflow of developing a proximity-service application.

proximity service should be considered by App developers. However, it is too complex and tedious to cover all these requirements every time. Proximity service development framework can help designers reduce these pressures by realizing the required functionalities in middle framework, which can be efficiently and stably reused by various ProSe developers.

Whether developing a proximity service application either directly or through a framework, it is imperative to analyze which functionalities are essential.

5.3 From ProSe Application Designers' View: Components, Requirements, and Existing Works

ProSe framework can be composed of three layers: physical layer, context layer, and application layer. In this subsection, requirements of each layer and cross-layer requirements are summarized and discussed. Then, according to the summarized dimensions, the existing proximity service solutions are compared.

5.3.1 Requirements in Developing ProSe Applications

As shown in Figure 5.4, the proximity service can be divided into three layers. The lowest layers are the physical layers (i.e., wireless link and sensors). They handle the creation and destruction of connections

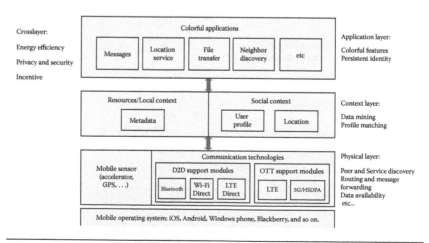

Figure 5.4 ProSe development framework and their corresponding requirements.

between devices, collect data from various sensors, and offer primitives such as neighbor discovery, message exchange, among other functionalities to the context layer.

Context layer can reuse the primitives offered by the lower layers to extract and manage the large amount of contexts from heterogeneous environments. By mining these contexts, higher level knowledge about users can be discovered such as interests of users and relationships with other people. This knowledge is significantly important for the construction and proactive service in application layer.

Finally, reusing the primitives offered by the two lower layers, the application layer will offer multiple proximity-specific applications and service to users.

There are some requirements that need to be considered among these three layers while designing and developing proximity-service applications. Besides these layer-specific requirements, there are also some other important issues across all the layers, which is called cross-layer requirements.

5.3.1.1 Physical Layer

1. *Peer and service discovery*: Peer discovery denotes the process of discovering physical peers in proximity and retrieving the content/service preference metadata from the peer [13]. The result of peer discovery may be processed by system or user to determine whether the peer is worth to connecting. It is a basic function for many ProSe use cases. During device discovery, a discovering device will periodically broadcast beacons to announce its existence, whereas other devices periodically scan and each may respond to this message once it receives the beacon [15].

2. *Routing and message forwarding*: A network abstraction is needed that performs the routing and the message forwarding between different networks to enable communication in proximity network [16]. In general, D2D link is single-hop communication, whereas multihop forwarding schemes are typically needed to extend the coverage. Current works are explicitly divided into delay-tolerant networking (DTN) (asynchronous) and wireless mesh networking (synchronous).

In DTN, when two devices encounter, they exchange the information they are carrying, store the information, and forward it to next encountered devices, thus increasing the reach of shared information to distant users or large group of users [6]. In wireless mesh network (WMN), each of the multiple devices (*nodes*) communicates directly with their neighbors, and messages from one node to another are forwarded through the mesh via intermediate nodes [11].

3. *Data availability*: Users may join or leave the network at any time, but their profiles must remain available from other online nodes. Data availability describes the possibility to get access to data of a user, not only when he/she is online but also when he/she has already left the network. Data availability is one of the most important factors. Proximity service must thus provide mechanisms to ensure high availability and topic-connectivity, even for unpopular users and under high churn rates. Topic-connectivity measures the fraction of users that can be reached by all their friends in the P2P overlay [17,18].

5.3.1.2 Context Layer

1. *Data mining*: There are many full-fledged sensors in mobile devices such as GPS, accelerator, and gravity sensor. The data collected from sensors and received by D2D communication may generate data redundancy due to the spatial correlation between sensor observations and obtained content, which inspires the techniques for in-network data aggregation and mining [19]. Data mining techniques may be utilized to provide high quality, useful, and real-time context information, hence improving the quality and efficiency of proximity-service applications and services. There are two main approaches to using data mining: one is through the Internet; and the other is through distributed mobile devices and their surrounding environments.

With the advent of cloud computing, a recent strategy is to use *network analysis software* on the Internet to automatically extract social information from the online social networking sites, including the users' identities, interests, and relationships with others. The computation power and the rich resources on the cloud servers enable fast processing of large amount of data. However, most of the data on websites are not available to the browsers, but are hidden in forms, databases, and interactive interfaces. Although many web servers provide application programming interfaces (APIs) that enable easy access to the hidden data, they normally require some form of authentication in order to be used.

Data mining can also be performed on the distributed mobile devices, which focuses on the context and interaction of users with their environments and surrounding people. Data mining software should be able to access and analyze the data stored on the mobile devices of users, such as contact lists and location history, which reflect the closest social contacts or real friends of the users [10].

2. *Profile matching*: Proximity-service applications can enable more tangible face-to-face social interactions in public places such as parks, stadiums, and train stations. However, the first step that chooses whom to interact with is difficult because it is crucial for user to select those who can lead to the most meaningful social interactions. A naive method is to choose those with valid public-key certificates, which nevertheless can only establish the identities of the certificate holders. This method is thus insufficient if Alice has multiple candidate neighbors who are all strangers and each have a valid public key certificate. A more viable method is for Alice to select those whose social profiles mostly match hers. Widely known as profile matching, this method is rooted in the social fact that people normally prefer to socialize with others having similar interests or background over complete strangers. In proximity service, application should be able to automatically find the persons around with similar profiles [20].

5.3.1.3 Application Layer

1. *Colorful features or functions*: The users are at the terminal ends of MSNs. They are only concerned with the functionalities. Thus, number of novel and useful features and

functions should be extended to end users in future MSNs. For instances: (1) *Interactive communications*, users can send social messages to their friends. (2) *Location service*, obtain location information from their smartphones in several ways, such as through GPS, the Internet, or cellular networks. This service not only enables mobile users to get their current location information, but also it enables the users to inform their friends of this information. (3) *Messaging/file transfer*, enabling users without Internet connectivity to access the Internet via the connectivity of peers who are willing to provide relaying services by receiving, storing, and forwarding messages or files through their mobile devices. (4) *Neighbor discovery*, the user study brought up that following nearby users (friends and unknown) has been the most interesting use case for local social application. (5) *Notifying a user when something happens*, the method of notifying a user depends on the relevancy of the information for the user, which derives from context (e.g., time, place, and other users in proximity). A user can have the control to select (and teach) the application notification according to personal preferences. Less important events can be collected to a log that can be browsed at a later time. (6) *Group/broadcast communications*, in public safety networks, providing services like police, fire and ambulance, and D2D group/broadcast communications are required features [10,15].

2. *Persistent identity*: Usually, there are no centralized servers to store information about a user. The identifier of one user is required to identify him/her not only within the context it is created, but also in all of his or her future transactions over the network. Transferred data are also linked to the original owner [17].

5.3.1.4 Crosslayer

1. *Energy efficiency*: During this process of communication, the battery consumption should also be considered and a solution for energy saving becomes necessary. Usually, the power consumption of a smartphone arises from: (1) the processor

and display/touch screen; (2) radio interfaces such as Wi-Fi, Bluetooth, and 3G; and (3) sensing devices such as GPS, accelerometer, proximity sensor, and camera.

A proximity-service application should provide mechanisms to efficiently use the resources of the battery and ensure that the application performs well on mobile devices with limited resources. Nevertheless, there is still a trade-off between the quality of user experience and power consumption, as users are more concerned about the effectiveness of ProSe applications, which tends to increase with power consumption in some situations. Thus, it is challenging to develop applications for MSNs that provide a favorable user experience without significantly reducing the battery life of smartphones [10,21].

2. *Privacy and security*: It is challenging for proximity communication to ensure the privacy of personal profiles, which often contain highly sensitive information related to gender, interests, political tendency, health conditions, and so on [20]. Thus it is important for proximity-service applications to take actions to ensure the security while communicating and generating reliable privacy strategy or users' profile.

On one hand, ProSe application should make use of cryptography to ensure communications' security. The application should have a method to verify the identity of other users, whether it is their real identity or the identity used in the social network [6]. A confidential and secure group communication support in a mobile environment also should be provided for applications that require it, such as meetings or collaborative editing of documents. As no central security authority is used, a distributed identification management for secure deployment and distribution of user's identifications is required [16].

On the other hand, restrictions on who can see what should be supported to preserve user's privacy. For example, a user may be agreeable for particular friends to see precisely where(s) he is, but other users should see only an approximation of the user's location. Similarly, rather than presenting the particular activity a user is engaged in, they may prefer that their activity status is simply shown as *busy* [22].

3. *Incentive*: In an autonomous proximity service without any infrastructure, communications between smartphones or PCs may need another device as a relay to form a two-hop or multihop connection. Therefore, the success of such networks which is called *user provided network* (UPN) relies on encouraging users to participate as a relay node since both demand (clients' requests) and provision (hosts' availability) depend on users' participation.

Nevertheless, more often than not, the participants of clients and hosts have conflicting interests. For example, clients would prefer to receive services at low cost, while hosts would prefer to charge high prices. Moreover, mobile devices have tight energy budgets that lead to stringent energy consumption constraints for forwarding messages. Clearly, it is of paramount importance to design incentive mechanisms for reconciling the objectives of all the participants.

Such incentive mechanisms are responsible to determine how much resources each user needs to contribute, in terms of energy and Internet bandwidth, in order to maximize the service capacity (i.e., the aggregate amount of data delivered to users). Accordingly, it dictates how this capacity will be shared by the different clients, as each one of them should receive service in accordance to her contributions. More specifically, such an incentive scheme needs to address the following issues:

- *Efficiency*: The mechanism should ensure the efficient allocation of the resources, that is, maximize the user-perceived service performance by taking into account the Internet access offered to each user, as well as the energy and monetary cost the user incurs.
- *Fairness*: The scheme should satisfy a fairness rule that accounts for the different resources (energy and bandwidth) that each user contributes and consumes (due to her multiple roles).
- *Decentralized implementation*: The mechanism should be amenable to distributed execution for networks with a large number of nodes. Users should be able to decide their routing and Internet access strategies independently based only on local information and minimum signaling from their one-hop neighbors.

- *Indirect reciprocation*: A user being served by another user may not be able to return the favor immediately by offering similar services. Hence, a resourceful user may be reluctant to help other less resourceful users. The incentive mechanism needs to induce cooperation among users even in these cases.
- *Future provision*: Some users may not have communication needs in a certain time period, and therefore may not be willing to participate. This may deteriorate the overall service performance. The mechanism should manage to encourage users to participate even if they currently have no communication needs [23].

Recently, many efforts have been done on incentive mechanism of proximity services from academia. Yufeng Wang et al. consider the incentive mechanism in autonomous networks as a system rule instead of using prices and payments as candidates to ensure the efficiency and fairness, whose goal is to influence participants to behave in a certain manner (so-called rule-based incentive mechanisms) [24]. To address issues on indirect reciprocation and future provision, a virtual currency system is introduced by Iosifidis et al. [25], which encourages users who currently do not have communication needs, to participate and serve other users, so as to collect virtual money (that can be used later when they have needs). Similarly, it enables users with poor Internet connections to utilize the service by paying other users. Clearly, this system increases the number of users who are willing to participate in the service and are able to cooperate with each other.

Expect encouraging users to provide their networks. Incentives also play a pivotal role in stimulating user-generated content (UGC), which includes users' preferences, points-of-interests (POIs), and comments. Acquiring these data, proximity-service system can provide customized services. The challenges to realize such incentive mechanism are same like UPN. While in UGC, guarantee of the quality of users' content is also an important issue. Nonfinancial social incentives such as virtual currency mentioned above may stimulate participants to make more UGC contributions, but its quality is questionable [26]. As a result, a reputation system is essential for incentive mechanism in UGC.

However, currently proximity services that are smoothly integrated with incentive mechanism in either UPN or UGC are rare, low attentions are paid on such mechanisms while designing proximity services. It is noticed that after addressing all the key challenge in forming a proximity service, reliable incentive mechanism will be the next field of proximity service that attract many researchers' focus.

5.3.2 Comparison of Existing ProSe Applications and Solutions

Several platforms or solutions have been proposed to meet some properties that we think they are important factors in proximity-service applications. These platforms are listed in Table 5.2 and discussed below:

Bluetooth mesh network (BMN) framework is a typical mesh network, based on Bluetooth low energy, a hallmark feature of Bluetooth core specification v4.0 featuring ultralow power consumption, low latency, and enhanced range. With the adoption of Bluetooth v4.1, a BLE device can be a master in one network and can act as a slave in another network, thus allowing the possibility for the formation of a mesh network

The architecture of BMN is composed of three layers: Android BLE layer, Mesh service layer, and application layer. Using directed acyclic graph (DAG), the BMN formed nodes into a mesh network, making multihop message-transferring possible. Although the BMN team only focuses on implementation of Mesh net, they had not considered more ProSe feature like profile match or service discovery. In brief, BMN is a good model of multihop network, which is the core composition of ProSe [27].

STARS is a decentralized approach to building social networking based on ad-hoc P2P architecture in order to remove the bound of a centralized provider and provide a flexible way of connecting people. By introducing a verification agents mechanism, a user is only allowed to join and communicate to other users if he/she is qualified by all of the existing verification agents. STARS allows user to build interest-based network but there is a set of permissions associated with it in order to control how users interact with received data. These permissions bring the ownership of user data back to the owner. In addition, two modes of sharing control are taken into account to balance the

Table 5.2 Properties of Several Proximity Solutions or Platforms

NAME	PHYSICAL/NETWORK	FEATURES THAT THE SOLUTION SUPPORT		
		SERVICE/CONTEXT	END USER	CROSS-LAYER
BMN [27]	Peer discovery: Bluetooth broadcast. Routing and message forwarding: DAG (Directed Acyclic Graph)	None	Persistent identity: Mac Address of Bluetooth	Energy efficiency: Less power and low latency
STARS [17]	Peer discovery: Ad-hoc discovery. Data availability: Two modes of sharing control	None	Persistent identity: A global unique identifier	Security: Verification agents. Privacy: Data controlled by users
My-direct [21]	Peer discovery: Wi-Fi Direct	None	None	Energy efficiency: Intelligent Energy Module (IEM). Privacy: A privacy layer definition
Social index [28]	Peer discovery: Ad-hoc discovery	Data mining: Social interest graph. Profile matching: Anonymized profile called Index	Colourful application: Share contact card or chat	Security: Encrypt message with a hash
SuperNova [29]	Peer discovery: Super-peer. Data availability: StoreKeepers	Profile matching: Friend list	None	None
Haggle [30]	Peer discovery: Push-based search dissemination. Routing and message forwarding: Store-carry-forward (the combining of content share with delegation)	Profile matching: Search resolution	Colourful application: Open-source and thus many applications are developed on it	Security: Built-in mechanism
Contrail [31]	Data availability: Cloud relays	Profile matching: Sender-side filters	Colourful application: API offered	Security: Encrypted data exchange on cloud relay. Privacy: Data filtering on trusted devices

availability of data and a global unique identifier combined by e-mail, password, and MAC address are used for account registration [17].

My-Direct make use of Wi-Fi Direct as a solution to improve the connectivity between the nodes of social network. It defines a privacy layer that serves to identify users and verify their degree of friendship. My-Direct uses a tuple containing the MAC address of the device, the name that identifies the device during connection, and the bond between users. The schema of the tuple can be seen as: (MAC, DEVICE NAME, BOND). This work also aims to create an intelligent mechanism for the management of mobile device resources to provide a reduction in power consumption. The intelligent energy module (IEM) has two agents: one for the management of wireless networks (network agent) and another for managing the display (display agent). When the My-Direct is active, the display agent will periodically check the battery level of the mobile device and the current sets the screen brightness according to this information. Meanwhile, the network agent will turn off or on the wireless networks, with the exception of Wi-Fi Direct, according to preestablished rules, to minimize battery consumption [21].

Social index is a content discovery application that focuses on finding interesting new content in the proximity. In social index, interests are mined from the globally unique social objects found from the mobile phone to form a social interest graph and then stored in database. The presentation for the interests is independent from the source to make processing the information easier.

These interests are shared through the ad-hoc communication, and an anonymized profile (called an index) is evaluated automatically according to these interests, current context, and user feedback.

When social index engine finds another user or new shared content, it uses an evaluation algorithm based on the index to determine how interesting the event might be to the user while preserving privacy.

The evaluation results in a numeric social index value, which can also be presented to the user. The user provides feedback of the discoveries to the engine through the user interface, and the feedback is used to refine the index to provide more interesting discoveries. The user can interact with the new content via the social index user interface, for example, share a contact card or chat with other users [28].

SuperNova proposes a architecture for realizing decentralized online social networks with the introduction of super-peers. Any node may become a super-peer by providing his services to the system in general. Super-peers provide storage to the new nodes that do not have enough friends in the network for some initial period; they also maintain and manage different types of services for proximity network by cooperating among themselves. To keep the availability of data and also help new user on communication, super-peers can act as storekeepers. Storekeepers are list of users who have agreed to keep a replication of another user's data, so that users can store their content in case they do not have adequate friends to do so, or if their friends are already overloaded. In addition, by maintaining a directory of users, SuperNova can also build a community around a user based on interests.

In fact, super-peers are volunteer agents. Users may want to become super-peers out of altruism, for the sake of the reputation (e.g., being an influential member for an interest-based community) as well as potentially to monetize their special roles (e.g., run advertisements) [29].

Haggle is a system for ad-hoc content sharing. Like SuperNova, Haggle can bridge disconnected or never encountered devices by a push-based dissemination. It disseminates a node's interests as a persistent search query that can propagate to nodes never encountered directly. This persistent search allows content to be pushed back to the interested nodes, leveraging content delegates that carry content, although they have no interest in it. This altruistic delegation can be achieved by using existing forwarding algorithms, increasing the likelihood that content items reach interested parties that are only connected via nodes with no interest in the items. Meanwhile, with content delegation, Haggle enables store-carry-forward on behalf of others based on interests.

Moreover Haggle foregoes traditional naming and addressing and instead uses search to resolve a content item's interest group by matching the interests of users against the item's metadata (or vice versa). Although similar content-sharing systems typically value every content item the same, Haggle uses a ranked search to judiciously decide which content to exchange, and in which order. The search matches a device's locally stored content against the interests of other users that the device has collected, prioritizing relevant content when contacts are time limited and resources scarce. Thus, search enables dissemination of content in order of how strongly users desire it, offering delay and resource savings by exchanging the content that matters. Ranked searches, combined with delegation, allow Haggle to balance the short-term benefit of exchanging a content item between two nodes against the long-term benefit to the network as a whole.

Haggle's source code is available online through an open-source license, and provides a readily available platform for researchers to build and study mobile applications for ad-hoc content sharing. A wide variety of applications have already been built on top of Haggle: photo-share shares pictures taken with a mobile phone's camera, MailProxy allows e-mailing without infrastructure, MobiClique and Opportunistic Twitter enable ad-hoc *Facebooking* and *tweeting*, whereas Haggle-ETT provides an electronic triage tag for disaster areas. It is believed that more applications like these will be implemented in the future [30].

Different from Haggle which is open source, so that others can make applications on the top of it, *contrail* offers its API for application running on the mobile device that makes it easy for developers to build social network applications that are decentralized yet efficient including: location notification application, real-time interactive, content sharing, sensor aggregation, and status uploading sharing like Facebook.

Contrail is a communication platform that allows decentralized social networks to overcome challenges like data availability, privacy, and energy using content filters and cloud-based relays. At the heart of contrail is a simple cloud-based messaging layer that enables basic connectivity between smartphones, allowing them to efficiently and securely exchange encrypted data with other devices. Contrail's cloud layer consists of stateless application servers and a persistent storage tier. When devices connect to the cloud, they interact with one of these application servers, which are called *proxy*. If a device uploads a message meant for an offline recipient, its proxy stores the message in the storage tier. When the recipient device comes online, its proxy checks the storage tier for any messages meant for it and transfers them. On the other hand, if the recipient is online and connected to some other application server, the two proxies interact directly to transfer the message, without the storage tier in the critical path. To provide multicast efficiency, the cloud layer allows senders to specify multiple recipients on a message.

Over this messaging layer, contrail implements a novel form of publish/subscribe that uses sender-side content filters to minimize bandwidth and energy usage while preserving privacy. Contrail filter is simply an application-defined function that accepts some unit of data as input and returns true or false. Once a user installs content filters on her friends' devices that express her interests, she will subsequently receive new data generated by her friends that match the filters. Contrail's content filters give us privacy, since the filtering of data occurs on trusted edge devices, not central servers. They also give us upload and download efficiency: a device only uploads data matching a filter installed on it by another device. Conversely, it only downloads data matching a filter installed by it on another device.

In summary, the combined edge-based content filters and cloud-based relaying layer let contrail offer all the properties we were seeking: social network applications built using contrail are privacy aware (the cloud-based relay has knowledge of the communication pattern but does not see the data), can work across devices decoupled in space and time, and are naturally efficient in terms of energy and bandwidth [31].

5.4 From ProSe Development Framework Builder View: Benefits and Challenges

In the earlier section, we overview the requirement in building proximity service applications and list some typical examples. Their contributions only focus on completing an efficient D2D proximity service or applications. To be specific, these works mainly consider about discovery, forwarding strategies, context management, privacy, and so on. However, these solutions, applications, or architecture cannot facilitate and make the development process of ProSe applications more efficient. Thus, it is important to develop a reusable framework that allows application developers to exploit the potential advantages of proximity service with minimal knowledge of the underlying technologies and connectivity. Figure 5.5 illustrate such a framework. The proximity-service application development framework can make development process easier. Have been provided with the property of easy to develop and zero configuration, development framework will be able to help application developer to explore proximity, mobility,

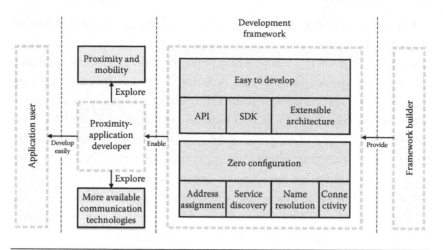

Figure 5.5   Relationships among framework builder, development framework, application developer, and application user.

and more communication technologies. As a result, ProSe developers can quickly leverage existing solution to solve basic issues in different layers discussed in Section 5.3, and put emphasis on providing more specific functions or services in various unique scenarios.

In the next subsection the benefits of introducing a proximity service development framework are discussed, including easy to develop and ease-of-configuration will be introduced. However, realizing these two functions or benefits is not easy, challenges to realize them is given afterward.

5.4.1 Benefits of Introducing Development Framework

A development framework can help developers to address these low-level but complex problems instead of doing these by developers themselves. Nowadays there have been several development framework proposed by academies or industries (will be introduced in detail in next section); most of these frameworks' contributions or benefits can be divided into two categories: one is easy to develop and the other is zero configuration. Easy to develop includes providing interfaces to developers that have solved basic requirements in proximity-service applications. In addition, zero configuration means that this framework can choose the best net configuration while developers would not worry about any network problems. As a result, ProSe application developers can focus on the core functionality in their own applications, such as particular application in specific scenario.

- *Easy to develop*: Developers face quite numbers of barriers in developing proximity-service applications. While the framework can simplify the process of developing proximity-service applications. The simplification of this process is obtained by providing high-level abstractions with lightweight interfaces to mobile application developers. All the technology-specific actions can be abstracted away from the upper layers, and only a small set of interfaces is made available. For example, a set of technology-independent interfaces should be provided to handle communication requirements such as neighbor discovery, connection and disconnection, and message delivery. While the actual implementation of these interfaces is hidden from the developers. Facilitated application integration and reuse must also be functions of the framework [21,32].
- *Ease-of-configuration*: There may be complex configurations for users to set before they can discover and use services. Although developers can consider these problems and solve them for users, developers in turn have to deal with these troubles. As a result, framework builders need to enable the proximity-service application development framework to reduce network configuration to zero (or near zero) in proximity networks, which is also called zero configuration (Zeroconf) [33]. This ease-of-configuration will make the framework both user- and developer-friendly.

5.4.2 Challenges in Building ProSe Application Development Framework

Although proximity-service application developers benefit greatly from development framework, many challenges or future works still exist in building such a framework.

- *Challenges in easy to develop*: Besides providing interfaces to developers, frameworks should also enable new technologies or strategies be easily and smoothly plugged. What is more, a software development kit (SDK) combined with these features will be better if possible.

- *Extensible architecture*: Mobile devices nowadays are capable of communicating with several communication technologies, such as Bluetooth, Wi-Fi, Wi-Fi Direct, NFC, or 3G/4G. However, it is difficult to rely on more than one communication technology and to adapt it at runtime according to availability or nonfunctional requirements such as latency or throughput. In that direction, an important contribution would be to create a communication infrastructure that permits applications to communicate using any available communication technology. That infrastructure would have to handle the complexities mentioned before and permit application developers to focus solely on the functional requirements of their applications. The outcome of using such communication framework is a simpler way to build applications that explore physical proximity [32].

- *Development kits*: A SDK is typically a set of software development tools that allows the creation of applications for a certain software development platform. In general, the SDK will implement APIs in the form of some libraries to interfaces to a particular programming language, and provides easy-to-use mechanism for extending new technologies. Besides, it can also include some sample code and supporting technical notes or other supporting documentation to help clarify points made by the primary reference material [34]. Particularly in developing proximity applications, providing a SDK can further help developers to quickly understand where to start.

- *Challenges in zero configuration*: In general, zero configuration includes automatic address assignment, service discovery, and name resolution. While in proximity network, connectivity is also an important factor. Connectivity means that proximity systems must ensure with high probability that all user friends remain connected. For example, when the Wi-Fi is not available, the public communication services such as the cellular network can be used to accomplish the delivery of information. On the other hand, when the public network is destroyed in an emergency, the personal range network is still available, which gives an opportunity to share information based on the opportunistic network

[35]. To guarantee the connectivity among different communication services, an efficient mechanism should be considered while building proximity-service application development frameworks.

5.5 Summary of Existing ProSe Application Development Frameworks

As it stands today, some industries or communities have introduced several frameworks for proximity-service application development. Among startups, commotion wireless project* builds a customized firmware to enable Wi-Fi access points and other devices to form mesh networks, with a focus on ease-of-deployment [11]; The Serval project† aims to bring infrastructure-free mobile communication to people in need, such as during crisis and disaster situation when vulnerable infrastructure like phone cell tower and mains electricity are cut-off [36]; AllJoyn‡ is an open source project, initially developed by Qualcomm Innovation Center, Inc., San Diego, California and hosted by the AllSeen Alliance, which provides a universal software framework and core set of system services that enable interoperability among connected products and software applications across manufacturers to create dynamic proximal networks. The range of consumer products enabled by AllJoyn is very wide: from the mobile devices consumers always have with them, to the appliances and media equipment in their homes, to the electronics in their cars, and the office equipment in their workplaces. The ambitious vision of AllJoyn is to enable the Internet of everything near in proximity area [37].

Some academics also present their own frameworks for MSNP. Proxima is a framework for the Android platform that employs ad-hoc device-to-device connections and proactive mesh routing for a decentralized topology with solely proximity-based rich content dissemination [38]; crossroads is a proximity-based framework that brings a set of expressive APIs and proximity-optimized services to balance app development overhead and developers' expressiveness [1]; USABle is a communication framework for ubiquitous systems to help developers to implement communication primitives [32]; in addition,

* Commotion Wireless: https://commotionwireless.net/
† The Serval Project: http://www.servalproject.org/
‡ AllJoyn: https://www.alljoyn.org/

CAMEO is a light-weight context-aware middleware platform for mobile devices designed to support the development of real-time mobile social network (MSN) applications [39].

Table 5.3 summarizes and compares these ProSe development frameworks.

5.5.1 Commotion Wireless Project

Commotion is an open source *device-as-infrastructure* communication platform integrating users' existing cell phones, Wi-Fi enabled computers, and other wireless-capable personal devices to create metro-scale peer-to-peer (mesh) communications networks [41].

The main purpose of commotion is to provide lower-cost access to communications for any number of barriers that may interfere with that access. In some scenarios, although it is not possible to access outside Web pages without the use of an Internet service provider, commotion wireless can handle files, and communication among those users connected.

Commotion is software that is installed on wireless routers. It enables the creation of P2P (mesh) communications networks. The commotion project's goal is to provide an easy-to-assemble package of software and documentation that makes building mesh networks accessible for a wide audience. It is reported that commotion can be used in places where there is no connectivity, where connectivity has gone down, or where there may be surveillance [42,43].

The features of commotion platform are mainly about the following:

- Allows existing Wi-Fi enabled devices (e.g., laptops, smart-phones, and home wireless routers) to network directly to form a distributed (wireless mesh) communications infrastructure
- Supports encrypted and anonymous data and voice communications transit throughout the network
- Provides local communications even if Internet connectivity is disrupted or severed
- Allows existing, unmodified GSM cell phones to connect and exchange anonymous calls, text messages, and other information with other devices on the network

Table 5.3 Summary of the Frameworks for Proximity-Service Application Development

FRAMEWORK	EASY TO DEVELOP		ZERO CONFIGURATION				PROXIMITY-SPECIFIC PRIMITIVES
	WELL-DEFINED APIS	SDK	ADDRESS ASSIGNMENT	SERVICE DISCOVERY	NAME RESOLUTION	CONNECTIVITY	
Commotion wireless project	Yes	No	Yes	Yes (multicast DNS service discovery)	Yes (use the router itself to resolve host name lookups)	No	None
Serval project	Yes	No	Yes	Yes (called Cooee)	Yes (name resolution from user's Android contacts)	No	Installed phone-by-phone
AllJoyn	Yes	Yes	Yes	Yes	Yes	Yes	None
Proxima [38]	Yes	No	Yes	Yes	Yes	No	No
Crossroads [1]	Yes	No	Yes	Yes	Yes	Yes	Group dissemination
USABle [32]	Yes	No	Yes	No (to be implemented in the future)	Yes	Yes (based on the management of redundant links)	Permit communication among devices that are not within their communication range by multihop and carry-and-forward strategies
CAMEO [40]	Yes	No	Yes	Yes	Yes	No	Disseminate content to interested users that could never be directly connected with the source of that information (with *travel nodes*)

- Enables any Internet uplink (i.e., any device in the network that is connected to the Internet) to share Internet access to every other device on the network, regardless of the connection type used (e.g., satellite, dial-up modem, mobile phone, fiber, and DSL/Cable)

5.5.2 Serval Project

The Serval project aims to provide infrastructure for direct connections between cellular phones through their Wi-Fi interfaces, without the need of a mobile phone operator. The project allows for live voice calls whenever the mesh is able to find a route between the participants. Text messages and other data can be communicated using a store and forward system called Rhizome, allowing communication over unlimited distances and without a stable live mesh connection between all participants [44]. In addition, when the Serval Mesh software is installed on a handset, it offers itself for installation onto other phones. In this way, the mesh itself can be deployed in a crisis situation, instead of having to be deployed beforehand. This represents a compelling solution to a common logistical problem in provision of services in crisis situations.

The architecture of Serval software is about four parts, as detailed below:

1. *Text messaging and file distribution*: The Serval project software suite currently includes the ability to make mesh-based telephone calls, send SMS-like short messages (MeshMS), and distribute, share, and exchange files, including Serval Mesh software updates. It also comprises a maps application that shares map tiles and POIs (Serval Maps). Already there is also an integration with Open Data Kit tools for the collection of forms and other structured data from the field (Serval SAM). Each of these components is complemented by a robust security framework that offers strong protection of private communications.
2. *Field data collection and visualization*: Serval maps and Serval SAM can source map tiles, surveys, and other data collection and visualization elements available on an isolated mesh

network and perform rudimentary visualization of that data, demonstrating the feasibility of infrastructure-free operation. The local-first dissemination behavior of single-hop bundle-based store-and-forward protocols is well suited with this environment, because local information will tend to be available where it is required. Combining the Serval project's maps application with the Open Data Kit integration offers the potential to create an integrated, distributed, and infrastructure-independent, but not infrastructure-ignorant, crisis-mapping platform that directly answers various humanitarian operational needs.

3. *Security framework*: Underlying these functions is strong cryptographic protection of communications through the Serval project's security framework. The Serval project uses a Diffie–Hellman shared-key agreement to encipher communications between pairs of parties, and to protect data through digital signatures for publicly distributed information. This provides considerably better security than standard cellular communications.

4. *Planned capabilities*: In addition to Ushahidi*-like crowd-sourced information systems, there is also great utility in text-messaging systems, such as SMS and Twitter. The Serval Mesh software offers prototype of mesh-based equivalents to these services. Serval MeshMS allows the sending of encrypted text messages on the mesh. MeshMS also offers a primitive Twitter-like service in the form of broadcast unencrypted MeshMS messages. By supplementing that feature with hashtag and sender filtering tools, it would be possible to create a distributed social networking platform that allows for local informal information sharing without dependence on any infrastructure. It would also be very simple to add a secure voice mail application. Although this would be useful in and of itself, when considered in the context of the already pictorial user interface, it would be especially valuable for providing accessibility to people with poor literacy skills [45].

---

* Ushahidi platform, http://www.ushahidi.com/

5.5.3 AllJoyn

Qualcomm recently introduced AllJoyn, an open-source, general networking framework that supports multiple direct networking technologies (Wi-Fi Direct, Bluetooth, etc.), and enables ad-hoc, proximity-based communication without the use of an intermediary server. Now, this open project is hosted by AllSeen Alliance (https://allseenalliance.org/), and many consumer brands have signed on. AllJoyn's vision is to enable the Internet of everything near users, which provides a software framework and a set of services that enable interoperability among connected products and software applications, across manufacturers, to create dynamic proximal networks. The range of consumer products enabled by AllJoyn is very wide: From the mobile devices, consumers always have with them, to the appliances and media equipment in their homes, to the electronics in their cars, and the office equipment in their workplaces.

Technically, the most basic abstraction of AllJoyn system is AllJoyn bus. It provides a fast and lightweight way to move marshaled messages around the distributed system. In many cases, the connections to the bus can be thought of as being coresident with specific processes. AllJoyn bus is typically extended across devices, to provide interprocess communication between components attached to the bus, and communicate without having to deal with the details of the underlying mechanisms. Another important abstraction is AllJoyn router, in AllJoyn-based proximity-service applications; each device runs an AllJoyn router. The AllJoyn router provides an abstraction layer that handles all of the transport mechanisms, message routing, and namespace management. AllJoyn router on each device communicates with routers on other devices, so apps never communicate directly because they talk through the bus. The communication protocol is transport-independent, currently supporting Wi-Fi and Bluetooth. New transports will be added to AllJoyn over time, but app developers do not need to modify their apps to handle them because AllJoyn does it for developers [37].

5.5.4 Proxima

Proxima is a framework for the Android platform by Salmon et al. in 2014, which employs ad-hoc D2D connections and proactive mesh routing for a decentralized topology with solely proximity-based rich content dissemination.

The Proxima framework provides a fully asynchronous thread-safe API and can be used by multiple client applications on a single device. It acts like a *neighbors and resources finder* server. Communication with the framework is done using asynchronous method calls with user-supplied callback functions. Users are sent broadcast intents when changes occur, such as a new device coming into proximity. As a mobile application developer, to use the Proxima framework, all you need to do is to register the application with the framework as a client at the beginning. Then, the framework will send the devices found in the neighborhood to you, and any changes in the network will be updated via self-defined callback function. The test of developing a real-life application called TuneSpy strongly demonstrates the ease-of-writing proximity-based applications with Proxima.

With its unique semantics of purely local information dissemination, Proxima could potentially allow application developers to incorporate this proximity-based functionality into their existing applications, or develop entirely new classes of unique mobile applications. In addition to this, Proxima also provides a nearly zero-configuration interface. The only aspect that needs to be configured is the device/user name (similar to that of the Bluetooth device name), which can be achieved programmatically with a simple API call. What is more, the framework is very lightweight at only 6 MB (including all necessary native binaries and compiled code), representing a small overhead for addition into applications.

In brief, Proxima can take care of the transport mechanics, device specifics, and configuration issues, leaving the application developer to focus solely on implementing their application [38].

5.5.5 *Crossroads*

Crossroads is a proximity-based social interaction (PSI) framework, which brings a set of features to balance the development overhead and developer expressiveness [1]. Crossroads has three main differentiators:

1. Crossroads adopts the approach of passing application-level hints via APIs to aggregate and schedule pending transmissions while matching the app requirement with system

resources. At the extremes, the framework can either infer the traffic semantics with a generic model for all apps, or expose all the low-level network functionalities to apps. However, both do not balance between developers' burden and intention expressiveness.

2. In network topology management, crossroads combines star topology and interval-decaying beacons to better achieve link robustness and device efficiency. In addition, this framework exposed the notion of persistent virtual links, if the current physical link deteriorates, crossroads can switch to another radio interface without interrupting the apps.

3. Recognizing group communication as an important PSI primitive, we designed a group dissemination protocol that addresses the problem of load hotspots in many existing solutions.

5.5.6 USABle

USABle framework is divided into three layers—connection-aware layer, network layer, and application layer—aiming to improve modularity and adaptability of the framework.

The lower layer is connection-aware layer, which is responsible for creating links between two devices using the available technologies. The main features implemented in this layer are (1) neighbor discovery and connection, (2) message exchange, and (3) neighbor disconnection. These features are exposed to the upper layers as interfaces. Each communication technology available is encapsulated as a different module in the connection-aware layer, plugged into the framework, and used by the upper layers to route and send messages between devices.

Using the interfaces offered by the lower layer, the network layer is responsible for routing messages to their destinations because the lower layer only sends one-hop messages to physically colocated devices. The goal of the network layer is to permit communication among devices that are not within their communication range. Since no supposition about a centralized routing service should be made, messages are sent using multihop or carry-and-forward strategies and all devices forming the network are responsible for forwarding messages.

Reusing the functionalities form the two previous layers, the application layer offers communication primitives to applications, enabling applications send messages to devices, not necessarily directly connected.

The main contributions of USABle are that (1) it provides a communication framework that routes messages between different technologies, (2) it provides a higher transparency since abstracts from developers have the complexity of dealing with the communication technologies, (3) it permits routing strategies that uses concepts such as multihop or carry-and-forward to be more easily implemented, and (4) it offers to application developers implementations of two well-known multihop routing strategies and three carry-and-forward strategies [32].

5.5.7 CAMEO

CAMEO is a light-weight context-aware middleware platform for mobile devices, designed to support the development of real-time MSN applications. CAMEO allows personal mobile devices, which occasionally meet in a physical location, to automatically discover users' common interests, available services, and resources through opportunistic communications. To this aim, it implements optimized networking protocols, resource management mechanisms, and context data processing features.

As a modular software architecture, CAMEO consists of a single software package containing two subpackages:

- Local Resource Management Framework (LRM-Fw), aims at implementing features strictly related to the interaction with the local resources of the device, both hardware (e.g., embedded sensors) and software (e.g., communication primitives and programming libraries). It is also in charge of managing the interactions between the node and the remote sources (e.g., single external sensors, sensors networks, and centralized repositories).
- Context-Aware Framework (CA-Fw), represents the core of CAMEO, being responsible for the collection, management, and processing of all the context information (local, external,

and social) and the development of internal context- and social-aware services (e.g., forwarding protocols and resource sharing services).

In addition, CAMEO provides an API toward MSN applications and it directly interacts with an external module for the user's profile definition called User Profile Module [46].

In general, through CAMEO, applications like Tourist-MSN can provide the users with functionalities such as (1) identifying users in the social context interested in a specific content, post or discussion; (2) disseminating selected contents to interested users; (3) generating ratings of available contents depending on the local user's interests; (4) establishing discussion forums with other users; and (5) cooperatively annotating content and enriching it, thanks to multimedia editing functionalities contributed by devices available in the environment [40].

5.6 Conclusion

Proximity services are extremely promising. This chapter overviews the current situation and trend of proximity service. Existing technologies used to serve the proximity awareness can be broadly divided into OTT and D2D solutions, which are naturally complementary to each other, to facilitate to design and build efficient ProSe applications. However, there is still lack of the practical proximity-service applications or solutions. This chapter mainly focuses on helping researchers to overcome two main difficulties while studying how to implement ProSe applications. The first is lack of comprehensive understanding and detailed analysis of special requirements in designing a proximity-service application. The second is that easy-to-use, platform-independent, and implementation-compatible development frameworks are still absent.

This chapter thoroughly investigates the requirements that need to be considered while designing and implementing ProSe applications, together with introduction and comparison of existing proximity-service platforms. Next, benefits of introducing development frameworks and challenges to build such frameworks are discussed. Some typical frameworks are also summarized and compared to provide deep knowledge about these frameworks.

References

1. Liang, C.J.M., H. Jin, Y. Yang, L. Zhang, and F. Zhao. Crossroads: A framework for developing proximity-based social interactions. In: *Proceedings of the International Conference on Mobile and Ubiquitous Systems: Computing, Networking, and Services*, Tokyo, Japan, Springer Cham, December 2–4, 2013, pp. 168–180.
2. Tsai, Y.H., Y.T. Lin, K. Loa, T.Y. Tsai, C.C. Chien, D.C. Huang, and S.T. Sheu. Proximity-based service beyond 4G network: Peer-aware discovery and communication using E-UTRAN and WLAN. In: *Proceedings of the 12th IEEE International Conference on Trust, Security and Privacy in Computing and Communications*, Melbourne, Australia, July 16–18, 2013, pp. 1345–1350.
3. Wang, Y., L. Wei, Q. Jin, and J. Ma. Device-to-device based mobile social network in proximity (MSNP) on smartphones: Architecture, challenges and prototype. *Future Generation Computer Systems (FGCS)*, 2016. doi:10.1016/j.future.2015.10.020.
4. GIGAOM RESEARCH. Proximity-based mobile social networking: Outlook and analysis. Available online: http://research.gigaom.com/report/proximity-based-mobile-social-networking-outlook-and-analysis/. Access date: April 18, 2017.
5. Bostanipour, B., B. Garbinato, and A. Holzer. Spotcast: A communication abstraction for proximity-based mobile applications. In: *Proceedings of the 11th IEEE International Symposium on Network Computing and Applications (NCA)*, Cambridge, MA, August 23–25, 2012, pp. 121–129.
6. Kulmala, J. Developing local social applications on mobile devices. Master of Science Thesis, Tampere University of Technology, 2013. Available online: https://dspace.cc.tut.fi/dpub/handle/123456789/21631, Access date: April 18, 2017.
7. Feng, J. Device-to-device communications in LTE-advanced network, networking and internet architecture. Doctor Thesis, Telecom Bretagne, University of Southern Brittany, 2013. Available online: https://hal.inria.fr/file/index/docid/983507/filename/2013telb0296_Feng_Junyi.pdf. Access date: April 18, 2017.
8. Lin, X., J.G. Andrews, A. Ghosh, and R. Ratasuk. An overview of 3GPP device-to-device proximity services. *IEEE Communications Magazine*, 2014; 52(4): 40–48.
9. Gardner-Stephen, P. The serval project: Practical wireless ad-hoc mobile telecommunications. Rural, Remote & Humanitarian Telecommunications Fellow, Flinders University and Founder, Serval Project Inc, 2011. Available online: http://developer.servalproject.org/files/CWN_Chapter_Serval.pdf. Access date: April 18, 2017.
10. Hu, X., T.H. Chu, V.C. Leung, E.C.H. Ngai, P. Kruchten, and H.C. Chan. A survey on mobile social networks: Applications, platforms, system architectures, and future research directions. *IEEE Communications Surveys & Tutorials*, 2015; 17(3): 1557–1581.

11. Hasan, S., Y. Ben-David, G. Fanti, E. Brewer, and S. Shenker. Building dissent networks: Towards effective countermeasures against large-scale communications blackouts. In: *Proceedings of the 3rd USENIX Workshop on Free and Open Communications on the Internet*, Washington, DC, August 13, 2013.

12. Thilakarathna, K., A.C. Viana, A. Seneviratne, and H. Petander. Mobile social networking through friend-to-friend opportunistic content dissemination. In: *Proceedings of the Fourteenth ACM International Symposium on Mobile Ad hoc Networking and Computing*, Bangalore, India, July 29–August 1, 2013.

13. Chang, C. Service-oriented mobile social network in proximity. *Ph.D. dissertation, Caulfield School of Information Technology Monash University*, 2013. Available online: http://kodu.ut.ee/~chang/ServiceOrientedMSNP.pdf. Access date: April 18, 2017.

14. Wang, Y. and J. Ma. *Mobile Social Networking and Computing: A Multidisciplinary Integrated Perspective*. Boca Raton, FL: CRC Press, 2014.

15. Lin, X., R. Ratasuk, A. Ghosh, and J.G. Andrews. Modeling, analysis, and optimization of multicast device-to-device transmissions. *IEEE Transactions on Wireless Communications*, 2014; 13(8): 4346–4359.

16. Gabler, J., R. Klauck, M. Pink, and H. Konig. uBeeMe: A platform to enable mobile collaborative applications. In: *Proceedings of the 9th International Conference on Collaborative Computing: Networking, Applications and Worksharing (Collaboratecom)*, Austin, TX, October 20–23, 2013, pp. 188–196.

17. Trieu, Q.L. and T.V. Pham. STARS: Ad-hoc peer-to-peer online social network. In: *Proceedings of the 4th International Conference on Computational Collective Intelligence*, Ho Chi Minh City, Vietnam, Springer Berlin Heidelberg, November 28–30, 2012, pp. 385–394.

18. Olteanu, A. and G. Pierre. Towards robust and scalable peer-to-peer social networks. In: *Proceedings of the 5th Workshop on Social Network Systems*, Bern, Switzerland, April 10, 2012.

19. Data monitoring and mining in Mobile ad hoc network. Available online: http://en.wikipedia.org/wiki/Mobile_ad_hoc_network#Data_monitoring_and_mining. Access date: April 18, 2017.

20. Zhang, R., J. Zhang, Y. Zhang, J. Sun, and G. Yan. Privacy-preserving profile matching for proximity-based mobile social networking. *IEEE Journal on Selected Areas in Communications*, 2013; 31(9): 656–668.

21. Luiz, M.D.A., M.A. Nunes, and A. Ribeiro. A middleware architecture for mobile social networking with intelligent energy saving. In: *Proceedings of the 2013 Federated Conference on Computer Science and Information Systems*, Kraków, Poland, Polish Information Processing Society (PTI), September 8–11, 2013, pp. 57–62.

22. Brooker, D., T. Carey, and I. Warren. Middleware for social networking on mobile devices. In: *Proceedings of the 21st Australian Software Engineering Conference*, Auckland, New Zealand, April 6–9, 2010, pp. 202–211.

23. Iosifidis, G., L. Gao, J. Huang, and L. Tassiulas. Incentive mechanisms for user-provided networks. *IEEE Communications Magazine*, 2014; 52(9): 20–27.
24. Wang, Y., A. Nakao, and A.V. Vasilakos. Heterogeneity playing key role: Modeling and analyzing the dynamics of incentive mechanisms in autonomous networks. *ACM Transactions on Autonomous and Adaptive Systems (TAAS)*, 2012; 7(3): 31.
25. Iosifidis, G., L. Gao, J. Huang, and L. Tassiulas. Enabling crowd-sourced mobile internet access. In: *Proceedings of the 33rd IEEE Conference on Computer Communications (INFOCOM)*, Toronto, Canada, April 27–May 2, 2014, pp. 451–459.
26. Ji, Q., H. Shen, Y. Mao, and Y. Zhu. Knowledge barter-auctioning: An incentive for quality of user-generated content in online communities. In: *Proceedings of the 2014 International Conference on Social Computing*, Beijing, China, August 8, 2014.
27. Sirur, S., P. Juturu, H.P. Gupta, P.R. Serikar, Y.K. Reddy, S. Barak, and B. Kim. A mesh network for mobile devices using Bluetooth low energy. In: *Proceedings of the 14th IEEE SENSORS Conference*, Busan, South Korea, November 1–4, 2015, pp. 1–4.
28. Kulmala, J., M. Vataja, S. Rautiainen, T. Laukkarinen, and M. Hännikäinen. Social index: A content discovery application for ad hoc communicating smart phones. In: *Proceedings of the European Conference on Service-Oriented and Cloud Computing*, Málaga, Spain, Springer, Berlin, Heidelberg, September 11–13, 2013, pp. 244–253.
29. Sharma, R. and A. Datta. Supernova: Super-peers based architecture for decentralized online social networks. In: *Proceedings of Fourth International Conference on Communication Systems and Networks (COMSNETS)*, Bangalore, India, January 3–7, 2012, pp. 1–10.
30. Nordström, E., C. Rohner, and P. Gunningberg. Haggle: Opportunistic mobile content sharing using search. *Computer Communications*, 2014; 48: 121–132.
31. Stuedi, P., I. Mohomed, M. Balakrishnan, Z.M. Mao, V. Ramasubramanian, D. Terry, and T. Wobber. Contrail: Decentralized and privacy-preserving social networks on smartphones. *IEEE Internet Computing*, 2014; 18(5): 44–51.
32. Maia, M.E.F., R.M.C. Andrade, C.A.B. de Queiroz Filho, R.B. Braga, S. Aguiar, B.G. Mateus, R. Nogueira, and F. Toorn. USABle: A communication framework for ubiquitous systems. In: *Proceedings of the 28th International Conference on Advanced Information Networking and Applications*, Victoria, Canada, May 13–16, 2014, pp. 81–88.
33. Siddiqui, F., S. Zeadally, T. Kacem, and S. Fowler. Zero configuration networking: Implementation, performance, and security. *Computers & Electrical Engineering*, 2012; 38(5): 1129–1145.
34. Software development kit. Available online: http://en.wikipedia.org/wiki/Software_development_kit. Access date: April 18, 2017.

35. Yu, Z., Y. Liang, B. Xu, Y. Yang, and B. Guo. Towards a smart campus with mobile social networking. In: *Proceedings of the 4th International Conference on Cyber, Physical and Social Computing, Internet of Things (iThings/CPSCom)*, Dalian, China, October 19–22, 2011, pp. 162–169.

36. The serval project. Available online: http://developer.servalproject.org/dokuwiki/doku.php?id=content:about. Access date: April 18, 2017.

37. Wang, Y., L. Wei, Q. Jin, and J. Ma. AllJoyn based direct proximity service development: Overview and prototype. In: *Proceedings of the 17th International Conference on Computational Science and Engineering (CSE)*, Chengdu, China, December 19–21, 2014, pp. 634–641.

38. Salmon, J.L. and R. Yang. A proximity-based framework for mobile services. In: *Proceedings of the IEEE 3rd International Conference on Mobile Services*, Anchorage, AK, June 27–July 2, 2014, pp. 124–131.

39. Arnaboldi, V., M. Conti, and F. Delmastro. CAMEO: A novel context-aware middleware for opportunistic mobile social networks. *Pervasive and Mobile Computing*, 2014; 11: 148–167.

40. Arnaboldi, V., M. Conti, and F. Delmastro. Implementation of CAMEO: A context-aware middleware for Opportunistic Mobile Social Networks. In: *Proceedings of the 12th IEEE International Symposium on a World of Wireless, Mobile and Multimedia Networks (WoWMoM)*, Lucca, Italy, June 20–24, 2011, pp. 1–3.

41. Commotion on Wikipedia. Available online: http://en.wikipedia.org/wiki/Commotion_Wireless. Access date: April 18, 2017.

42. FAQ of commotion. Available online: https://commotionwireless.net/about/faq/. Access date: April 18, 2017.

43. Commotion wireless: Free and open way to network. Available online: http://abcnews.go.com/Technology/commotion-wireless-free-open-network/story?id=18659257. Access date: April 18, 2017.

44. Serval project on Wikipedia. Available online: http://en.wikipedia.org/wiki/Serval_project. Access date: April 18, 2017.

45. Gardner-Stephen, P., R. Challans, J. Lakeman, A. Bettison, D. Gardner-Stephen, and M. Lloyd. The serval mesh: A platform for resilient communications in disaster & crisis. In: *Proceedings of Global Humanitarian Technology Conference (GHTC)*, San Jose, CA, October 20–23, 2013, pp. 162–166.

46. Arnaboldi, V., M. Conti, and F. Delmastro. Cameo software architecture and APIs: Technical specification. Available online: http://cnd.iit.cnr.it/fdelmastro/CAMEO.pdf. Access date: April 18, 2017.

PART II
FUNDAMENTAL ISSUES

6

SMARTPHONE-BASED HUMAN ACTIVITY RECOGNITION

6.1 Introduction

Nowadays, smartphones are ubiquitous and becoming more and more sophisticated, with ever-growing computing, networking, and sensing abilities. This has been changing the landscape of people's daily life and has opened the doors for many interesting applications, ranging from health and fitness monitoring, personal biometric signature, urban computing, assistive technology, and elder-care to indoor localization and navigation, and so on. Besides the inclusion of sensors, such as accelerometer, compass, gyroscope, proximity, light, global positioning system (GPS), microphone, camera, the ubiquity and unobtrusiveness of the phones and the availability of different wireless interfaces, such as Wi-Fi, 3G, and Bluetooth, make them an attractive platform for human activity recognition (HAR) [1–3].

Basically, human activities can be detected with the following four approaches [4]: wearable device, video-based device, ambience device, and smartphone-based pervasive device. The wearable device approach is to hold or wear some devices with embedded sensors to detect the posture and motion of the wearer's body and then use classifiers to identify activity. The wearable device provides an obtrusive way of monitoring. Moreover, the battery life and the weight of the device are major problems that should be solved in the designing of such standalone systems. Video-based approach is another method for activity recognition that uses camera as a vision sensor and is an unobtrusive way for activity monitoring. Due to the fixed nature of the camera, it is mostly used in indoor environment. Ambience device approach is to use multiple installed sensors to collect the data from a person when he or she is close to them. These systems can be used for indoor environments. They are not well suited for a normal living environment because the excessive noise in a normal living environment decreases the signal obtained from the sensors. Smartphone-based pervasive approach is classified as a new method for activity recognition which utilizes smartphones as a platform for detecting the activity. As a typical example, smartphone-based fall detection system can work almost everywhere with minimal requirement: the accelerometer sensor embedded inside smartphones.

HAR takes the raw senor reading as inputs and predicts a user's motion activity. These sensors have become a rich data source to measure various aspects of a user's daily life. The typical activities include walking, jogging, and sitting. Due to its unobtrusiveness, low or no installation cost, and easy-to-use properties, smartphones are becoming the main platform for HAR [5]. In fact, despite these advantages, activity recognition on smartphones also faces challenges such as the battery limitation of the phones, limitations in processing, storage compared to more powerful stations, and diverse human behavior, that is, various usage of mobile phones by different people, which need to be investigated for the realization and further adoption of smartphone-based activity recognition systems [6–8]. Sensors embedded in smartphones are the source for raw data collection in activity recognition, which will directly impact the accuracy in the recognition period and the applicability of the classification models. After collecting the raw data from different sensors, the next step is to preprocess it before performing any further calculation. Similar to the task of data mining, extracting the *right* features is critical to the final recognition performance. From data mining perspective, activity recognition can be regarded as a multiclass classification problem, and many existing classifiers can be plugged in.

The remainder of the chapter is organized as follows: Section 6.2 introduces the basic concepts about smartphone-based HAR, including sensors commonly used and various activities detected. Considering the methodology of activity recognition is generally composed of two main phases: training phase and activity-classifying phase. Section 6.3 focuses on the elemental processes of activity recognition in each phase and summarizes the typical performance metrics. Section 6.4 presents the typical applications of activity recognition including daily life monitoring, health care, indoor localization, and so on. Section 6.5 discusses the challenges and open issues related to smartphone-based activity recognition, including energy

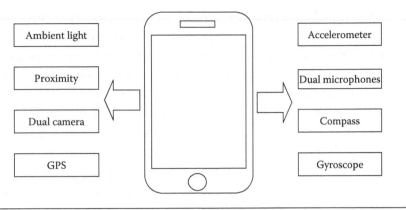

Figure 6.1 Typical sensors available on smartphone.

consumption, smartphone placement, training burden, and security and privacy. Finally, Section 6.6 briefly concludes this chapter.

6.2 Background

6.2.1 Common Sensors Embedded in Smartphones

As mobile phones have matured as a computing platform and acquired richer functionality, these advancements often have been paired with the introduction of new sensors, as shown in Figure 6.1. The basic function of common sensors is briefly described in Table 6.1.

Microelectromechanical sensors (*MEMS*) are made on a tiny scale and usually on silicon chips using techniques of computer-chip manufacturing. All physical sensors in smartphones are made using these

Table 6.1 Description of Sensors' Function

SENSOR	DESCRIPTION
Accelerometer	Measure the acceleration force that is applied to the device
Ambient temperature sensor	Measure the ambient room temperature
Gravity sensor	Measure the force of the gravity
Light sensor	Measure the ambient light level
Linear acceleration	Measure the acceleration force
Magnetometer	Measure the ambient geomagnetic field in three axes
Barometer	Measure the ambient air pressure
Proximity sensor	Measure the proximity of an object relative to the screen
Humidity sensor	Measure the humidity of ambient environment
Gyroscope	Measure the orientation of device in pitch, roll, and yaw

techniques, including pressure sensor, accelerometer, gyroscope, and the compass. Especially, from programming viewpoint, the sensors referenced through the Sensor class in Android may be of two types: *raw* sensors or *synthetic* (or *composite* or *virtual*) sensors. Raw sensors correspond to the actual physical components inside the Android device and gives raw sensing data. Synthetic sensors provide an abstraction layer between application code and low-level device components by either combining the raw data of multiple raw sensors or by modifying the raw sensor data to make it easier to handle. They may report a physical quantity by referring to two or three sensors (such as reporting orientation by referring to the compass that gives a north–south–east–west bearing and the accelerometer that gives tilt), or alternatively they may manipulate the sensor reading before reporting it, for example, by integrating the gyroscope data before using it in addition to magnetometer and accelerometer to get a better determination of orientation. Regardless of the sensor type, the programmer accesses any type of sensor in the same way using the sensor application programming interface (API).

6.2.1.1 Accelerometer The raw data stream from the accelerometer is the accelerations (in the units of g-force) of three axes, whose directions are predefined as shown in Figure 6.2.

Those raw data are represented in a set of vectors: $<A_x, A_y, A_z>$. A time stamp can also be returned together with the three axes readings. Most of the existing accelerometers provide a user interface to configure the sampling frequency so that the user could choose a best sampling rate through experiments.

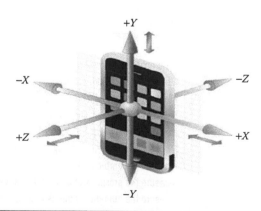

Figure 6.2 Illustration of three axes in accelerometer sensor of smartphone.

Due to the fact that it directly measures the subject's physiology motion status, accelerometer has been used heavily in smartphone-based activity recognition. For example, if a user changes his or her activity from walking to jogging, the activity will reflect on the signal shape of the acceleration reading along the vertical axis—there will be an abrupt change in the amplitude. Moreover, the acceleration data could indicate the motion pattern within a given time period, which is helpful in the complex activity recognition.

6.2.1.2 Compass Sensor Compass is a traditional tool to detect the direction with respect to the north–south pole of the earth by the use of magnetism. The compass sensor on smartphone works with a similar functionality. Figure 6.3 shows the compass reading display screen on a smartphone. It begins from the absolute north and the actual reading indicates the angle between current smartphone-heading direction and the absolute north in clockwise. For example, the reading of heading to absolute East is 90° and heading to absolute West is 270°. The raw data reading from a compass sensor is the float number between 0° and 360°. Compass reading can be used to detect the direction change in the user's motion such as walking.

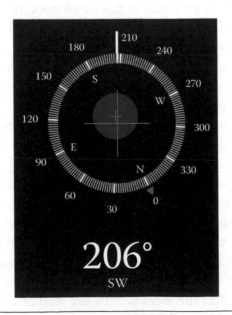

Figure 6.3 Illustration of compass reading display screen on a smartphone.

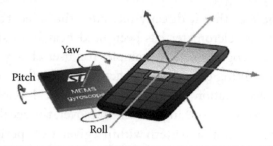

Figure 6.4 Three axes of gyroscope on smartphone.

6.2.1.3 Gyroscope Gyroscope measures the phone's rotation rate by detecting the roll, pitch, and yaw motions of the smartphone along the *x*, *y*, and *z* axis, respectively. The axes directions are shown in Figure 6.4. The raw data stream from a gyroscope sensor is the rate of the rotation in radian per second (rad/s) around each of the three physical axes on smartphones. Gyroscope is helpful in the navigation applications as well as some smartphone games. In activity recognition research, gyroscope is used to assist the mobile orientation detection.

6.2.2 Activities

In the previous subsection 6.2.1, several commonly available sensors on the mobile phone are listed. In this subsection, the aim is to give an overview about the activities that can be recognized by state-of-the-art activity recognition systems using those sensors.

In many cases, human activity is approximated with two proxies, namely location and motion, which correspond to two orthogonal types of activity recognition [9].

The best way of introducing locational activity recognition may be to compare it to the better known and closely related localization task (this would be discussed in Chapter 7). In the latter, the objective is to estimate the position of a device anywhere as accurately as possible using techniques such as assisted GPS or pedestrian dead reckoning (PDR). In contrast, locational activity recognition is interested in the meaning of the user's location rather than its coordinates. In other words, a locational activity recognition system distinguishes between locations only if it helps determine what the user is doing. Typical locations that reflect the user's activity are workplace, home, restaurant, shopping centre, and so on.

In motional activity recognition, the user's activity is abstracted as a motion state or a mode of transportation. Certain motion states can be identified from cellular signals even without any knowledge about the location of observed cell towers. For examples, fluctuations in GSM signals have been shown to be sufficient to determine with reasonable accuracy whether a mobile phone carrier is walking, driving in a motor car, or staying stationary. Accelerometers embedded in recent smartphones can serve the same purpose and further can distinguish between finer movements such as sitting, standing, or running.

Location- and motion-associated activity recognition are the two dominating types of activity recognition using mobile phones. Besides these, recent applications consider using mobile phones for more complex activities, for instance, in the field of sports such as outdoor bicycling, soccer playing, lying, walking, rowing with the rowing machine, running, sitting, standing, and walking using accelerometers or for daily activities such as shopping, using a computer, sleeping, going to work, going back home, working, and having lunch, dinner, or breakfast. Some recent applications also consider using mobile phones for detecting dangerous situations such as falls [10].

Activities recognized by the sensor's data can also be classified in terms of the complexity of activities. A simple locomotion could be walking, jogging, walking downstairs, taking elevator, and so on. The complex activities are usually related to a combination of a longer period of activities (e.g., taking bus and driving). The activities may only correspond to the movements of certain parts of the body (e.g., typing and waving hand). There are several healthcare-related activities, such as falling, exercise, and rehabilitations. Furthermore, smartphone-based mood/emotion detection (or mood, including happy, sad, active, and lazy) has emerged in the field of affective computing [11].

Table 6.2 summarizes the sample types of activities that are inferred in state-of-the-art activity recognition systems on mobile phones classified into five different categories according to their objectives.

6.3 Core Techniques

In general terms, activity recognition consists of associating sensor readings and other inputs to a label taken from a set of distinguishable activities. The task therefore involves (1) determining a set of

Table 6.2   Classification of Human Everyday Life Activities

CLASS	ACTIVITY TYPES
Locomotion	Walking, running, sitting, standing, still, lying
Daily activities	Shopping, using computer, sleeping, going to work, going back home, working, lunch, dinner, breakfast, in a conversation, attending a meeting
Exercise	Outdoor bicycling, soccer playing, biking on a fitness bike
Mode of transportation	Biking, traveling with a vehicle, riding a bus, driving
Health related activities	Falls, rehabilitation activities, following routines
Emotion detection	Happy, sad, active, and lazy, etc.

activity labels and (2) assigning sensor readings and other inputs to the appropriate activity labels. Activity labels are the output of an activity recognition system. Activity labels are assigned based on context information and background knowledge. Context information includes sensor data, time, and user inputs. Background knowledge can either be provided by experts or mined automatically. For example, the set of activities that may be performed at a specific location is an example of useful background knowledge. The general process of activity recognition is clearly demonstrated in Figure 6.5.

Basically, smartphone-based activity recognition includes two main phases: training phase and classifying phase. The main steps in both phases include (a) preprocessing of sensor data and segmentation and (b) feature extraction.

The preprocessing step contains noise removal and representation of raw data. The segmentation process is usually applied to continuous stream of sensor data to divide the signal into smaller time segments as

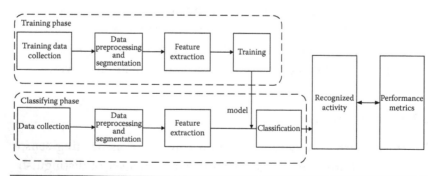

Figure 6.5   General framework of activity recognition on smartphones.

retrieving useful information from a continuous stream of sensor data is a difficult problem. For this purpose, different segmentation methods can be applied to time-series data that enable us to gather useful information from continuous stream of sensor data. The feature extraction phase includes the generation of abstractions that accurately characterize the sensor data. In other words, large input sensor data are reduced to a smaller set of features, called the feature vector, that represents the original data in the best way. Dimensionality reduction techniques such as principal component analysis and sequential forward feature selection can be applied to remove the irrelevant features to decrease the computational effort and memory requirements in the classification process. The aim of dimensionality reduction is to reduce the computational complexity and increase the performance of the activity recognition process.

After the previous steps, the generated feature vectors can be used directly in the classifying step. Finally, the classification phase includes mapping the sensor data (i.e., the extracted feature set) to a set of activities. The classification technique may involve a simple threshold-based scheme or a machine learning scheme based on pattern recognition or neural networks. Common pattern recognition algorithms include decision tables, decision trees, hidden Markov models (HMM), Gaussian mixture models, and support vector machines.

In brief, the functionalities of three main processes are as follows:

- *Data preprocessing and segmentation*: This phase aims to remove the noise from raw data and to divide the signal into smaller time segments for retrieving useful information from a continuous stream of data.
- *Features extraction and evaluation*: This phase aims at extracting and evaluating meaningful parameters able to univocally characterize each class, therefore enabling the classification process.
- *Classification*: This phase refers to the last stage that is the association of the input pattern to a specific class, according to specific decision rules. In this phase, usually, machine learning approaches are utilized to classify unknown activities, and various metrics are used to evaluate the classification performance.

6.3.1 *Raw Data Collection and Preprocessing*

The recognition process starts with collecting data from the sensors available on smartphones, including accelerometer sensor, GPS, and gyroscope. The way to collect the raw data will directly impact the recognition accuracy and the applicability of the classification models. It is shown that the recognition model trained from one subject's data has a lower accuracy in recognizing another subject's activities. Another subtle issue in raw data collection is the sampling rate. Almost all smartphones provide APIs to allow users to configure the sampling rate of sensors. Although data collected at a higher rate provide more information, it may also introduce more noise. Therefore, a higher sampling rate does not always lead to a higher accuracy. When reading sensor data, it is important to understand the types of errors encountered. Being familiar with why a sensor reading might return bad data may play a key role in developing algorithms to detect and process the erroneous data. Common causes of error in sensor data are discussed in the following subsections [12].

6.3.1.1 *Types of Errors*

- *Human error, systematic error, and random error*: Human errors are mistakes made by humans in making a measurement (such as incorrectly reading a value from a graph). Systematic errors are errors that affect the accuracy of a measurement—they are a constant offset from the true value (for instance, taking a measurement with the magnetometer with a magnet nearby). In some cases they can be predicted or removed by calibration or by changing the measurement scheme. On the other hand, random errors such as noise result in imprecise measurements and cannot be removed by these techniques.
- *Noise*: Noise is the random fluctuation of a measured value. Although noise can be categorized (brown noise, white noise, and so on) and statistically quantified, these details are not usually necessary for programming with Android sensors. Low-pass filters can help mitigate the effects of noise when necessary.
- *Drift*: Drift describes slow, long-term wandering of data away from the real-world value. Drift may occur due to the sensor

reading itself degrading over time. It can also occur if a sensor value is integrated. In such cases, a small offset will add up in each integration, which results in the drifting away from the real measurement.

- *Zero offset (or "offset," or "bias")*: If the output signal is not zero when the measured property is zero, the sensor has an offset or bias. For example, if the average accelerometer measurement when the device is flat on a table is not exactly $(0, 0, -9.80665 \text{ m/s}^2)$, the accelerometer has an offset. If a gyroscope does not measure exactly $(0, 0, 0)$ rad/s when stationary, even a small zero offset will show up as an integration error when the gyro data are integrated to find the angle.

- *Time delays and dropped data*: As Android is not a real-time operating system, some measured data values can be delayed, resulting in incorrect timestamps. Data may even sometimes be dropped when the device is busy.

- *Integration error*: The gyroscope reports angular rotation rate in radians per second, however it would be more useful in many applications to know the amount by which the device has rotated, which can be obtained by integrating the gyroscope's readings and finding a rotation angle in radians. However, the zero offset and drift in the gyroscope measurements mean that simple integration will give poor results. Over time, it will quickly explode to give large unphysical numbers even with no actual rotation, because offset and drift are accumulated under the integral, each time the integration executes. For example, the accelerometer was likely sampled irregularly because of the Android framework's implementation of the sampling mechanism. The Android API offers four abstract sampling rates for its accelerometer sensor (listed from fastest to slowest): Fastest, Game, Normal, and UI. The actual physical sampling capabilities of the accelerometers likely vary from device to device, so these sampling rate options are used more as guidelines than as actual physical sampling rates. Another reason is because of how application on the Android framework receives acceleration readings. The Android API only allows the acquisition of an acceleration sample on an "onSensorChanged()" event, which triggers whenever the Android OS determines that the accelerometer values have changed. As such, an acceleration sampling application cannot force a reading of the accelerometer at predetermined intervals. Combined with the fact that Android OS likely has varying loads of activity from moment to moment, the "onSensorChanged()" is not scheduled firmly, resulting in irregularly sampled values. To regularize sampled values, the holes in the signal were interpolated via a process called data linearization.

In brief, there are many sources of noise and uncertainty in mobile phone data. Devices can be turned off, not recharged, or forgotten. There are issues with radio communication such as poor indoor reception and fluctuating connections. Cell tower allocation obeys the operator's strategy, which is not known a priori. Bluetooth errors include detecting people who are not physically proximate through certain types of windows or doors. There is also a probability that Bluetooth will not discover other proximate devices.

6.3.1.2 Techniques to Address Error After collecting the activity data, the data are subjected to appropriate noise reduction techniques in the preprocessing step. Once filtered, the data are processed to extract various features [12].

- *Rezeroing*: If there is an offset present that is affecting your application, it may be useful to rezero the sensor measurements. This is as simple as storing a calibrated value (potentially stored when the user clicks a Calibrate button) and subtracting it from each measured value. For instance, the device may be placed flat on a surface and the *downward* direction as measured by the accelerometer can be calibrated. This is simple enough, but the trick is getting the user to actually perform the calibration, and know how and when to do so.
- *Filters*: Low-pass filters filter out any high-frequency signal or noise and have a *smoothing* effect on data. High-pass filters filter out slow drift and offset and just give the higher frequency changes. The *cutoff frequency* is the approximate transition frequency above or below which the data are filtered out. Band-pass filters reject both low-frequency and

high-frequency data and just keep the data in some frequency range of interest. For example:

Since most of the energy captured by the accelerations and angular rates associated to human movements is below 15 Hz [13], low-pass filter, for example, a 10th-order Butterworth filter with a 15 Hz cutoff frequency is often used to remove the high-frequency components of the noise. The signal-to-noise ratio taken on the samples above 15 Hz demonstrate that higher frequency components are primarily composed of noise and contain almost no actual contribution to the walking data [14]. This phase is necessary for extracting useful information from the low-cost sensor signals enabling the human motion recognition process. It is worth pointing out that high-frequency components could be exploited to identify other types of contexts, such as driving a car, and so on.

- *Sensor fusion*: Sensor fusion refers to using more than one sensor to take advantage of the strengths of each sensor and mitigate the effects of the weaknesses. For example, the accelerometer can give a relatively accurate measurement of the *downward* direction, but it has the disadvantage that it can never tell us the north-south-east-west yaw of the device. However, the compass can supplement that measurement to give yaw. A more complicated sensor fusion approach might also add integrated gyroscope data to give an app access to faster and lower noise changes than the accelerometer and compass can give. In effect, an app would primarily use the high-quality gyroscope data to get orientation information, but *nail it down* and prevent it from drifting by continually comparing it to the zero-drift accelerometer and compass data.

6.3.2 Feature Computation

After the collection of labeled data and preprocessing (noise removal and representation of raw data), feature extraction (abstractions of raw data to represent main characteristics) step is followed. As in any other data mining tasks, extracting the *right* features is critical to the final recognition performance. The feature extraction phase transforms raw data into a reduced representation called a feature vector. The accuracy of the system depends strongly on how data are represented. Features should help differentiate between activities.

Since mobile phones are generally energy constrained and extending phone battery life is an essential requirement, smartphone-based activity recognition requires features that are both light-weight (energy efficient) and accurate (possess high discriminating power) to preserve battery life and to ensure high accuracy. For activity recognition, features can be extracted in time domain, frequency domains, and discrete domain.

6.3.2.1 Time-Domain Features Time-domain features contain the basic statistics of each data segment and those of different segments [13]. Simple mathematical and statistical metrics can be used to extract basic signal information from raw sensor data. There are more than 70 features in time domain, such as average, maximum, minimum, standard deviation, variance, zero cross (count of change of sign of values), and time between peaks (TBP in the sinusoidal waves).

Reference 13 focuses on the selection of time-domain features for classification of daily activities. The main goal is to eliminate the unnecessary features and determine the most discriminative ones. It will lead to increase the success rate of classification, while saving energy of mobile phone due to less number of features. Experiments show that using only 20 features (even 10 features) result in the same success rate compared to the values obtained with 70 features. It is summarized that 15 features are the minimum number of features that give the best success rate of classification.

Intuitively, complicated machine learning solutions are not feasible for smartphone-based HAR, because the energy supply is exhausted quickly by those computation-consuming learning algorithms. A threshold-based approach is less complex, which helps to save battery power. An energy-efficient fall detection scheme on smartphone is presented in Reference 23, which detected the typical phases of falling behavior: free fall, impact, post impact, and stability by comparing the current sensor data with different threshold values.

6.3.2.2 Frequency-Domain Features Frequency-domain techniques have been extensively used to capture the repetitive nature of a sensor signal. This repetition often correlates to the periodic nature of a specific activity such as walking or running. A commonly used signal transformation technique is the Fourier transform (FT), which can represent in the frequency domain (or spectrum), the important characteristics of a time-based signal such as its average (or DC component, the first coefficient in the spectral representation of a signal, and its value is often much larger than the remaining spectral coefficients) and dominant frequency components. In this spectral representation, the main periods or repetition intervals of the signal are represented by non-zero values or coefficients at the corresponding frequency axis value. In addition to the fast Fourier transform (FFT) and its spectral representation, other frequency-based representation has been used. For example, the wavelet transforms represent a time-domain signal as a decomposition of a set of weighted orthonormal vector basis or coefficients. These transforms, although less common, provide computational advantages over the more established FFT computation [15].

The energy of the signal can be computed as the squared sum of its spectral coefficients normalized by the length of the sample window. The energy metric can be used to identify the mode of transport of a user with a single accelerometer, respectively, walking, cycling, running, and driving. The entropy metric can be computed using the normalized information entropy of the discrete FFT coefficient magnitudes excluding the DC component. Entropy helps to differentiate between signals that have similar energy values but correspond to different activity patterns. Together with the mean, energy, and correlation, entropy has been used in several activity recognition approaches.

In walk detection, Reference 16 extracted walking periods by thresholding the ratio between the energy in the band of walking frequencies and the total energy across all frequencies, where the energy was computed using continuous/discrete wavelet transform (CWT/DWT).

6.3.2.3 Discrete-Domain Features Accelerometer and other sensor signals are time series data, which can be transformed into strings of discrete symbols, called symbolic representation, which is one of the dimensionality reduction techniques in time series data mining. Symbolic representation of time series uses an alphabet *a* (usually finite) to reduce the dimensionality of the time series, which can be defined as a map: $[s_i \ldots s_n] \xrightarrow{f} a_k$, $a_k \in \boldsymbol{a}$, mapping time series data

$[s_i \cdots s_n]$ to symbol a_k by using map relation f, and a_k belong to alphabet $\boldsymbol{a}$. Symbolic representation of time series has attracted much attention, because this paradigm not only reduce the dimensionality of time series but also benefit from the numerous algorithms used in bioinformatics and text data mining.

The symbolic aggregate approximation method (SAX) is one of the most powerful methods of symbolic representation of time series. SAX is based on the fact that normalized time series have *high Gaussian distribution*, so by determining the breakpoints that correspond to the alphabet size, one can obtain equal sized areas under the Gaussian curve. SAX works as follows: (1) the time series are normalized; (2) the dimensionality of the time series is reduced by using piecewise aggregate approximation (PAA). In PAA the times series is divided into equal sized frames and the mean value of the points within the frame is computed. The lower dimensional vector of the time series is the vector whose components are the means of the successive frames; (3) the PAA representation of the time series is discretized. This is achieved by determining the number and location of the breakpoints. The number of breakpoints is related to the desired alphabet size (which is chosen by the user), that is, alphabet_size = number(breakpoints) + 1. Their locations are determined by statistical lookup tables so that these breakpoints produce equal sized areas under the Gaussian curve. The interval between two successive breakpoints is assigned to a symbol of the alphabet, and each segment of the PAA that lies within this interval is discretized by corresponding symbol.

Once signals have been mapped to strings, exact or approximate matching and edit distances are key techniques used to evaluate string similarity and thus either find known patterns or classify the user activity. Some typical metrics of evaluating string similarity are as follows:

- Euclidian-related distances between symbols are defined by the corresponding numeric distance between the signal values that correspond to each symbol in the string representation.
- Levenshtein edit distance determines the minimum number of symbol insertions, deletion, and substitutions needed to transform one string into the other.

- Dynamic time warping (DTW) is a metric for measuring similarity between two sequences that may vary in length and can thus correspond to different time basis. It can capture similarities of strings with distinct sampling period, but has a relatively high computational cost.

In brief, a key aspect of this transformation is discretization process, while there is potential for information loss because a symbol alphabet can lead to a substantial compression in the representation of a signal.

Discussion of the techniques for extracting this activity information from raw accelerometer data is presented in Reference 15. Time-domain metrics are dominated by the cost of normalization and none of them involve complex arithmetic operations such as trigonometric or logarithmic functions. Some of those simple metrics exhibit a very small number of multiplications; in some extreme cases (e.g., differences), only additions and comparisons are required. In general, any of these metrics can be a good candidate for implementation on mobile devices. Somewhat surprising is the good performance of some of the simpler time-based metrics making them a method of choice for embedded mobile devices where energy and storage are at premium.

For complex activity, time-domain features are less accurate than frequency-domain features. However, frequency-domain features require higher components to discriminate between different activities. Their calculation requires longer time windows, and thus they increase computational cost and are not suitable for real-time applications.

6.3.3 Classification Methods

The activity recognition process can be summarized as determining a target set of activities, collecting sensor readings, and assigning sensor readings to the appropriate activities. In other words, it is the process of how to interpret the raw sensor data to classify a set of activities.

Many of the activity recognition studies focus on the use of statistical machine learning techniques to infer information about the user activities from raw data. The learning phase can be supervised or unsupervised. Supervised techniques rely on labeled, that is, associated with a specific class or activity, sample observations to build classifiers, whereas unsupervised techniques do not rely on labeled data.

In general, a classification process can be considered as a mapping function that given an input pattern characterized by a set of d features, named as feature vector $f = \left[f_1, f_2 \ldots f_d \right]$, assigns each feature vector to one of the n possible classes $c = [c_1, c_2 \ldots c_n]$. Classifier algorithms are traditionally divided into two groups:

- *Supervised classifier*: The labeled data whose classes are known are used to train the classifier and then to assign unlabeled data to one of the known classes.
- *Unsupervised classifier*: Here the classes are not known a prior but are defined when the classification is completed, this is the case, for example, of the clustering classification.

Supervised learning methods are composed of two main phases: training and classifying. In the training phase, machine learning approaches utilize a given set of examples or observations, called the *training set*, to discover patterns from the sensor readings. These examples or observations should be associated with a specific class of activity. Labeling the data in the training phase is usually a tedious and complex process. The user can also label each activity being performed by using an automatic voice recognition system or utilizing video camera, in other words, the activities are automatically labeled by the system.

Models are learnt from training data in essentially two ways. Most activity recognition systems rely on supervised learning. In this type of learning, training data are collected by the sensing module and other inputs and then transformed into a set of training instances by the preprocessing module. Each training instance is then labeled for the user's true activity when data were collected. The supervised inference model generalized from training data is to infer the activity labels of unseen instances in test situations.

Unsupervised learning aims to relieve the user from the burden of labeling, usually by clustering data based on some distance measure. One of the most common clustering techniques is K-Means, which tries to minimize the total intracluster variance. In either case, devising an appropriate distance measure is a difficult problem and can be quite subjective.

6.3.3.1 Performance Metrics for Activity Recognition   In the classifying (classification) phase of an activity recognition system, the output classes should be compared with the ground truth, that is, what the user was actually doing in order to evaluate the success of a classification scheme. Specifically, performance metrics are usually calculated in three steps [17]. First, a comparison is made between the returned system output and what is known to have occurred (or an approximation of what occurred). From the comparison, a scoring is made on the matches and errors. Finally, these scores are summarized by one or more metrics, usually expressed as a normalized rate or percentage.

The methodology of evaluating machine learning (ML) algorithms is predominantly made through the statistical analysis of the models using the available experimental data. The vast majority of the studies use cross-validation with statistical tests to compare classifiers' performance for a particular dataset. The most common method is the confusion matrix $M_{n \times n}$, which allows representing the algorithm performance by clearly identifying the types of errors (false positives and negatives) and correctly predicted samples over the test data. The element m_{ij} in the matrix $M_{n \times n}$ is the number of instances from class i that were actually classified as class j. From it, various metrics can also be extracted such as model accuracy, sensitivity, specificity, precision, and F_1 – measure. In addition, it is also necessary to quantitatively compare various classifier algorithms, in terms of the number of activities recognized, prediction speed, memory consumption, and so on.

6.3.3.1.1 Confusion Matrix   The common method used to analyze the performance of classification in HAR field is the confusion matrix $M_{n \times n}$. Rows represent the actual activities and columns represent the predicted ones. Therefore, each matrix cell M_{ij} shows the number of instances of activity i that were predicted as activity j. It is clear then that all the values within the matrix diagonal are correct predictions and classification errors otherwise. From this matrix, we can visualize four different helpful values used to estimate the various statistical measures regarding the system performance. These are visible in the simplified confusion matrix of two classes (a and b) from Table 6.3. We take these values by assuming that class a is the class of interest

Table 6.3   The Simplified Confusion Matrix of Two Classes (*a* and *b*)

ACTUAL CLASS	PREDICTED CLASS	
	a	*b*
a	True positives (TP)	False negatives (FN)
b	False positives (FP)	True negatives (TN)

or positive condition. The following values can be obtained from the confusion matrix in a binary classification problem:

- *True positives* (*TP*): Actual samples of class *a* correctly predicted as class *a*.
- *True negatives* (*TN*): Actual samples of class *b* correctly predicted as class *b*.
- *False positives* (*FP*): Actual samples of class *b* incorrectly predicted as class *a*.
- *False negatives* (*FN*): Actual samples of class *a* incorrectly predicted as class *b*.

In addition, other metric can also be extracted from matrix $M_{n \times n}$. The accuracy is the most standard metric to summarize the overall classification performance for all classes and it is defined as follows:

$$\text{Accuracy} = \frac{TP + TN}{TP + TN + FP + FN}$$

The precision, often referred to as positive predictive value, is the ratio of correctly classified positive instances to the total number of instances classified as positive:

$$\text{Precision} = \frac{TP}{TP + FP}$$

The recall, also called true positive rate or sensitivity, is the ratio of correctly classified positive instances to the total number of positive instances. This is a measure of how good is the classifier to correctly predict actual positives samples:

$$\text{Recall} = \frac{TP}{TP + FN}$$

In contrast, the specificity measure, also called the true negative rate, shows the ability to correctly predict actual negative samples. It is formulated as

$$\text{Specificity} = \frac{TN}{TN + FP}$$

The F_1 – measure combines precision and recall in a single value:

$$F_1 - \text{measure} = \frac{2 \cdot \text{Precision} \cdot \text{Recall}}{\text{Precision} + \text{Recall}}$$

This accuracy gives an indication of how good the overall performance of a classifier is. Moreover, we can also use the error measure for expressing the opposite. It denotes the deviation of the measurement from the truth and it can be obtained in terms of the accuracy as follows:

$$\text{Error} = 1 - \text{accuracy}$$

6.3.3.1.2 Qualitative Criteria In order to select the most appropriate classification system for a particular application, it is also imperative to take into account a set of various qualitative criteria in addition to the statistical measures. These aspects can help to make important decision trade-off, during this selection process. Some of them are described in Table 6.4.

6.3.3.2 Taxonomy of Classifiers The core task of classifier is to classify the readings from multiple sensors (the so-called example) into correct activities. In the training phase, the classifier is built from training dataset that denotes the given examples or observations. The instances in the training set may not be labeled. In some cases, labeling data is not feasible because it may require an expert to manually examine the examples and to assign a label based upon their experience. This process is usually tedious, expensive, and time consuming in many data mining applications.

According to whether the training dataset may or may not be labeled, the classifiers can be divided into two categories: supervised and unsupervised. Many existing systems have already adopted

Table 6.4 Qualitative Criteria of Selecting Appropriate Classification System

CRITERION	DESCRIPTION
Online capability	Measures whether the system is able to perform activity recognition in real time
Recognition time	Is the time delay associated with the activity prediction process? For instance, the length of the time window related to each prediction and its CPU processing time
Battery consumption	Quantifies the energy expenditure of portable devices and how this affects their continuous operation time
Memory and CPU usage	Considering the limited memory and CPU abilities in smartphones, it is critical to share resources with others applications
Obstrusivity	Evaluates how comfortable is the system for the user (e.g., sensors location and weight, presence of wires, fitting time)
User's privacy	Examines whether the system protects the accessed personal data of its users from external sources
Number and type of classes	Be able to be classified by the system
Number and type of sensors	Be required for the recognition of activities
Modular design	Indicates whether or not the system allows its integration with others or the adaptation of new sensors and devices

supervised learning methods and yielded high classification accuracy. For example, some systems store sensor data in a nonvolatile medium while a person from the research team supervises the collection process and manually registers activity labels. Other systems feature a mobile application that allows the user to select the activity to be performed from a list [18]. In this way, each sample is matched to an activity label and then stored in the server.

Considering that supervised learning methods are popularly used in HAR field, this chapter mainly emphasizes on discussing some supervised learning-based classification schemes, including decision tree, KNN, HMM, and SVM. In an unsupervised learning method, the training data consist of only input vectors without their associated activity labels. It aims to find certain similarities or discover distinguishable structure within the input data (e.g., clustering). But unsupervised learning is seldomly employed in HAR field, due to its high complexity and computation overhead.

6.3.3.3 Decision Tree   Decision-tree induction algorithms are highly used in a variety of domains for knowledge discovery and pattern recognition. They have the advantage of producing a comprehensible classification/regression model and satisfactory accuracy levels in several application domains, such as business intelligence, health care, and biomedicine. Due to its low complexity in implementation and excellent interpretation, decision tree is adopted as the main classifier in many activity recognition researches. Decision trees are an efficient nonparametric method that can be applied either to classification or to regression tasks. They are hierarchical data structures for supervised learning whereby the input space is split into local regions in order to predict the dependent variable.

Normally, the decision tree consists of two parts, tree building and tree pruning. In the stage of tree building, the initial state of a classification tree, called the root node, is the first node to which all the patterns of the training set are assigned. If the training example consists of all the same class, then there is only a need for the root node. However, if the training examples at the root node consist of two or more classes, a test node is made that will split the training set into two subspaces, or secondary nodes. These can either become terminal nodes, in which a classification is reached, or another test node. The process is repeated until each branch results in a terminal node and a completely discriminating tree is obtained.

In decision tree, root and internal nodes hold a test over a given dataset attribute (or a set of attributes), and the edges correspond to the possible outcomes of the test. Leaf nodes can either hold class labels (classification), continuous values (regression), or even models produced by other machine learning algorithms. For predicting the dependent variable value of a certain instance, one has to navigate through the decision tree. Starting from the root, one has to follow the edges according to the results of the tests over the attributes. When reaching a leaf node, the information it contains is responsible for the prediction outcome. Decision trees can be regarded as a disjunction of conjunctions of constraints on the attribute values of instances. Each path from the root to a leaf is actually a conjunction of attribute tests, and the tree itself allows the choice of different paths, that is, a disjunction of these conjunctions. Other important definitions regarding decision trees are the concepts of depth and breadth. The average number of layers (levels) from the root node to the terminal nodes is referred to as the average depth of the tree. The average number of internal nodes in each level of the tree is referred to as the

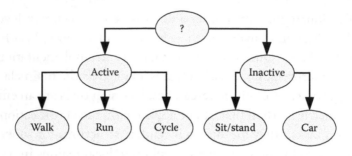

Figure 6.6 Illustration of decision tree used to recognize human activities.

average breadth of the tree. Both depth and breadth are indicators of tree complexity, that is, the higher their values are, the more complex the corresponding decision tree is [19].

Reference 20 adopted decision tree as the classifier in the two-stage activity recognition process presented in Figure 6.6. The work first classified the activity into two categories: active (e.g., walking, running, and cycling) and inactive (e.g., driving and idling), and then each activity was further classified within the first level category.

6.3.3.4 KNN K-nearest neighbor (KNN) algorithm is one of the analogical learning algorithms. Owing to its strong robustness and simplicity, KNN algorithm is widely used in various fields. KNN is a supervised learning algorithm where the result of new instance query is classified according to scale of KNN categories. The training examples are vectors in a multidimensional feature space, each with a class label. The training phase of the algorithm consists only of storing the feature vectors and class labels of the training samples. In the classification phase, K is a user-defined constant, and an unlabeled vector (a query or test point) is classified by assigning the label which is most frequent among the K training samples nearest to that query point. The degree of proximity between query instance and KNNs can be measured in terms of various distance metrics, for example, the simple Euclidean distance.

When it comes to processing massive high-dimensional datasets, one shortcoming of the traditional KNN is the time complexity of making classification, since the traditional KNN algorithm is a lazy learning method, which needs to store all the training samples before the classification. For a query instance, its distance to all training samples should be calculated, which may lead to considerable overhead when the training dataset is large. In addition, the algorithm is based on the assumption that all the classes of the samples have the same percentage in the training set, which is obviously not always consistent with actual situation. As class population is unbalanced, for example, one class has large sample size whereas others have small size in the training set, it would cause erroneous judgment. Fast or approximate nearest-neighbor search, for example, can be used to cut down the search complexity for real-time analysis.

A mobile application for iPhone has been developed for activity recognition and data collection [21]. The experimental results show that 1-NN classifier is a suitable candidate for the real-time implementation.

6.3.3.5 HMM A hidden Markov model (HMM) is a statistical Markov model in which the system being modeled is assumed to be a Markov process with unobserved (hidden) states.

The HMM focuses on analyzing correlation of spatiotemporal information and modeling them mathematically. There are four significant components in the structure of HMM: hidden states, observations, transition probability, and emission probability, respectively. Each transition has a pair of probabilities: a transition probability (which provides the probability for undertaking the transition) and an emission probability (which defines the conditional probability of emitting an output symbol from a finite alphabet, given a state). A common formal characterization of HMM is shown as:

- $\{s_1, s_2, s_3 \ldots s_n\}$—A set of N hidden states. The state at time t is denoted by the random variable q_t.
- $\{v_1, v_2, v_3 \ldots v_m\}$—A set of M distinct observations. The observation at time t is denoted by the random variable o_t. The observations correspond to the physical output of the system being modeled.
- $A = [a_{ij}]$—An $N \times N$ matrix for the state transition probability distributions, where a_{ij} is the probability of making a transition from state s_i to s_j: $a_{ij} = p(q_{t+1} = s_j | q_t = s_i)$.
- $B = [b_j(i)]$—is the output probability at time t in state s_j: $b_j(i) = p(o_t = v_i | q_t = s_j)$.
- $\pi = [\pi_i]$—The initial state distribution, where π_i is the probability that the state s_i is the initial state: $\pi_i = p(q_1 = s_i)$.

Usually, using the compact notation $\lambda = (A, B, \pi)$ to represent an HMM. The specification of an HMM involves the choice of the number of states N, the number of discrete symbols M, and the specification of the three probability densities A, B, and π.

HMMs can be considered as a set of states which are traversed in a sequence hidden to an observer. The only thing that is visible is a sequence of observed symbols, emitted by each of the hidden states while they are traversed. Related to HMM, the following several efficient algorithms are used for learning and recognition:

- Forward–backward algorithm is used for determining the probability that an emission sequence was generated by a given HMM. The forward pass computes the probability of being in a state at a particular time, given the observation sequence up to that time, so summing over all states at the end is required.
- Baum–Welch algorithm is used for estimating transition and emission probabilities of a HMM, given an observation sequence and initial *guesses* for these values. As an expectation-maximization algorithm, it uses an iterative search for the parameters with the highest likelihood.
- Viterbi algorithm (it is the training algorithm, not to be confused with Viterbi decoding algorithm) is used to find the most likely sequence of hidden states, given a HMM and an observation sequence. It computes the recursive likelihood of being in each state at the next time step until the end of the sequence, at which point, the algorithm backtracks to give the most likely state sequence.

HMMs are trained on a sequence of features, and cannot improve the accuracy, as the temporal correlations can be captured. Due to its lesser computational cost and better overall accuracy, HMM was used for walk detection and step counting in Reference 16. In walk detection, a two-state (walk/idle) HMM with Gaussian emissions was trained in an unsupervised fashion. In step counting, an HMM was trained to distinguish between different phases of the gait period such as heel-off, toe-off, heel strike, and foot stationary. The experiments showed that a simple HMM with two hidden states produced accurate step-counting results.

6.3.3.6 SVM Support vector machines (SVMs), including support vector classifier (SVC), and support vector repressor (SVR), are among the most robust and accurate methods in all well-known data mining algorithms. SVMs, which were originally developed by Vapnik in the 1990s, have a sound theoretical foundation rooted in statistical learning theory and require only as few as a dozen examples for training, and are often insensitive to the number of dimensions. The effectiveness of SVM models makes them particularly appealing in several and heterogeneous real-world problems, including applications on smartphones.

Three SVM algorithms were explored in Reference 18, including linear (L1 SVM, conventional L2 linear SVM) and nonlinear (L2 Gaussian SVM) approaches on the HAR dataset, and the selection criterion was subject to prediction speed and the possibility of applying them in devices with limited resources to provide less computational complexity and energy consumption. Activities that have been recognized are *single* activities that can be expressed with one verb, such as sitting, walking, holding a mobile phone, and throwing a ball. The ability to locate periods of walking, as well as walking speed, can be used for several applications. Walking speed is essential in providing additional context when, for example, monitoring disease progress in different patient populations. A data fusion approach was proposed to estimate the walking speed using smartphone, where SVM is selected as classifiers [22]. Table 6.5 summarizes the advantages and disadvantages of various classification methods.

6.3.4 Data Analysis Tools

With the continued exponential growth in data volume, large-scale data mining and machine learning experiments have become a necessity for many researchers without programming or statistics backgrounds. Waikato Environment for Knowledge Analysis (WEKA) is a gold standard framework that facilitates and simplifies this task by allowing specification of algorithms, hyperparameters, and test strategies from a streamlined Experimenter GUI.

Nowadays, WEKA is recognized as a landmark system in data mining and machine learning. It has achieved widespread acceptance

Table 6.5 Classification Methods and Its Advantage and Disadvantage

CLASSIFICATION METHODS	ADVANTAGES AND DISADVANTAGES
Decision tree	Advantage: • Simple to understand and interpret • Have value even with little hard data • Can be combined with other decision techniques Disadvantage: • Calculations can get very complex particularly if many values are uncertain and/or if many outcomes are linked
KNN	Advantage: • Very simple implementation • Robust with regard to the search space • Classifier can be updated online at very little cost as new instances with known classes are presented Disadvantage: • Expensive testing of each instance • Sensitiveness to noisy or irrelevant attributes • Sensitiveness to very unbalanced datasets
HMM	Advantage: • The HMM can be used for generating alignments • Separate HMM built for recognizing particular structures can be merged to create HMM that recognize sequences of structures Disadvantage: • The Viterbi algorithm is expensive • HMM needs to be trained on a set of seed sequences and generally requires a larger seed than the simple Markov models
SVM	Advantage: • It has a regularization parameter, which makes the user think about avoiding over-fitting • SVM is defined by a convex optimization problem (no local minima) for which there are efficient methods Disadvantage: • Biggest limitation of the support vector approach lies in choice of the kernel, sadly kernel models can be quite sensitive to over-fitting the model selection criterion

within academia and business circles and has become a widely used tool for data mining research. It contains implementations of a number of learning algorithms, and it allows to easily evaluating them for a particular dataset using cross validation and random split, among others. WEKA also offers a Java API that facilitates the incorporation of new learning algorithms and evaluation methodologies on top of the pre-existing framework.

In brief, WEKA would not only provide a toolbox of learning algorithms but also a framework inside which researchers could implement new algorithms without having to be concerned with supporting infrastructure for data manipulation and scheme evaluation.

WEKA's basic characteristics are as follows [23]:

- *Data preprocessing*: Similar to a native file format (ARFF), WEKA is compatible with several other formats (for instance, CSV, ASCII Matlab files) and database connectivity through JDBC.
- *Classification*: Classification is done with about 100 methods. Classifiers are divided in *Bayesian* methods, lazy methods (closest neighbor), rule-based methods (decision charts, OneR, Ripper), learning threes, and learning based on diverse functions and methods. On the other hand, WEKA includes metaclassifiers, such as bagging, boosting, stacking, multiple instance classifiers, and interfaces implemented in Groovy and Jython.
- *Clustering*: Unsupervised learning supported by several clustering schemes, such as EMbased Mixture model, k-means, and several hierarchy clustering algorithms.
- *Selection attribute*: The set of attributes used is essential for the performance classification. There are several selection criteria and search methods available.
- *Data visualization*: Data may be visually inspected representing attribute values against class or against other attribute values. There are visualization-specialized tools for specific methods.

The Java Data Mining Package (JDMP) is an open-source Java library for data analysis and machine learning ("The Java Data Mining Platform," http://www.jdmp.org/). It facilitates the access to data sources and machine learning algorithms (e.g., clustering, regression, classification, graphical models, optimization) and provides visualization modules. JDMP provides a number of algorithms and tools but also interfaces to other machine learning and data mining packages (WEKA, LibLinear, Elasticsearch, LibSVM, Mallet, Lucene, Octave, etc.).

6.4 Application

Activity recognition is a core building block behind many interesting applications. Those applications can be classified into the following categories [11], according to their targeted beneficial subjects: (1) application for the end users such as fitness tracking, health monitoring, fall detection, behavior-based context-awareness, home and work automation, and self-managing system; (2) applications for the third parties such as targeted advertising, research platforms for the data collection, corporate management, and accounting; and (3) applications for the crowds and groups such as social networking and activity-based crowdsourcing. These types of applications are not mutually exclusive; an application that benefits groups will invariably benefit end users. In this section, we briefly review some representative applications.

6.4.1 HAR for End Users

6.4.1.1 Daily Life Monitoring Applications in daily life monitoring usually aim to provide a convenient reference for the activity logging or assisting in exercise and healthy life styles. In commercial market, there exist some popular products (e.g., gadget). These devices are equipped with the embedded sensors such as accelerometer, gyroscope, GPS, and they track people's steps taken, stairs climbed, calorie burned, hours slept, distance travelled, and quality of sleep, and moreover, an online service is provided for users to review data tracking and visualization in reports. Compared with smartphone sensors, these devices are more sophisticated, since their sensors are designed specifically for the activity detection and monitor. The drawback is that, in a sense, they are intrusive instruments for human being.

Smartphone applications with activity recognition techniques have been shown up in recent years as an alternative solution. These applications usually have similar roles as the specialized devices. They track users' motion logs such as jogging route, steps taken, and sleeping time. By mining the logged data, they may offer the user a summary on his or her life style and report the sleeping quality. Such applications include personal life log system, the popular iphone application Nike + iPod (Apple: Nike + iPod application, http://www.apple.com/ipod/nike/), and sleep cycle (Sleepcycle, http://www.sleepcycle.com/). Compared to the devices mentioned above, these applications are easier to use and less or zero cost because they are installed on smartphone and do not need extra sensors or devices. Social interactions play an important role in the daily life. Sensing capabilities of smartphones can play a significant role in automatic detection of social interactions. Reference 24 described a method of detecting interactions between people, specifically focusing on interactions that occur in synchrony, such as walking. Walking together between subjects is an important aspect of social activity and thus can be used to provide a better insight into the social interaction patterns. They rely on sampling smartphone accelerometer and Wi-Fi sensors only. The results show that from seven days of monitoring using seven subjects in real-life setting, it achieves 99% accuracy, 77.2% precision, and 90.2% recall detection rates.

6.4.1.2 Health Care The World Health Organization (WHO) reported that 28%–35% of people aged 65 years and more fall each year and the rate increases to 32%–42% for those over 70 years of age [25]. There is a growing need in elderly care (both physically and mentally), partially because of the retirement of the baby boomer generation. Fall is a major problem associated with old age. Fall can force elderly people to depend on others, severely affecting their quality of life. Therefore, it is important to develop a technology that can monitor gait of elderly people that looks for precursors to falls. Those applications could help prevent harms, for example, to detect older people's dangerous situations. Activity recognition and monitor sensors could help elders in a proactive way such as life routine reminder (e.g., taking medicine), living activity monitoring for a remote robotic assist. Moreover, with the assistance of activity recognition, elderly people (and caregivers/medical personnel) can keep track of the level of activities being performed by them on a regular basis. A smartphone-based application, smart fall detection, was designed in Reference 4, which detects the falls through the trained multilayer perception (MLP) neural network utilizing the accelerometer sensor data and GPS on smartphones. Data from the accelerometer are evaluated with the MLP to determine a fall. When neural network detects the fall, a help request will be sent to the specified emergency contact using SMS, and subsequently whenever GPS data are available, the exact location of the fallen person will be sent.

The advantage of using smartphones as a pervasive fall detection system is that the user is more likely to carry the phone throughout the day because of its necessity in daily living, whereas users may forget to wear special sensors. Easy installation, short training times, low price, and being standalone make smartphone-based pervasive devices well suited for the elderly as a reliable and ubiquitous technology. In comparison with wearable device, video-based and ambience devices that are not applicable for everyone because of some limitations in installation, price, and so on, this approach is a trustable, standalone, and portable method for fall detection that helps the elderly to live independently. On the other hand, the youth care is another field that benefits from the activity recognition research. Applications include monitoring infants' sleeping status and predicting their demands on food or other stuff. Activity recognition techniques are also used in children's autism spectrum disorder (ASD) detection [25].

6.4.2 HAR for the Third Parties

6.4.2.1 Corporate Management and Accounting
Activity recognition can also assist with employee management and accounting for employee time. Reference 26 demonstrated that mobile activity recognition systems can be used to build pervasive, context- and activity-aware networks for the monitoring of hospital staff, whose whereabouts and activities are important information for colleagues. In other contexts, where the location and activities of employees are not critical to health care, this technology certainly raises questions regarding employer ethics. However, voluntary AR has proven successful in other contexts. For example, the Progressive® insurance company runs a voluntary program where drivers are given a device that monitors the acceleration and duration of their driving. If a driver is deemed to drive safely, he or she is given a discount on his or her insurance payments (Snapshot, Progressive. http://www.progressive.com/auto/snapshot-privacy-statement). Similarly, a company or organization could utilize AR on a voluntary basis, perhaps in combination with incentives.

6.4.2.2 Targeted Advertising
A user's contextual history, their location in particular, is analyzed to determine the user's interests. For example, if a user frequently goes to Chinese restaurants, then a

system may determine that the user prefers Chinese food to other types of cuisine, and the inference would be useful for certain advertisers, such as the owners of a newly opened local restaurant. However, this fact may not be easily determined just through online behavioral analysis, HAR could be greatly complementary to detect the fact.

6.4.3 Applications for Crowds and Groups

6.4.3.1 Activity-Based Social Networking   Social networking sites, such as Facebook and Twitter, provide users a medium to communicate their daily activities and thoughts with a broad audience. Applications like Foursquare allow users to advertise their current location, so broadcasting a user's current activity may be a valuable social tool. Although fitness tracking technologies exist that provide daily or weekly fitness updates to social networking profiles, they do not provide responses to a user's current activity. In fact, nearly all facets of social networking rely on manual user operations. With AR, an automated aspect to social networking can be introduced allowing users to automatically update their profiles based on their current activity, location, and proximity to other users.

Beyond the simple posting of current activities, activity recognition can be used to generate social networks. After identifying users who are in proximity of one another and share activity patterns, an application can suggest a sphere of friends who fit similar activity qualities. This methodology creates a new basis on which to build a network, one that is based on more than just proximity or interests and one that instead links individuals based on their physical actions.

6.4.3.2 Activity-Based Crowdsourcing   Some areas will naturally have high concentrations of specific activities (such as running at a track or sitting in a stadium). By analyzing the activity of many people in the same area, applications can learn and tag places and times as popular for biking or other recreation. Once these patterns are discovered, applications can detect abnormal behavior. If, for instance, a large number of people are suddenly running where they would normally sit or walk, the system could generate a notification of a possible emergency or disaster.

6.4.4 Vehicle-Related Applications

Activity recognition by smartphone not only can recognize the activities of human but also can be applied to vehicles. As vehicle manufacturers continue to emphasize on safety, smartphones can record and analyze various driver behaviors and external road conditions that could potentially be hazardous to the health of the driver, the neighboring public, and the automobile. Effective use of these data can educate a potentially dangerous driver on how to safely and efficiently operate a vehicle. With real-time analysis and auditory alerts of these factors, they can increase a driver's overall awareness to maximize safety.

LaneQuest is proposed by Aly et al. [27], which leveraged the ubiquitous and low-energy inertial sensors available in commodity smartphones to provide an accurate estimate of the car's current lane. Reference 28 addressed the problem of distinguishing the driver from passengers using a fusion of embedded sensors (accelerometers, gyroscopes, microphones, and magnetic sensors) in a smartphone. The system only utilizes naturally arising driver motions, that is, sitting down sideways, closing the vehicle door, and starting the vehicle to determine whether the user enters the vehicle from left or right and whether the user is seated in the front or rear seats.

6.5 Challenges and Open Issues

Although, the research and application on activity recognition are beneficial from the mobile sensors' unobtrusiveness, flexibility, and many other advances and witnessed great development and deployment, there exist several challenges in this filed. In this section, some typical challenges and potential solutions are discussed.

6.5.1 Energy Consumption and Battery Life

Activity recognition applications require continuous sensing as well as online updating for the classification model, both of which are energy consuming. For the online updating, it might also require significant computing resources (e.g., mobile phone memories). Lower sampling frequency means less work time for the heavy-duty sensor. However, whether the low sampling frequency is feasible for detecting human activities is still an open question. On the other hand, low sampling frequency may result in the loss of sampling data, reducing the recognition rate with low-resolution sensory data. So there is a trade-off between energy consumption and recognition rate.

Energy efficiency can be improved in many ways, such as using an adaptive scheme that does not poll the sensors as frequently when the user (i.e., phone) appears to be stationary. Based on the observation that the required sampling frequency differs for different activities, the adaptive accelerometer-based activity recognition (A3R) strategy is proposed in Reference 6, which adaptively makes the choices on both sampling frequency and classification features, to reduce both energy and computing resource cost. Reference 7 adopted adaptive sampling and duty cycling of smartphone accelerometer to achieve the balance in the trade-off between the context inference accuracy and power consumption caused during the context-inferring process of smartphone sensors.

6.5.2 Quality of Smartphone Sensors

It remains doubtful whether the qualities of built-in smartphone sensors are adequate to produce fall detection and prevention systems with acceptable performance. For example, the usual dynamic ranges of the built-in accelerometers in smartphones are insufficient for accurate fall incident detection. Usually, acceptable dynamic ranges for accelerometers from ±4 g to ±16 g have been mentioned in previous publications. Smartphone typically contains accelerometers with dynamic ranges of ±2 g or less. Therefore, in smartphone-based HAR-related applications (e.g., fall detection), adequate attention should be paid to the quality of the sensors. Specifications of the sensors should satisfy the minimum requirements of the applications. Or alternatively, multiple sensor data should be appropriately integrated to enhance the data quality provided by smartphone sensors.

6.5.3 Burden

Since the classification of activities is mostly based on the use of statistical machine learning techniques, a learning phase is required. Mostly, supervised or semisupervised learning techniques are utilized and such techniques rely on labeled data, that is, associated with a specific class or activity. Labeling the data in the training phase is usually a tedious and complex process. In most of the cases, the user is required to label the activities and this, in turn, increases the burden on the user. Hence, user-independent training and activity recognition are required to foster the use of HAR systems where the system can use the training data from other users in classifying the activities of a new subject. Besides the user-independent recognition, it is also important that HAR algorithms should work on different mobile phone platforms in a device-independent manner for the acceptance of such systems by masses.

6.5.4 Smartphone Placement and Use Habits

Challenges also stem from the different uses of the phones by different people and the differences in the way people perform activities. For instance, the phone context problem [3] arises from the human behavior when the phone is carried in an inappropriate position relative to the event being sensed. Especially with HAR using inertial sensors, location where the phone is carried, such as in the pocket or in the bag, and the orientation of the phone affect the activity recognition performance. For example, some people, mainly women, rarely carry their phones in their pockets. This impacts the utility of activity recognition service. It would be addressed by making recognition models more flexible, so that they can adapt to different body locations.

6.5.5 Security and Privacy Threats

It is shown that the location of screen taps on smartphones can be identified from the readings of motion sensors, such as by the accelerometer and gyroscope, which could cause security problems. Using this information, an attacker can monitor user's inputs, such as keyboard presses and icon taps. Hence, it is an open field how to address the threats arising from the unrestricted access to motion sensors by the activity recognition systems. Similarly, privacy emerges as a big concern in activity recognition applications due to the potential of collecting personal data, particularly if the data reveal a user's location and speech. This raises the requirement for developing suitable privacy-preserving mechanisms.

6.6 Conclusion

Smartphones are ubiquitous and becoming more and more sophisticated. Built-in inertial sensors, open-source operating systems, state-of-the-art wireless connectivity, and universal social acceptance make smartphone a very good alternative to conventional dedicated activity recognition tools. This has been changing the landscape of people's daily life and has opened the doors for many interesting data mining applications. HAR is a core building block behind these applications. It takes the raw sensors' reading as inputs and predicts a user's motion activity. This chapter presents a comprehensive survey of the recent advances in activity recognition with smartphone sensors. First, the basic concepts of activity recognition (such as sensors, activity types, etc.) are overviewed. Then, the core data mining techniques behind the main stream activity recognition algorithms are discussed by analyzing their major challenges and by summarizing a variety of real applications enabled by activity recognition. The activity recognition based on smartphone sensors leads to many commercial applications. However, the performance and usability of current systems still need to be improved because of some existing issues, including limited smartphone resource, especially energy, placement and sensor quality, and security and privacy.

References

1. Su, X., H. Tong, and P. Ji. Activity recognition with smartphone sensors. *Tsinghua Science and Technology*, 2014; 19(3): 235–249.
2. Yang, H. and S. Fong. Improving the accuracy of incremental decision tree learning algorithm via loss function. In: *Proceedings of the IEEE 16th International Conference on. Computational Science and Engineering (CSE)*, Sydney, Australia, December 3–5, 2013.
3. Incel, O.D., M. Kose, and C. Ersoy. A review and taxonomy of activity recognition on mobile phones. *BioNanoScience*, 2013; 3(2): 145–171.
4. Mokaram, S., K. Samsudin, and A.R. Ramli. A pervasive neural network based fall detection system on smart phone. *Journal of Ambient Intelligence and Smart Environments*, 2015; 7(2): 221–230.
5. Susi, M., V. Renaudin, and G. Lachapelle. Motion mode recognition and step detection algorithms for mobile phone users. *Sensors*, 2013; 13(2): 1539–1562.
6. Liang, Y., X. Zhou, Z. Yu, and B. Guo. Energy-efficient motion related activity recognition on mobile devices for pervasive healthcare. *Mobile Networks and Applications*, 2014; 19(3): 303–317.

7. Yurur, O., C.H. Liu, X. Liu, and W. Moreno. Adaptive sampling and duty cycling for smartphone accelerometer. In: *Proceedings of the IEEE 10th International Conference on Mobile Ad-Hoc and Sensor Systems (MASS)*, Hangzhou, China, October 14–16, 2013.

8. Lane, N.D., E. Miluzzo, H. Lu, D. Peebles, and T. Choudhury. A survey of mobile phone sensing. *IEEE Communications Magazine*, 2010; 48(9): 140–150.

9. Yan, Z., V. Subbaraju, D. Chakraborty, A. Misra, and K. Aberer. Energy-efficient continuous activity recognition on mobile phones: An activity-adaptive approach. In: *Proceedings of the 16th annual International Symposium on Wearable Computers (ISWC)*, Newcastle, UK, June 18–22, 2012.

10. Mehner, S., R. Klauck, and H. Koenig. Location-independent fall detection with smartphone. In: *Proceedings of the 6th International Conference on PErvasive Technologies Related to Assistive Environments (PETRA)*, Island of Rhodes, Greece, May 29–31, 2013.

11. Lockhart, J.W., T. Pulickal, and G.M. Weiss. Applications of mobile activity recognition. In: *Proceedings of the 14th ACM International Conference on Ubiquitous Computing (Ubicomp)*, Pittsburgh, PA, September 5–8, 2012.

12. Milette, G. and A. Stroud. *Professional Android Sensor Programming*. Birmingham: Wrox, 2012.

13. Buber, E. and A.M. Guvensan. Discriminative time-domain features for activity recognition on a mobile phone. In: *Proceedings of the IEEE Ninth International Conference on Intelligent Sensors, Sensor Networks and Information Processing (ISSNIP)*, Singapore, April 21–24, 2014.

14. Cheng, Q., J. Juen, Y. Li, V. Prieto-Centurion, J.A. Krishnan, and B.R. Schatz. GaitTrack: Health monitoring of body motion from spatio-temporal parameters of simple smart phones. In: *Proceedings of the International Conference on Bioinformatics, Computational Biology and Biomedical Informatics (BCB)*. Washington, DC, September 22–25, 2013.

15. Figo, D., P.C. Diniz, D.R. Ferreira, and J.M.P. Cardoso. Preprocessing techniques for context recognition from accelerometer data. *Personal and Ubiquitous Computing*, 2010; 14(7): 645–662.

16. Brajdic, A. and R. Harle. Walk detection and step counting on unconstrained smartphones. In: *Proceedings of the 2013 ACM International Joint Conference on Pervasive and Ubiquitous Computing (UbiComp)*, Zurich, Switzerland, September 8–12, 2013.

17. Ward, J.A., P. Lukowicz, and H.W. Gellersen. Performance metrics for activity recognition. *ACM Transactions on Intelligent Systems and Technology (TIST)*, 2011; 2(1): 6.

18. Anguita, D., A. Ghio, L. Oneto, X. Parra, and J.L. Reyes-Ortiz. Training computationally efficient smartphone-based human activity recognition models. In: *Proceedings of International Conference on Artificial Neural Networks (ICANN)*, Sofia, Bulgaria, September 10–13, 2013.

19. Barros, R.C., A.C.P.L.F. de Carvalho, and A.A. Freitas. *Automatic Design of Decision-Tree Induction Algorithms*. Springer International Publishing, Germany, 2015.

20. Siirtola, P. and J. Röning. Recognizing human activities user-independently on smartphones based on accelerometer data. *International Journal of Interactive Multimedia & Artificial Intelligence*, 2012; 1(5): 38–45.

21. Thiemjarus, S., A. Henpraserttae, and S. Marukatat. A study on instance-based learning with reduced training prototypes for device-context-independent activity recognition on a mobile phone. In: *Proceedings of 2013 IEEE International Conference on Body Sensor Networks (BSN)*, Cambridge, MA, May 6–9, 2013.

22. Altini, M., R. Vullers, and C.V. Hoof. Self-calibration of walking speed estimations using smartphone sensors. In: *Proceedings of the First Symposium on Activity and Context Modeling and Recognition (ACOMORE)*, Budapest, Hungary, March 24, 2014.

23. Navas, P.C.J., Y.C.G. Parra, and J.I.R. Molano. Big data tools: Haddop, MongoDB and Weka. In: *Proceedings of the International Conference on Data Mining and Big Data (DMBD)*, Bali, Indonesia, June 25–30, 2016.

24. Garcia-Ceja, E., V. Osmani, A. Maxhuni, and O. Mayora. Detecting walking in synchrony through smartphone accelerometer and Wi-Fi traces. In: *Proceedings of European Conference on Ambient Intelligence (AMI)*, Eindhoven, the Netherlands, November 11–13, 2014.

25. Albinali, F., M.S. Goodwin, and S. Intille. Detecting stereotypical motor movements in the classroom using accelerometry and pattern recognition algorithms. *Pervasive and Mobile Computing*, 2012; 8(1): 103–114.

26. Favela, J., M. Tentori, L.A. Castro, V.M. Gonzalez, E.B. Moran, and A.I. Martinez-Garcia. Activity recognition for context-aware hospital applications: Issues and opportunities for the deployment of pervasive networks. *Mobile Networks and Applications*, 2007; 12(2–3): 155–171.

27. Aly, H., A. Basalamah, and M. Youssef. LaneQuest: An accurate and energy-efficient lane detection system. In: *Proceedings of the 13th IEEE International Conference on Pervasive Computing and Communications (PerCom)*, St. Louis, MI, March 23–27, 2015.

28. Park, H., D.H. Ahn, M. Won, S.H. Son, and T. Park. Poster: Are you driving? Non-intrusive driver detection using built-in smartphone sensors. In: *Proceedings of the 20th Annual International Conference on Mobile Computing and Networking (MobiCom)*, Maui, HI, September 7–11, 2014.

7

INDOOR LOCALIZATION AND TRACKING SYSTEMS

7.1 Introduction

With the rapid development of the mobile Internet, the demand of positioning and navigation is increasing gradually, especially in complex indoor environments, in which the positions of the immobile/mobile objects, for example, individuals, facilities, and goods, are generally needed to be determined in real-time for provisioning of various smart services. These complex indoor environments include airport halls, exhibition halls, warehouses, supermarkets, libraries, underground car parks, and other environments. In general, the indoor positioning systems aim to locate and track an object within buildings or closed environment by exploiting various radio waves, optical tracking, magnetic field or ultrasonic technology, and so on. Positioning and tracking system is the basis for many applications on monitoring and activity recognition. For example, in the supermarket, by obtaining the relevant position information of the consumers and target commodities, the supermarket is able to provide the services of the route guidance and the intelligent shopping guide. Moreover, the indoor positioning system can be specially employed for guiding the rescuers to search the building and to rescue the trapped personnel quickly when a sudden disaster happens. In addition, the position technology can be used in hospital, that is, the patient's monitoring and the management of the medical equipment.

Basically, global positioning system (GPS) is the most widely used satellite-based positioning system, which can provide accurate location of objects with GPS component. However, GPS is not suitable for indoor positioning due to a variety of obstacles. For solving the indoor positioning problems, many solution technologies are proposed in literature [1,2]. In this chapter, some popular indoor positioning technologies and systems are presented, including radio frequency identification (RFID), Bluetooth technology, Wi-Fi fingerprinting, and pedestrian dead reckoning (PDR).

The remainder of this chapter is organized as follows: Section 7.2 summarizes the background of indoor localization including several measuring principles and some commonly used performance metrics. In Section 7.3, wireless communication-based localization technologies (Bluetooth; ultra-wideband, ultra-wide band (UWB); RFID; IR; ultrasound; etc.) are discussed. Especially, Wi-Fi fingerprint and dead-reckoning localization technologies are mainly introduced and compared. In Section 7.4, open research issues and potential solution are presented. Finally, we briefly conclude this chapter.

7.2 Background

This section introduces the background of indoor localization, including several performance metrics like accuracy, precision, complexity, and several kinds of basic measuring principles like triangulation, proximity, scene analysis, and so on.

7.2.1 Performance Metrics

There are mainly six performance metrics in indoor positioning systems. These metrics are accuracy, precision, complexity, robustness, scalability, and cost [3].

1. *Accuracy*: Accuracy (or localization error) is the most important requirement of positioning systems. Usually, mean distance error is adopted as the performance metric, which is the average Euclidean distance between the estimated location and the true location. Accuracy can be considered to be a potential bias or systematic effect/offset of a positioning system. The higher the accuracy, the better the system; however, there is often a trade-off between accuracy and other characteristics. Some compromise between *suitable* accuracy and other characteristics is needed.
2. *Precision*: Accuracy only considers the value of mean distance errors. However, location precision considers how consistently

the system works, that is, it is a measure of the robustness of the positioning technique as it reveals the variation in its performance over many trials. Some schemes regarded the location precision as the standard deviation in the location error or the geometric dilution of precision (GDOP); others defined it as the distribution of distance error between the estimated location and the true location. In the latter, the cumulative probability functions (CDF) of the distance error are usually used for measuring the precision of a system. When two positioning techniques are compared, if their accuracies are the same, the system with the CDF graph reaching high probability values faster is always preferred, because its distance error is concentrated in small values. In practice, CDF is described by the percentile format. For example, one system has a location precision of 90% within 2.3 m (the CDF of distance error of 2.3 m is 0.9), and 95% within 3.5 m; another one has a precision of 50% within 2.3 m and 95% within 3.3 m. The former system is preferred, because of its higher precision.

3. *Complexity*: Complexity of a positioning system can be attributed to hardware, software, and operation factors. Software complexity, that is, computing complexity of the positioning algorithm is emphasized. If the computation of the positioning algorithm is performed on a centralized server side, the positioning could be calculated quickly due to the powerful processing capability and the sufficient power supply. If it is carried out on the mobile unit side, the effects of complexity could be evident. Most of the mobile units lack strong processing power and long battery life, so positioning algorithms with low complexity under the constraints of accuracy and precision are ideal. Usually, in practice, it is difficult to derive the analytic complexity formula of different positioning techniques, thus the computing time is considered. Location rate is an important indicator for complexity. The dual of location rate is location lag, which is the delay between a mobile target moving to a new location and reporting the new location of that target by the system.

4. *Robustness*: A positioning technique with high robustness should function normally even when some signals are not

available, or when some of the received signal strength (RSS) value or angle character are never seen before. Sometimes, the signal from a transmitter unit is totally blocked, so the signal cannot be obtained from some measuring units. The only information to estimate the location is the signal from other measuring units. Sometimes, some measuring units could be out of function or damaged in a harsh environment. The positioning techniques have to use this incomplete information to compute the location.

5. *Scalability*: A location system may need to scale on two axes: geography and density. Geographic scale means that the area or volume is covered. Density means the number of units located per unit geographic area/space per time period. As more area/space is covered or units are crowded in an area/space, wireless signal channels may become congested, more calculation may be needed to perform location positioning, or more communication infrastructure may be required. Another measure of scalability is the dimensional space of the system. The current system can locate the objects in 2D or 3D space. Some systems can support both 2D and 3D spaces.

6. *Cost*: The cost of a positioning system may depend on many factors. Important factors include money, time, space, weight, and energy. The time factor is related to installation and maintenance. Mobile units may have tight space and weight constraints. Measuring unit density is considered to be a space cost. Sometimes, we have to consider some sunk costs. For example, a positioning system layered over a wireless network may be considered to have no hardware cost if all the necessary units of the network have already been purchased for other purposes. Energy is an important cost factor of a system. Some mobile units (e.g., electronic article surveillance, EAS, tags and passive RFID tags) are completely energy passive. These units only respond to external fields and, thus, could have an unlimited lifetime. Other mobile units (e.g., devices with rechargeable battery) have a lifetime of several hours without recharging.

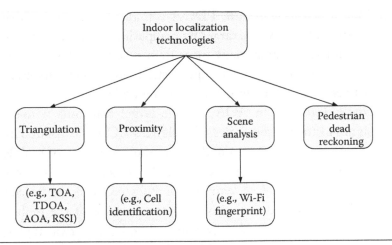

Figure 7.1 Category of indoor localization principles.

7.2.2 Positioning Principles

There are mainly four principles used in building indoor positioning systems. These principles are triangulation, scene analysis, proximity, and PDR as shown in Figure 7.1.

7.2.2.1 Triangulation

Triangulation uses the geometric properties of triangles to estimate the target location. It has two derivations: lateration and angulation. Lateration estimates the position of an object by measuring its distances from multiple reference points. So, it is also called range measurement techniques. The typical technologies are mainly time of arrival (TOA), time difference of arrival (TDOA), angle of arrival (AOA), and received signal strength indication (RSSI). As triangulation-based indoor localization needs a lot of beacon nodes, its deployment cost is very high.

1. *Time of arrival*: The principle of TOA is to measure the arrival time between the mobile target and the at least three known beacon nodes, and multiplies the signal speed to calculate the distances between the target and beacon nodes. Then taking the beacon nodes as the centers of circles, the distances between target and beacon nodes as the radius of circles, the intersection of circles is the coordinate of the mobile target.

2. *Time difference of arrival*: TDOA is to determine the relative position of the mobile target by computing time difference of the signal arrival at many measurement units. TDOA positioning method does not require strict time synchronization between the mobile target and beacon nodes.

3. *Angle of arrival*: In AOA, mobile target transmits wireless signal to the beacon nodes. The signal's AOA is measured by antennas on the beacon nodes, and the coordinate of mobile target is calculated by the principle of geometry. In two-dimensional space, at least two signal's AOA of the beacon nodes are needed for calculating the position of the mobile target. Based on the line direction of the signal arrival angles, the target position is the intersection of them.

4. *Received signal strength indication*: Principle of RSSI positioning method is that the signal receiver measures the received power, and the distance between nodes is calculated by using a transmission loss formula.

7.2.2.2 Proximity Proximity algorithms provide symbolic relative location information. Usually, it relies upon a dense grid of antennas, each having a well-known position. When a mobile target is detected by a single antenna, it is considered to be collocated with it. When more than one antenna detects the mobile target, it may be considered to be collocated with the one that receives the strongest signal. This method is relatively simple to be implemented over various types of physical media.

In particular, the systems using infrared (IR) radiation and RFID are often based on this method. Another example is that the cell identification (Cell-ID) or cell of origin method. This method relies on the fact that mobile cellular networks can identify the approximate position of a mobile handset by knowing which cell site the device is using at a given time. The main benefit of Cell-ID is that it is already in use today and can be supported by all mobile handsets.

In brief, proximity-based indoor localization has low complexity, but its accuracy is also low.

7.2.2.3 Scene Analysis Scene analysis techniques [4] involve examination and matching a video/image or electromagnetic characteristics (such as Wi-Fi fingerprint) viewed/sensed from a target object.

The analysis of electromagnetic *scene* sensed by a target object defined by electromagnetic signals and their strengths from different transmitters provides the determination of location using a pattern matching, radio map technique. Using video cameras, a positioning system can detect significant patterns in a video data stream to determine the user's location. If users wear badges with certain labels, they can be detected in video images and thus localized and tracked in indoor environment covered by a camera. At the other extreme are techniques involving the matching of perspective video images of the environment captured by a camera, worn by a person or mounted on a mobile robot platform, to 3D models stored in an image/video database of the mobile device.

7.2.2.4 Pedestrian Dead Reckoning The PDR techniques [4] estimate successive positions of a moving pedestrian starting from a known position through estimations of traveled distance and direction of movement. In indoor environment, such systems use accelerometers to obtain human velocity rate information through step detection and step length estimation. Also, digital compass and gyroscope measurements are used for direction and angular rate information. All measured data are processed to continuously calculate the location, direction (bearing), and velocity of a moving object without the need to preinstall beacon nodes in the building.

Table 7.1 compares the four typical positioning principles in terms of the following metrics: accuracy, complexity, and cost.

In brief, as triangulation-based indoor localization suffers from interference and multipath problems, and proximity has low accuracy, these two principles are not popular these years. On the contrary, scene analysis and PDR are the two most popular principles in indoor localization, for scene analysis is of high accuracy, and PDR is of low complexity, just using the sensors on mobile terminals.

Table 7.1 Comparison among Four Typical Positioning Principles

POSITIONING PRINCIPLE	ACCURACY	COMPLEXITY	COST
Triangulation	Medium	High	High
Proximity	Low	Medium	Medium
Scene analysis	High	High	High
Pedestrian dead reckoning	Medium	Low	Low

This chapter mainly focuses on Wi-Fi fingerprinting and PDR localization technologies.

7.3 Wireless Technologies Used in Indoor Localization

7.3.1 Existing Wireless Technologies for Indoor Localization

1. *Radio frequency identification*: In general, RFID-based indoor localization systems are composed of three main parts: RFID tags, readers, and miniature antennas between the tags and the readers [5]. The reader detects the vicinity of a tag and retrieves the data stored in that tag. Therefore, the absolute location of the tag is not known but the RFID system is aware that a tag is placed at a certain range that depends on the type of the system used, either active or passive RFID. There are several methods for performing accurate positioning using active RFID technology. These methods employ techniques such as AOA, TDOA, and RSSI that achieve accuracy in the range of 1–5 m, or even below 1 m, depending on the density of tag deployment and RFID reading ranges.

2. *Bluetooth*: Bluetooth is a wireless technology that can be used for localization and tracking, mainly indoors. Bluetooth positioning systems have similar working principles as the self-localization schemes of sensor networks. The operation principle of both types of systems is based on obtaining the range information to anchor devices or access points (APs) and exploring unknown device locations using various algorithms. The majority of the available research and commercial systems are based on trilateration using the RSSI for calculating distances between Bluetooth devices, although several reported systems have explored proximity, cell-based approach.

3. *UWB*: UWB communication technology has emerged, providing better positioning accuracy than Wi-Fi. It is suitable for high-precision real-time positioning using TOA, TDOA, AOA, and fingerprinting. UWB signals are less sensitive to multipath distortion and environment than conventional RF-based positioning systems, so they can achieve higher

accuracy. At bandwidths of at least 500 MHz and high-time resolution in the order of nanoseconds, it is possible to obtain accuracy of ranging and localization at cm-level.

4. *Infrared positioning*: In IR-based systems, each tracked person is wearing a small IR device that emits a unique pulse signal representing its unique identifier. The signals are detected by at least one particular IR sensor in the vicinity. A location server estimates IR device location by aggregating data obtained from fixed IR sensors deployed within the indoor environment.

5. *Ultrasound*: Ultrasound location systems are composed of ultrasonic tags and readers. An example is the Active Bats system developed at AT&T Cambridge [4]. The Active Bats system uses physical devices such as ultrasonic tags sending information to a receiver mounted each square meter. Based on the data received, the system computes the user's location. The location is computed using time of flight (TOF) triangulation based on the receivers mounted on the ceiling. The receivers are connected using a wired network, which enable to reset the receivers at any time.

7.3.2 Wi-Fi Fingerprint-Based Indoor Localization

7.3.2.1 Basic Concepts about Wi-Fi Fingerprint Wi-Fi is one of the most popular wireless network technologies and widely deployed in almost every building. Therefore, it could be naturally used to estimate the location of mobile devices within this network. Indoor positioning system based on Wi-Fi fingerprint consists of three main physical entities: wireless network terminal, wireless local area networks (WLANs) hotspots (i.e., AP) in fixed-location, and position platform. As Figure 7.2 shows, Wi-Fi fingerprint-based indoor localization can be divided into two phases: training phase (to construct offline Wi-Fi fingerprint database) and real-time positioning phase.

In offline training stage whose goal is to construct the fingerprint database, the main work is to determine the sampling points according to certain distance in the WLAN signal coverage area and to forming a network with evenly distributed sample points. At each sampling point all visible hotspots' signal strength values, media

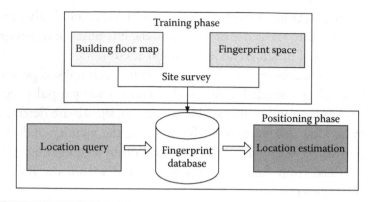

Figure 7.2 Typical architecture of Wi-Fi fingerprint-based localization system.

access control (MAC) address and the coordinate are measured, and these information as a record are saved in a database. The database information corresponding to these sampling points is called the location fingerprint.

In real-time positioning stage, the wireless network terminal is used to measure the signal strength of the visible WLAN hotspots, which is compared with the recorded data location in the fingerprint database, to consider the sampling points with maximum similarity as the positioning result. From a machine learning perspective, location fingerprinting can be viewed as training the computer to learn the rule between signal strength and location, and then to carry out the reasoning and judgment.

7.3.2.2 Improvements to Traditional Wi-Fi Fingerprint-Based Systems In general, there mainly exist two aspects to be improved in indoor localizations, including localization accuracy and efficient site survey.

7.3.2.2.1 Improving Localization Accuracy Wi-Fi-based localization suffers from the RSS variance problem, that is, accuracy is degraded when the RSS vectors observed in the localization phase are different from the ones collected during the training phase. The RSS variance problem is caused by differences in device type, user direction, and environmental changes between the two phases. The problem becomes serious in a pervasive environment, where users may have diverse types of smartphones and carry their smartphones in different places such as in a pocket or bag, or in their hand.

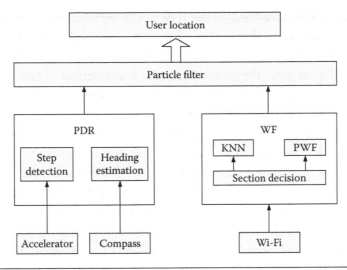

Figure 7.3 Framework of the proposed tracking system. (From Kim, Y. et al., Smartphone-based Wi-Fi tracking system exploiting the RSS peak to overcome the RSS variance problem, *Pervasive and Mobile Computing*, 9, 406–420, 2013.)

Reference 14 proposed to exploit the location of the maximum RSS, that is, when a peak RSS signal from a particular AP is detected, the current position can be estimated accurately by matching the peak location of the AP in the training data. Since a RSS peak is detected intermittently, the work combined the peak-based location estimation with PDR.

Figure 7.3 illustrates the structure of the proposed tracking system. Wi-Fi fingerprint (WF) is combined with PDR by using a particle filter. The PDR used in the system detects steps and the heading orientation by reading the accelerometer and digital compass. Since PDR provides relative location information, the absolute location is updated with WF. WF in the proposed system consists of two components: the K-nearest neighbor (KNN)-based method and the peak-based Wi-Fi fingerprinting (PWF) method. KNN is always available but the accuracy is not reliable when the RSS variance problem occurs. PWF provides highly accurate location estimation, but the scheme is hardly applicable in the places, such as a large hall, where explicit peak location does not exist. Therefore, a radio map is divided into several sections according to the availability of PWF in the training phase. In the localization phase, the system first decides the section where the user is located. A fine-grained localization is then performed with

PWF or KNN, according to the type of the section. The experiments in real environments demonstrated that the proposed system improves positioning accuracy by overcoming the RSS variance problem.

It is shown that the main causes of localization errors in RSS fingerprint-based localization schemes lie in the following aspects: (1) APs' diverse discrimination for fingerprinting a specific location. Intuitively, faraway APs may lead to large location estimation errors while close ones can help mitigate the location uncertainty. (2) RSS is inconsistent caused by signal fluctuations and human body blockages. (3) RSS measurements may be outdated due to hardware and software limitations of commodity wireless devices. In other words, the latest reported RSS values could be duplicates of previous scans performed several seconds ago. Considering user mobility, the outdated RSS could in fact be measurements done at a previous location.

Inspired by these insights, first, distinct AP's discriminatory ability with regard to a specific location is quantitatively differentiated. APs with stronger ability are emphasized with more weights in fingerprint matching, whereas others are de-emphasized. Second, noting fingerprint inconsistency, a robust regression technique in fingerprint matching is applied, in the hope of bounding the impact of RSS outlier values and ensuring accuracy under noisy measurements. Finally, multiple fingerprints in the fingerprint database are incorporated to deal with the outdated RSS values. Combining these techniques in a unified solution, DorFin, a novel scheme of fingerprint generation, representation, and matching, is proposed [7]. Extensive experiments demonstrated that DorFin achieves mean error of 2 m and more importantly bounds the 95th percentile error under 5.5 m; these are about 56% and 69% lower, respectively, compared with the state-of-the-art schemes.

7.3.2.2.2 Efficient Site Survey One of the key components in Wi-Fi fingerprint-based indoor localization schemes is the site survey, in which the specific location of sampled point should be explicitly given by engineers and be correlated with the Wi-Fi RSS fingerprint. Although site survey is time-consuming, and labor-intensive, it is inevitable for fingerprinting-based approaches, since the database is constructed by Wi-Fi RSS fingerprints labeled with specific location from on-site records.

An indoor localization system called WicLoc was developed in Reference 9, which collected Wi-Fi signals with user motions using crowdsourcing and designed a model to get Wi-Fi fingerprints of each interested location. Specifically, WicLoc utilized a weighted KNN algorithm to assign different weights to APs and achieve room-level localization. To obtain the absolute coordinate of users, a novel multi-dimensional scaling (MDS) algorithm called multidimensional scaling with calibrations (MDS-C) was designed to calculate the coordinates of the corridor and rooms, and used anchor points to match with the corresponding points in the map for calibration. Experimental results show that this system can achieve a competitive localization accuracy compared with state-of-the-art of Wi-Fi fingerprint-based methods while avoiding the labor-intensive site survey.

A wireless indoor logical localization approach, WILL proposed in Reference 8, which did not require the site survey process used in traditional approaches. WILL requires no prior knowledge of AP locations, and users are not required for explicit participation to label measured data with corresponding locations, even in the training phase. Its design rationale is that human motions can be applied to connect previously independent radio signatures under certain semantics. Specifically, the penetrating-wall effect of wireless signals of APs is a good starting point for characterizing different rooms or functional areas. Moreover, WILL exploits various sensors integrated on smartphones to obtain user movements, which will be further utilized to assist localization. In brief, WILL leveraged user motions to construct radio floor plan that is previously obtained by site survey. Fingerprints are partitioned into different virtual rooms, and a logical floor plan is accordingly constructed. Localization is achieved by finding a matching between logical and ground-truth floor plan. The experiment results show that WILL achieves an average room-level accuracy of 86%, which is competitive to existing designs that adopt time and labor-consuming site survey. Future research in physical floor plan construction, sophisticated floor plan mapping, and user behavior detection should be thoroughly conducted to make WILL practical in real life.

7.3.3 Pedestrian Dead Reckoning–Based Indoor Localization

PDR localization technologies use motion sensors to determine device's location without the need for infrastructure, with the scope limited by the accumulated error. Among various sensors, inertial measurement units (IMUs) are one of the most widely adopted instruments to measure mobility. Smartphone IMUs typically include three major types of sensors: accelerometer, gyroscope, and compass [10].

System framework: The dead-reckoning localization system is based on the PDR algorithm, which is recognized as a relative positioning algorithm, as shown in Figure 7.4. Notations E and N denote the east and north directions, respectively.

Starting from the user's initial position (x_0, y_0), we can calculate the next position, notated as (x_1, y_1), by utilizing the heading angle θ_1 and the displacement d_1. Based on the iterative process of position calculation, the coordinates of the user's k_{th} position are calculated by Equation 7.1:

$$x_k = x_0 + \sum_{i=1}^{k} d_i \sin \theta_i$$

$$y_k = y_0 + \sum_{i=1}^{k} d_i \cos \theta_i$$

(7.1)

where θ_i and $d_i (i = 1,...,k)$, respectively, stand for the heading angle and the stride length of step i. Hence, the user's position coordinates

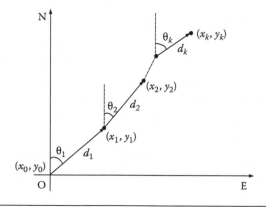

Figure 7.4 Basic model of PDR algorithm.

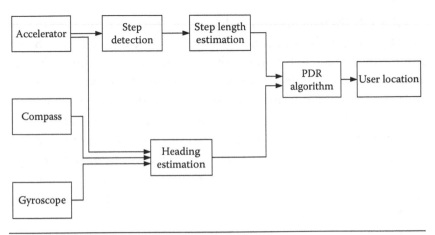

Figure 7.5 Architecture of pedestrian dead reckoning (PDR).

can be calculated by Equation 7.1 as soon as the parameters d_i and θ_i are estimated.

The system architecture is shown in Figure 7.5. In general, the filtered three-axis accelerometer data are utilized to detect the user's steps for location updating. Second, some empirical models are applied to estimate the pedestrian stride length as the displacement between every two adjacent positions. Third, a stable heading angle for each step is obtained. Finally, the PDR algorithm is performed to calculate the user's locations in a real-time manner.

7.3.3.1 Step Detection A complete single step of a pedestrian can be divided into two stages: the first stage is from the foot separating from the ground to the foot reaching the highest point, the acceleration at this stage starts to increase until it reaches a maximum value; the second stage is from the highest point of the foot to the foot hitting the ground, at this stage, the acceleration drops from the maximum value to minimum value.

Step detection is a basic module in most inertial based pedestrian localization and navigation systems. The physical underpinning is to search for cycles in acceleration traces to capture the repetitive movements during walking. When an acceleration trace was given as an input, step detection algorithms slice and label the trace into steps exploiting the repetitive patterns of walking, and the labels are then summed into step counts. Roughly these algorithms can be grouped as follows:

7.3.3.1.1 Temporal Analysis The cyclic property of walking is directly reflected in the acceleration trace in the time domain. Since heel strikes tend to introduce sharp changes, numerous schemes were proposed to detect thresholds, magnitude peaks, local variance peaks, local minima, and zero-crossings from the low-pass filtered acceleration trace [11]. Autocorrelation is a more robust means to magnify periodicity in the time domain regardless of the absolute amplitude of acceleration [12]. Steps can also be recognized by matching with a stride template either linearly (e.g., by cross-correlation) or nonlinearly (e.g., by dynamic time warping, DTW, [13]), yet at a higher cost.

The following subsections mainly introduce two typical schemes: threshold-based detect detection and DTW-based step detection.

1. *Threshold–based step detection*: Based on the physiological characteristics of the pedestrian, the waveform for the three-axis accelerometer modulus values can be obtained for the formation of cyclical changes. Therefore, the cyclical and characteristic values can be used to detect the steps. The three-axis accelerometer modulus values can be calculated as Equation 7.2:

$$a_{\mathrm{mag}} = \sqrt{a_x^2 + a_y^2 + a_z^2} \qquad (7.2)$$

where a_x, a_y, and a_z are the output data of triaxial accelerometer in the X, Y, and Z directions, respectively. Then, a single peak curve of accelerometer modulus can be obtained by using a digital low-pass filter, the peak point can be accurately detected, and the number of steps can be consequently calculated. Some small jitters might be produced during walking when the pedestrian is holding the phone. On this basis, the peaks appear in the output waveform of accelerometer modulus values, as shown in Figure 7.6.

According to the value of a_{mag}, a candidate step at time t_k, where k denotes the index of steps, is identified by following criteria:

- The total acceleration magnitude a_{mag} has to cross the threshold g from negative to positive.

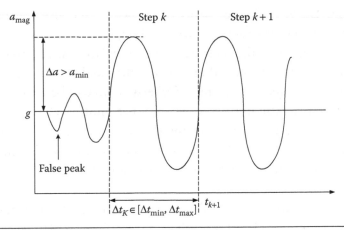

Figure 7.6 Identification of candidate steps.

- The time interval Δt between two consecutive steps must be within the interval threshold: $\Delta t \in \left[\Delta t_{min}, \Delta t_{max}\right]$. As the normal walking frequency of people is 0.5 Hz~5 Hz, so the lasting time of a step is 0.2 s~2 s. Consequently, those parameters are usually set as: $\Delta t_{min} = 0.2$ s, $\Delta t_{max} = 2$ s.
- The difference Δa between peaks of a_{mag} during a step phase and the threshold g has to be larger than a_{min}, usually a_{min} is set as 0.3.

As depicted above, the computational cost of threshold detection is low. However, its step detection error rate is relatively high, about 2%. Steps can also be recognized by matching with a stride template nonlinearly (e.g., by DTW), its step detection error rate is smaller than 2%, yet at a higher computational cost.

2. *Dynamic time warping*-based step detection: DTW has been widely used as an efficient way to measure the similarity between two discrete waveforms. A lower DTW distance denotes a higher similarity. For step detection, considering in normal walking when a user alternates her left and right foot, it can be expected similar waveform will repeat at every other step. Based on this observation, a DTW validation algorithm is designed to determine whether a step detected via peaks is a real step [13].

The DTW validation works as follows. Suppose the peak detection yields a series of detected steps $\{S_1, S_2, \ldots, S_n\}$. Then, for S_i, the algorithm calculates the DTW distance between S_i and S_{i-2}. If it is lower than a given threshold, both S_i and S_{i-2} are considered as real steps. Otherwise, S_i will be temporally labeled as false until S_{i+2}, when S_i has another chance to be validated after computing the distance between S_{i+2} and S_i. The distance is calculated over normalized accelerometer waveforms to reduce the influence of amplitude.

7.3.3.1.2 Spectral Analysis When the acceleration trace contains at least two walking cycles, it is possible to identify the repetitive walking patterns in the frequency domain. The rationale is that walking movements would generate dominant frequencies around 2 Hz, a unique spectral characteristic compared with other human activities. Short-term Fourier transform (STFT) [14] and wavelet transforms [15] have been employed to extract dominant frequency, and steps are counted as the sum of the transform coefficients of the walking frequency.

7.3.3.1.3 Feature Clustering Besides the above pattern analysis approaches, research literature also resorts to machine learning techniques to classify walking steps via features from acceleration traces. Various features have been explored, including statistics, entropy, as well as temporal correlation and Fourier transform coefficients [16]. Despite its high computational cost, feature clustering-based schemes are more general and are often applied to classify multiple human activities beyond walking.

Table 7.2 summarizes some recent smartphone-based step detecting schemes in terms of techniques, cost, and performance [10].

Table 7.2 Recent Smartphone-Based Step Detection Summary

TECHNIQUES	COMPUTATIONAL COST	STEP DETECTION ERROR RATE
Threshold detection [11]	Low	About 2%
Dynamic time warping [13]	Medium	<2%
Normalized autocorrelation [12]	Medium	0.6%
Spectral analysis [14]	Medium	<2%
Two-state hidden Markov model (HMM) clustering [16]	High	About 1.3%

Temporal analysis-based schemes are the most intuitive in concept and facilitate physical explanation on the extracted feature metrics. The primary drawback is that the cyclic walking patterns are mixed with other noises in the time domain.

Spectral analysis-based approaches offer an orthogonal domain to distinguish frequencies of walking and other noises, yet are less intuitive. For example, it is difficult to distinguish fine-grained walking patterns such as heel-up and heel-down directly from the signal spectrum. In addition, the accuracy of spectral analysis improves with the amount of signal periods contained in the input signal. Hence spectral analysis often requires more signal samples than temporal analysis.

Despite its high computational cost, feature clustering-based schemes are more general and are often applied to classify multiple human activities beyond walking. Feature clustering can also provide higher accuracy, given sufficient training efforts.

7.3.3.2 Step Length Estimation Knowing stride length is necessary when converting steps into distance traversed. Pedestrians may exhibit different stride lengths due to variety in height, walking speed, and style. Some pioneer efforts in stride length estimation are summarized as follows:

7.3.3.2.1 Constant Model The simplest method to estimating step length is to assume it as constant. Pedestrians have a surprisingly constant step length when walking. An intuitive way to estimate user-specific stride length is to divide a known walking distance by measured step counts. However, since the step length may be adjusted to different walking patterns such as rushing, ambling, or walking with others, this approach may induce bias on dominant walking patterns [17].

7.3.3.2.2 Linear Frequency Model Research in the medical community has shown a tight coupling between the step frequency and the walking speed. Although the precise relationship is nontrivial, it is common to fit a linear relationship [18].

Model 1: Reference 17 set up a system to collect over 4000 steps from 23 users with diverse physical characteristics and illustrated that the step length has a linear relation to the walking frequency. Based on that, they choose a frequency model as a step model, represented as

$$L = a \cdot f + b$$

Then, the next to do is to determine the value of the coefficients a and b for each person. As long as the coefficients a and b are determined, then step length L could be represented. Despite its small computational cost, the step length estimation accuracy is relatively low.

Model 2: A more complete model based on step frequency and the user height has recently been proposed and showed a 5.7% error over the travelled distance for 10 test subjects [19]:

$$L = h \cdot (a \cdot f + b) + c$$

where:

L is the estimated step length

h is the user height

f is the step frequency

a, b, and c are individual parameters

The experiment result indicates that this model improves the step length estimation accuracy as computational cost is still low.

Model 3: In Reference 11, the step length is estimated using a linear combination of step frequency and acceleration variance through the following equations:

$$L = \alpha \cdot f + \beta \cdot \upsilon + \gamma$$

where:

f is the step frequency

υ is the acceleration variance during one step

α and β are the weighting factors of step frequency and acceleration variance

γ is the constant

For different pedestrians, the model parameters α, β, and γ are different and required by offline training.

Table 7.3   Recent Smartphone-Based Step Length Estimation Summary

TECHNIQUES	COMPUTATIONAL COST	ACCURACY
Assumption of constant step length [17]	Low	Low
Linear relation between step length and step frequency [13]	Low	Low
Linear relation between step length and the combination of step frequency and the pedestrian height [19]	Low	Medium
Linear relation between step length and the combination of step frequency and acceleration variance [11]	Medium	Medium
A back propagation (BP) neural network [20]	High	High

7.3.3.2.3 Empirical Model   Reference 20 adopted the empirical model to estimate the stride length.

$$L = C \cdot \sqrt[4]{a_{max} - a_{min}}$$

where a_{max} and a_{min} stand for the maximum and minimum of modulus values of accelerometer, which are obtained from the step detection. C is the proportionality coefficient. Since the step lengths are determined by height, attitude, and frequency, the value of C is not constant and is significantly influenced by the pedestrian height and stride frequency. For the sake of predicting the value C accurately and real timely, a back propagation (BP) neural network was used to obtain this nonlinear mapping.

Table 7.3 summarizes some existing smartphone-based step length estimation schemes in terms of techniques, cost, and performance.

7.3.3.3 Heading Estimation   The direction of steps during walking is usually obtained by phone gyroscope and compass (magnetometer), and compass directly measures the absolute orientation (heading) of the phone with respect to the magnetic North. The main hurdles for accurate heading estimation via phone compass lies are twofold.

7.3.3.3.1 Magnetic Offset   Metal and conducting material nearby can disturb the perceived north of phone compass, thus leading to offset in heading estimation. Magnetic offset is location specific and thus unpredictable beforehand. Figure 7.7 illustrates the magnetic offset β between the original North (N) and the perceived North (N′).

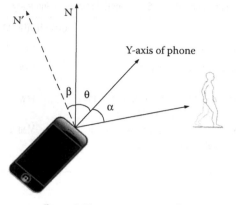

α: Placement offset   θ: Phone orientation   β: Magnetic offset

Figure 7.7 Illustration of magnetic offset and placement offset.

7.3.3.3.2 Placement Offset The compass measures the orientation of the phone (represented as the angle θ between the North and the direction of phone's Y axis), while the phone's heading may not be aligned with the moving direction of the user. Thus placement offset α refers to the difference between the phone's orientation and the moving direction of the user.

Many existing schemes attempted to fuse multiple sensors to improve estimation accuracy of heading direction. According to the additional sensory modalities utilized, we categorize these schemes as follows:

7.3.3.3.3 Inertial Verification Since multiple inertial sensors are integrated on a single smartphone, they tend to perceive similar movements during walking. For example, compass value is probably valid if the readings of phone compass and gyroscope experience correlated trend, which assists to discard compass values containing severe magnetic offset. Acceleration traces, on the other hand, can be utilized to identify the time when the phone placement is the same to that when the phone is first put into a pocket, which helps to accurately infer the moving direction of the user, given the initial phone placement offset.

For the case where the phone is put in a user's pant pocket, a heading estimation method was developed in References 11 and 13.

Step 1: Assume the initial phone placement offset is known. This assumption is reasonable as it is prerequisite to initialize the relation between user heading and phone heading so that the user heading can be tracked based on the sensor tracking of the phone heading. An example is when a user is standstill and looks at her phone, with the phone pointing in the direction of her walking, at this time, the phone heading is θ_1, also is user heading; then puts the phone in the pocket, at this time, the phone heading is θ_2, so the initial phone placement offset is $\alpha = \theta_2 - \theta_1$.

Step 2: During the walking, the phone heading is relatively stable with the leg movement. This is also quite reasonable as the phone moves with the leg in the pocket, exhibiting a periodic pattern in heading change. A special point, the so-called inference point, can be identified during each step, in which the relative orientation of the phone to the user body is the same as in the original state, that is, when the phone was just put into the pocket. At inference point, the placement offset between sensor's yaw and user heading is approximately the same as the moment when user is at standstill. So, at the inference point of every step, the user heading can be inferred, based on the current phone orientation θ and the initial phone placement offset α, as $\alpha + \theta$.

Note that a novel approach was proposed by Deng et al. [21] for user heading estimation, which did not use compass, and therefore can avoid magnetic field disturbance. First, the smartphone is initially held and gazed at by the user for a few seconds, whereas the smartphone's forward axis Y is aligned with the user's walking direction. Then angular velocity measured by gyroscope can be used to derive a rotation matrix that project acceleration data from accelerometer into user's coordinate system. Finally, principal component analysis (PCA) of projected acceleration signals is applied for walking direction vector extraction. User's heading direction can be inferred by combining the change in walking direction and the initial orientation.

7.3.3.3.4 Visual Reference Since modern buildings are mostly rectangular, the straight ceiling edges offer an orthogonal reference to determine heading information. A zero-configured heading sensing system for indoor mobile devices, Headio was proposed in Reference 22. Headio takes advantage of the observation that, as the development of suspended acoustical ceiling systems in modern building constructions, various ceiling objects, such as beams, panel grids, tube lamps, and ventilation ducts, are generally mounted in such a way that their straight edges are either parallel or perpendicular to the orientation of the building, to retain aesthetic neatness and to facilitate construction. Headio detected these visual patterns on ceilings, and used their straight edges to provide directional references. Therefore, when a user wants to determine his or her phone's current heading, Headio would capture ceiling images using the front-facing camera of the mobile phone, and automatically would compute the heading by integrating the directions of the detected ceiling edges and the device's own magnetometer readings.

To determine accurate device headings, Headio senses multimodal contexts with zero-configurations from the user. First, the system uses geolocation sensors on mobile devices to identify locations of users with building-level accuracy. This location information is then used in the detection of building orientations through online map services (e.g., Google Maps or Bing Maps). Second, since users hold mobile devices, such as smartphones, with arbitrary poses, Headio estimates the users' phone poses using the gravity sensors of the mobile devices and rectifies perspective distortions to minimize potential heading errors. Finally, to provide energy-efficient heading sensing, Headio collects various ambient contexts, such as phone placements and ambient indoor luminance, to assess the probability of getting effective ceiling images, and dynamically schedules sensing tasks. The system achieved average heading precision of 1° with arbitrary phone-holding poses. Although the heading estimation accuracy was improved by dozens than inertial schemes, the computational overhead, energy consumption, and the perquisite to take photos, impede its viability.

Table 7.4 compares some representative heading estimation approaches.

In summary, while compass directly provides the absolute directions of phones, the magnetic offset and placement offset considerably deviate compass readings from the actual moving direction. Recent advances fuse extra sensors with compasses to provide robust heading estimation, yet accuracy still remains a bottleneck for inertial based

Table 7.4 Representative Heading Estimation Approaches

TECHNIQUES	SENSORS USED	ACCURACY
Identify inference points [11]	CP+A	Medium
Estimate offset range via spectral analysis [12]	CP+G+A	N/A
Human walking pattern analysis and magnetic interference localization and isolation [23]	CP+G+A	High
Detect visual patterns on ceilings and integrate with compass readings [22]	CP+CA+A	High
Vector projection; PCA [21]	G+A	High

CP—compass, G—gyroscope, A—accelerometer, CA—camera.

indoor localization and navigation systems. Reference 23 reduced heading estimation error to less than 6° by in-depth video-based human walking pattern analysis and magnetic interference localization and isolation. The primary hurdle for this bottleneck is that inertial based heading estimation schemes exploit sensors to perceive the relatively unconstrained human behaviors, making precise walking direction a micromotion that requires subtle identification [23]. In contrast, visual reference-based approaches [22] leveraged some static landmarks such as ceiling edges, yet improving estimation accuracy at the cost of computation and energy consumption.

7.3.4 Comparison of Wi-Fi Fingerprint and PDR-Based Indoor Localizations

PDR and Wi-Fi fingerprinting are well-known approaches for indoor localization but each has its own advantages and limitations. PDR-based schemes naturally enable continuous location tracking and have low cost; location error accumulating over time is a major concern. Wi-Fi fingerprinting-based localization approach is an attractive alternative as it can leverage the widely existing Wi-Fi infrastructure (i.e., a commonplace nowadays in most indoor environments) and the ubiquitous presence of Wi-Fi interfaces on smartphones. Furthermore, it has high location accuracy with no accumulating error. But the Wi-Fi fingerprinting approach is not suitable for continuous location tracking of a mobile user because Wi-Fi scanning operations are relatively quite power hungry. The applicability and effectiveness of Wi-Fi fingerprinting are dependent on a number of factors including Wi-Fi AP density, spatial differentiability, and temporal stability

Table 7.5 Comparison of Wi-Fi Fingerprint and PDR

INDOOR LOCALIZATION TECHNOLOGIES	ADVANTAGE	LIMITATION
Wi-Fi fingerprinting	• Having high location accuracy without accumulating error. • Leveraging the common Wi-Fi infrastructure.	• Power hungry for continuous location tracking. • Training phase is time-consuming and expensive.
Pedestrian dead reckoning	• Enabling continuous location tracking. • Having low deployment cost.	• Location error accumulates over time. • Low location accuracy.

of the radio environment. Moreover, the training phase can be quite time-consuming and expensive. The advantages and disadvantages of those two popular indoor localization paradigms are summarized in Table 7.5.

7.4 Improvements to PDR and Wi-Fi Fingerprint-Based Indoor Localization Schemes

HiMLoc was proposed by Radu et al. [24], which synergistically uses PDR and Wi-Fi fingerprinting to exploit their positive aspects and limit the impact of their negative aspects. Specifically, HiMLoc combined location tracking and activity recognition using inertial sensors on mobile devices with location-specific weighted assistance from a crowdsourced Wi-Fi fingerprinting system via a particle filter. The integration of dead reckoning with Wi-Fi fingerprinting is based on the observation that some spaces in a building are more accurately localizable with Wi-Fi fingerprinting than others, which is a consequence of differences in spatial differentiability of the Wi-Fi environment among these spaces due to building structure and radio signal propagation effects. To exploit this observation, HiMLoc associates a weight for the Wi-Fi fingerprinting component in a particle filter that influences the extent to which it is relied on in the hybrid localization. This weight in turn inversely proportional to similarity is a metriccomputed by comparing a run-time Wi-Fi fingerprint with fingerprint database—smaller similarity area results in a higher weight and vice versa.

The system framework of HiMLoc is illustrated in Figure 7.8. Phone's sensors (accelerometer, compass, and Wi-Fi card) collect

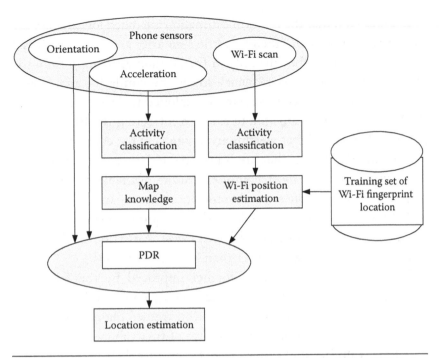

Figure 7.8 Schematic of HiMLoc hybrid localization mechanism. (From Radu, V. and M.K. Marina, HiMLoc: Indoor smartphone localization via activity aware Pedestrian Dead Reckoning with selective crowdsourced Wi-Fi fingerprinting, in *Proceedings of the 4th International Conference on Indoor Positioning and Indoor Navigation* (*IPIN*), Montbeliard-Belfort, France, October 28–31, 2013.)

sensor data (acceleration, orientation, and Wi-Fi scans) to be used as direct input to the system. The activity classification component determines what activity the user is performing within a short interval of time by sampling the acceleration data. If the estimated activity can be performed in just a very limited number of places inside a building, like going up and down the stairs or taking an elevator, then Map Knowledge can assist to determine these possible locations. Acceleration and orientation are used in the PDR component to track the continuous movement. Finally, if a Wi-Fi scan is available, it is used to extract a runtime Wi-Fi fingerprint. Such a fingerprint is compared with those in a fingerprint database (created via crowdsourcing). Estimations of these components are merged by the particle filter to obtain a single estimation for the whole system. At the end of this process, if Wi-Fi scan information is available, it is annotated with the estimated location and used to update the fingerprint database.

It is observed that the errors of estimation are much lower when the similarity area is small. Therefore, the similarity area can be utilized as a metric to judge the reliability of the location estimation from the Wi-Fi fingerprinting component—lower the similarity area, lower the error. The similarity area metric is exploited to decide when to use the Wi-Fi location estimation to assist in correcting the PDR via the particle filter, and higher weights are assigned to the estimations with a low-similarity area as they are considered to be more accurate.

By using just the most common sensors available on the large majority of smartphones (accelerometer, compass, and Wi-Fi card) and offering an easily deployable method (requiring just the locations of stairs, elevators, corners, and entrances), HiMLoc is shown to achieve median accuracies lower than 3 m in most cases.

However, HiMLoc can be further improved by the following aspects: balance the trade-off between energy consumption and system performance and deal with the unconstrained phone placement.

Probably the most important resource of smartphones is their battery. Long running applications have to consider their impact on this resource and to reduce energy consumption as much as possible. But there is always a trade-off between low energy consumption and system's performance. It is the case with localization on smartphones as well. Continuous use of Wi-Fi scans consumes more energy than continuous sampling of acceleration data. HiMLoc aims to reduce the number of Wi-Fi scans but also it keeps providing some Wi-Fi scans to reduce the error accumulation in the PDR caused by drift and noisy readings. In this trade-off between location accuracy and battery consumption, it was found that one Wi-Fi scan every 20 seconds is ideal for the cause. This was empirically determined based on analysis of the time between landmarks in one of the buildings investigated by HiMLoc. It may attribute to the feature imposed by the building, as well as the application's purposes. If the application requires high location accuracy, then this frequency should be increased to provide assistance for the PDR more often or decreased if the application constraints are not very strict. In brief, the energy-accuracy trade-off should be deeply investigated in future, according to the detailed requirements of specific application.

For the following two observations: an initial investigation of how people carry their phone and also the need to have the phone in a static position relative to the user's body so that the phone can detect user's movements more accurately, HiMLoc considered two cases of carrying phones: in the front pocket of trousers and in hand in front of the user. However, some people may prefer to carry their phones differently, like in a bag or purse, but these cases are very hard for location systems that use inertial sensors because their position is not fixed and the acceleration detected by the phone is a mixture between the bag's movement and the free movement of the phone inside the bag. Some possible work around this problem would be to learn the patterns of movements on the way, with more assistance from independent references, like landmarks or Wi-Fi, knowing that people tend to keep a uniform motion when walking.

The UnLoc system [25] combined the use of inertial sensors (accelerometer, compass, gyroscope) with the notion of natural and organic landmarks that are learnt over time for indoor navigation, whereas the use of Wi-Fi fingerprinting in UnLoc is limited to identifying organic landmarks based on radio environment.

The key observation is that certain locations in an indoor environment present identifiable signatures on one or more sensing dimensions. An elevator, for instance, imposes a distinct pattern on a smartphone's accelerometer; a corridor corner may overhear a unique set of Wi-Fi APs; a specific spot may experience an unusual magnetic fluctuation. It is reasonable to hypothesize that these kinds of signatures naturally exist in the environment and can be envisioned as internal landmarks of a building. Mobile devices that *sense* these landmarks can recalibrate their locations, while dead-reckoning schemes can track them between landmarks. The rationale of using landmarks lies in that the locations of these landmarks serve as restarting locations for pedestrians, hence dividing a user's long trajectories into multiple short trajectories and significantly reducing the accumulative measurement errors from inertial sensors. Figure 7.9 illustrates the system framework of PDR-based indoor localization assisted with landmark.

The landmarks investigated in this work include turns, elevators, escalators, stairs, and doors. The sensors involved in the identification of these landmarks are the accelerometer, magnetometer, gyroscope, barometer, and Wi-Fi.

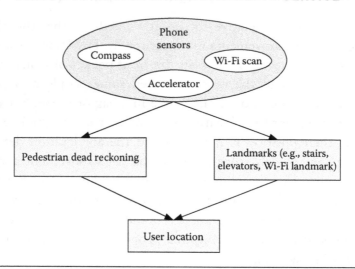

Figure 7.9 Architecture of landmark-assisted indoor localization. (From Wang, H. et al., No need to war-drive: Unsupervised indoor localization, in *Proceedings of the 10th ACM International Conference on Mobile Systems, Applications, and Services* (*MobiSys*), Low Wood Bay, Lake District, June 25–29, 2012, pp. 197–210.)

Turns: Due to the topology constraints in indoor environments, turns can be identified using angular or direction-related sensors. The most direct way to recognize turns is to use the magnetometer reading. However, the magnetometer output will be affected by electronic devices, which makes it unstable for the recognition of turns. An alternative way to do this is by leveraging the gyroscope information, which measures angular acceleration without the influence of any equipment. After smoothing the gyroscope output, turns can be easily distinguished. In addition, based on the direction of the pulse, left or right turns can be separated. In real situations, many turns can be detected. If the distance of two turns is smaller than localization accuracy, we may make a wrong decision, which will lead to a bias in the system. In order to avoid this problem, we only consider turns that are unique in the path as landmarks.

Elevators: Based on the unique pattern of vertical acceleration data, taking elevators can be easily identified. There will be a hypergravity and a hypogravity process, shown as a positive impulse and a negative impulse of vertical acceleration, respectively. Moreover, the length between the impulses reflects how many floors the pedestrian goes up or down, which can be important for multifloor localization.

Escalators: The acceleration pattern of taking escalators is similar to being stationary. In order to distinguish them, the magnetometer data are used. In the stationary condition, the variance of magnetometer data is small. However, since the motors in the escalators will dramatically influence the magnetic field, the variance of the magnetometer reading will be large when taking escalators.

Stairs: Going upstairs or downstairs has a similar pattern of acceleration as normal walking. An effective way to distinguish these two activities is to use the barometers in smartphones. It is known to all that the higher the elevation, the lower the ambient air pressure. Therefore, there will be a decrease in the ambient air pressure when going upstairs and an increase in the ambient air pressure when going downstairs.

Doors: The landmarks of doors contain two phases of acceleration: a low value for opening the door and a periodic pattern for walking out. If only based on this phenomenon, many false positive events will be produced. Another prominent property of passing through a door is the big change of the received signal strength of Wi-Fi. Based on this property, doors can be easily identified.

Results from three different indoor settings, including a shopping mall, demonstrate median location errors of 1.69 m. But, the weakpoints of UnLoc are as follows. Similar to HiMLoc, UnLoC deals with two cases of phone orientation: in pocket and in hand. Handling arbitrary phone orientations and their effect on the reported sensor values should be deeply investigated. As gyroscopes are becoming more available on new phones, they can be leveraged to map arbitrary orientation to a specific frame of reference. Meanwhile, orientation-independent features, such as the magnitude of the acceleration, should be contemplated. In order to limit the energy consumed from sensing, UnLoc simply avoids using sensors that have a high-energy footprint, for example, light and sound. However, more advanced energy-efficient technologies should be utilized. For instance, it is suitable to adaptively turn on sensors, perhaps depending on the density of landmarks available in the environment.

7.5 Conclusion

This chapter systematically overviews the current indoor positioning techniques and systems. First, some basic concepts of indoor localization are introduced, including performance metrics and positioning principle like triangulation, proximity, scene analysis, PDR, and so on. Especially, Wi-Fi fingerprint (i.e., a special kind of environmental scene) and PDR are two popular indoor localization paradigms. Basically, Wi-Fi fingerprint-based indoor localization can be divided into two stages: offline training phase (i.e., site survey to build fingerprint database) and online positioning phase. PDR localization technologies use motion sensors (accelerometer, digital compass, gyroscope, etc.) to determine location at the device without the need for infrastructure, which mainly consists of three modules: step detection, step length estimation, and heading estimation. Typical Wi-Fi fingerprint and PDR indoor localization schemes are summarized and compared, and the improved approaches integrating both paradigms are also discussed.

References

1. Deak, G., K. Curran, and J. Condell. A survey of active and passive indoor localization systems. *Computer Communications*, 2012; 35: 1939–1954.
2. Cruz, C.C., J.R. Costa, and C.A. Ferandes. Hybrid UHF/UWB antenna for passive indoor identification and localization systems. *IEEE Transactions on Antennas and Propagation*, 2013; 61: 354–361.
3. Liu, H., H. Darabi, P. Banerjee, and J. Liu. Survey of wireless indoor positioning techniques and systems. *IEEE Transactions on Systems, Man, and Cybernetics, Part C: Applications and Reviews*, 2007; 37(6): 1067–1080.
4. Stojanović, D. and N. Stojanović. Indoor localization and tracking: Methods, technologies and research challenges. *Facta Universitatis, Series: Automatic Control and Robotics*, 2014; 13(1): 57–72.
5. Zhang, L. Modeling of distance related range error of TDOA based UWB RFID indoor positioning system. 2014. Available online: https://dr.ntu.edu.sg/handle/10220/24261.
6. Kim, Y., H. Shin, Y. Chon, and H. Cha. Smartphone-based Wi-Fi tracking system exploiting the RSS peak to overcome the RSS variance problem. *Pervasive and Mobile Computing*, 2013; 9(3): 406–420.
7. Wu, C., Z. Yang, Z. Zhou, and M. Liu. DorFin: WiFi fingerprint-based localization revisited. *arXiv preprint*, 2013; arXivID:1308.6663.
8. Niu, J., B. Wang, L. Cheng, and J.J.P.C. Rodrigues. WicLoc: An indoor localization system based on WiFi fingerprints and crowdsourcing. In: *Proceedings of IEEE International Conference on Communications (ICC)*, London, UK, June 8–12, 2015.

9. Wu, C., Z. Yang, Y. Liu, and W. Xi. WILL: Wireless indoor localiza-tion without site survey. *IEEE Transactions on Parallel and Distributed Systems*, 2013; 24(4): 839–848.
10. Yang, Z., C. Wu, Z. Zhou, X. Zhang, X. Wang, and Y. Liu. Mobility increases localizability: A survey on wireless indoor localization using inertial sensors. *ACM Computing Surveys*, 2015; 47(3): 54.
11. Qian, J., J. Ma, R. Ying, P. Liu, and L. Pei. An improved indoor local-ization method using smartphone inertial sensors. In: *Proceedings of the 4th International Conference on Indoor Positioning and Indoor Navigation (IPIN)*, Montbeliard-Belfort, France, October 28–31, 2013.
12. Rai, A., K.K. Chintalapudi, V.N. Padmanabhan, and R. Sen. Zee: Zero-effort crowdsourcing for indoor localization. In: *Proceedings of the 18th ACM Annual International Conference on Mobile Computing and Networking (Mobicom)*, Istanbul, Turkey, August 22–26, 2012, pp. 293–304.
13. Li, F., C. Zhao, G. Ding, C. Liu, and F. Zhao. A reliable and accurate indoor localization method using phone inertial sensors. In: *Proceedings of the 2012 ACM Conference on Ubiquitous Computing (UbiComp '12)*, Pittsburgh, PA, September 5–8, 2012, pp. 421–430.
14. Brajdic, A. and R. Harle. Walk detection and step counting on uncon-strained smartphones. In: *Proceedings of the 2013 ACM International Joint Conference on Pervasive and Ubiquitous Computing (UbiComp '13)*, Zurich, Switzerland, September 8–12, 2013, pp. 225–234.
15. Wang, J.H, J.J. Ding, Y. Chen, and H.H. Chen. Real time accelerometer-based gait recognition using adaptive windowed wavelet transforms. In: *Proceedings of 2012 IEEE Asia Pacific Conference on Circuits and Systems (APCCAS)*, Kaohsiung, Taiwan, December 2–5, 2012.
16. Mannini, A. and A. Sabatini. A hidden Markov model-based technique for gait segmentation using a foot-mounted gyroscope. In: *Proceedings of IEEE International Conference of Engineering in Medicine and Biology Society (EMBS)*, Boston, MA, August 30–September 3, 2011.
17. Cho, D.K., M. Mun, U. Lee, W.J. Kaiser, and M. Gerla. Autogait: A mobile platform that accurately estimates the distance walked. In: *Proceedings of the 2010 IEEE International Conference on Pervasive Computing and Communications (PerCom)*, Mannheim, Germany, March 29–April 2, 2010.
18. Harle, R. A survey of indoor inertial positioning systems for pedes-trians. *IEEE Communications Surveys & Tutorials*, 2013; 15(3): 1281–1293.
19. Renaudin, V., M. Susi, and G. Lachapelle. Step length estimation using handheld inertial sensors. *Sensors*, 2012; 12(7): 8507–8525.
20. Tian, Z., Y. Zhang, M. Zhou, and Y. Liu. Pedestrian dead reckon-ing for MARG navigation using a smartphone. *EURASIP Journal on Advances in Signal Processing*, 2014; 1: 65.
21. Deng, Z.A., G. Wang, Y. Hu, and D. Wu. Heading estimation for indoor pedestrian navigation using a smartphone in the pocket. *Sensors*, 2015; 15(9): 21518–21536.

22. Sun, Z., S. Pan, Y.C. Su, and P. Zhang. Headio: Zero-configured heading acquisition for indoor mobile devices through multimodal context sensing. In: *Proceedings of the 2013 ACM International Joint Conference on Pervasive and Ubiquitous Computing (UbiComp '13)*, Zurich, Switzerland, September 8–12, 2013, pp. 33–42.

23. Roy, N., H. Wang, and R.R. Choudhury. I am a smartphone and I can tell my users walking direction. In: *Proceedings of the 12th ACM International Conference on Mobile Systems, Applications, and Services (MobiSys)*, Bretton Woods, NH, June 16–19, 2014.

24. Radu, V. and M.K. Marina. HiMLoc: Indoor smartphone localization via activity aware Pedestrian Dead Reckoning with selective crowdsourced WiFi fingerprinting. In: *Proceedings of the 4th International Conference on Indoor Positioning and Indoor Navigation (IPIN)*, Montbeliard-Belfort, France, October 28–31, 2013.

25. Wang, H., S. Sen, A. Elgohary, M. Farid, M. Youssef, and R.R. Choudhury. No need to war-drive: Unsupervised indoor localization. In: *Proceedings of the 10th ACM International Conference on Mobile Systems, Applications, and Services (MobiSys)*, Low Wood Bay, Lake District, June 25–29, 2012, pp. 197–210.

8

MOBILE CROWDSOURCING WITH BUILT-IN INCENTIVES

8.1 Introduction

The perspective of human computation argues that combining humans and machines can help tackle increasingly hard problems [1]. As typical evidence, crowdsourcing has emerged as an effective way to perform tasks that are easy for humans but remain difficult for computers. Note that the term crowdsourcing, a form of *peer production* was coined by Howe [2] to refer to the approach that outsources works to a large group of common people. Crowdsourcing is deemed particularly useful for tasks requiring a large number of viewpoints and problem solvers and, at the same time, it is deemed easy to perceive by the general public [3]. Actually, crowdsourcing platforms, such as Amazon Mechanical Turk (AMT), have enabled the construction of scalable applications for vast tasks ranging from product categorization and photo tagging to audio transcription and translation, and so on.

Nowadays, smartphones are ubiquitous and widely used around the world, in which rich sensors (e.g., global positioning system, GPS; accelerometer; camera) are built into, and multiple radios (e.g., Bluetooth, Wi-Fi, Cellular) are provided. This trend enables individuals with smartphones to sense, collect, process, and distribute data around people at any time and place. Naturally, the integration of smartphone-based mobile technologies and crowdsourcing offers a significant flexibility and leads to a new computing paradigm called mobile crowdsourcing (MCS), which can be fully explored for real-time and location-sensitive crowdsourced tasks.

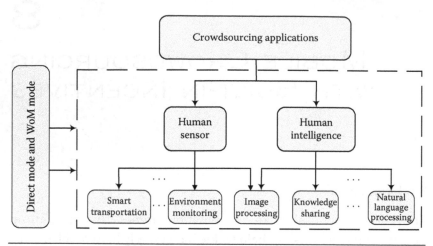

Figure 8.1   Taxonomy of the mobile-crowdsourcing applications.

MCS presents an exciting opportunity and is likely to enable a new generation of applications capable of bringing about positive changes in lives of ordinary citizens at the societal scale over the next several years. It has a broad spectrum of applications including but not limited to environmental monitoring, intelligent transportation, personalized medicine, and epidemiological investigation of disease vectors. As shown in Figure 8.1, most of the current MCS deployments fall in the following two categories: human intelligence and human sensor. The category of human intelligence utilizes human wisdom (wisdom of crowd) to perform tasks that are easy for humans but remain difficult for computers, such as knowledge sharing and natural language processing (NLP). The category of human sensor incorporates a concept of *human-as-a-sensor* into the MCS system to collect human observations (in addition to sensor measurements from the owners of mobile devices), which can be used in various fields, such as smart transportation and environment monitoring. As described above, the paradigm of crowd participation is explicitly divided into the direct and WoM modes.

Specifically, as a main application filed of MCS, crowdsensing (or mobile crowdsourced sensing) is an emerging area of interest for research and applications, through which the physical world can be sensed without deploying a sensor network by exploiting the billions of users' mobile devices/phones as location-aware data collection instruments for real-world observations. Mobile crowdsourced sensing offers a number of advantages over traditional sensor networks, which entails deploying a large number of static wireless sensor devices, particularly in urban areas. First, since participatory sensing leverages existing sensing (mobile phones) and communication (cellular or Wi-Fi) infrastructure, the deployment costs are virtually zero. Second, the inherent mobility of the phone carriers provides unprecedented spatiotemporal coverage and also makes it possible to observe unpredictable events (which may be excluded by static deployments). Third, using mobile phones as sensors intrinsically affords economies of scale. Fourth, the widespread availability of software development tools for mobile phone platforms and established distribution channels in the form of App stores makes application development and deployment relatively easy. Finally, by including people in the sensing loop, it is now possible to design applications that can dramatically improve the day-to-day lives of individuals and communities [4].

Crowdsensing applications can be broadly classified into two categories, personal and community sensing, based on the type of phenomena being monitored. In personal sensing applications, the phenomena pertain to an individual, for example, the monitoring of movement patterns (e.g., running, walking, exercising) of an individual for personal record-keeping or healthcare reasons. Another example of personal sensing is one that monitors the transportation modes of an individual to determine his or her carbon footprint. On the other hand, community sensing pertains to the monitoring of large-scale phenomena that cannot easily be measured by a single individual. For example, intelligent transportation systems may require traffic congestion monitoring and air pollution level monitoring. These phenomena can be measured accurately only when many individuals provide speed and air quality information from their daily commutes, which are then aggregated spatiotemporally to determine congestion and pollution levels in cities [5].

Community sensing is also popularly called participatory sensing or opportunistic sensing. In participatory sensing, the participating user is directly involved in the sensing action, for example, to photograph certain locations or events. This approach includes people into significant decision stages of the sensing systems, deciding actively what application requests to accept. Through participatory sensing, people who are carrying everyday mobile devices act as sensor nodes and form a sensor network with other such devices. A large number of mobile phones, PDAs, laptops, and cars are equipped with sensors and GPS receivers, which are potential candidate devices for participatory sensor nodes. The collection and dissemination of environmental sensory data by ordinary citizens are made possible by using participatory sensing through devices such as mobile phones. It does not require any preinstalled infrastructure.

In opportunistic sensing the user is not aware of active applications. He is not involved in making decisions, instead the mobile phone or smartphone itself makes decisions according to the sensed and stored data. Opportunistic sensing shifts the burden of supporting an application from the custodian to the sensing system, automatically determining when device can be used to meet application requests. When we consider different aspects of a person and his or her social setting like where he or she is and where he or she is going, what he or she is doing, what he or she is seeing, what he or she eats and hears, what are his or her likes and dislikes, his or her health-related conditions, and so on while developing mobile phone sensing systems, this is called people-centric urban sensing. In a people-centric sensing system, the focal point of sensing is humans instead of buildings or machines, and the visualization of sensor-based information is for the benefit of common citizens and their friends, rather than domain scientists or plant engineers [6].

Crowd members use their smartphones to contribute to the problem's solution by generating, processing, or sensing data of interest, which are in turn collected by the server, which results in a win–win situation in which both the open-call publisher and the mobile crowd are rewarded [7]. In this chapter, we propose a generic mobile-crowdsourcing framework that consists of multiple functional modules that are independent of specific applications and can accommodate multimodal data sources [8,9].

Although there are many applications and numerous crowdsourcing websites, most of them are based on voluntary participation. While participating in a crowdsourcing task, smartphone users consume their own resources that incur a heterogeneous variety of costs, such as data processing and transmission, charges by the mobile operator for the bandwidth needed for transmitting data, mobile device battery energy cost for sensing, processing power cost, or discomfort due to manual effort to submit data. We classify such sensing costs into two categories: (1) energy consumption for generating, processing, and transmitting the sensed data, and (2) local storage occupied by the sensed data. Users may exhibit various levels of tolerance to different categories of sensing costs according to the specific sensing task, environmental context, and system resource conditions. Such heterogeneity hence makes it challenging to maximize user benefits and reduce their reluctance against mobile-sensing tasks via modeling, analyzing, and fulfilling users' demands and preferences. On the other hand, users also expose themselves to potential security threats by sharing their sensed data with location tags. Therefore a user would be unwilling to contribute his or her smartphone's local resources, unless he or she receives a satisfying reward to compensate his or her resource consumption and potential security breach.

Without adequate user participation, it is impossible for crowdsourcing applications to implement good service quality, since sensing services depend crucially on users' sensed data. Thence it is plenty important for the survivability of the service to have suitable incentive mechanisms to motivate users to participate in crowdsourcing. In addition, ill-defined crowdsourcing tasks that do not provide workers with enough information about the tasks and their requirements can also lead to low-quality contributions from the crowd. This chapter will investigate the challenges of MCS and will propose a quality-aware incentive framework for MCS, QuaCentive, which, pertaining to all components in MCS, can motivate crowd to provide high-quality sensed contents, can stimulate crowdsourcers to give truthful feedback about quality of sensed contents, and can make platform profitable.

This chapter is organized as follows: In Section 8.2, we introduce several typical crowdsourcing applications including human sensor, image processing, knowledge sharing and NLP, word of mouth (WoM), and so on. Section 8.3 proposes a generic MCS framework. The key challenges of MCS are presented in Section 8.4, involving the task management, incentive, quality control, security, and privacy. Then, a quality-aware incentive framework QuaCentive is provided in Section 8.5. Finally, we briefly conclude this chapter.

8.2 MCS Framework and Typical Applications

8.2.1 MCS Enabling Modes

As shown in Figure 8.2, in general, three stakeholders are involved into MCS system: task publishers (i.e., crowdsourcer), crowdworkers, and crowdsourcing platform. Specifically, to crowdsource a task, its owner, crowdsourcer submits the task to a crowdsourcing platform, and optionally, after receiving the data/solutions provided by crowdworkers, rates the quality; crowdworkers choose to work on those tasks and attempt to submit their solutions as feedbacks; an intermediation platform (i.e., crowdsourcing platform) builds a link between the crowdsourcers and workers, which serves as a crowdsourcing enabler and has some rules for the whole life cycle of crowdsourcing, such as the skill-set, certification level, due date, expected outcomes, and payments for the crowdworkers.

In general, there exist two paradigms to mobilize users to participate in and contribute data to MCS system: direct and WoM mode. Direct mode adopts the centralized control, in which the MCS platform accepts an uploaded task from a crowdsourcer, divides it into microtasks, and coordinates crowdworkers when they conduct the tasks iteratively or in parallel. In other words, humans passively or actively select tasks and contribute to them, but do not generally interact and collaborate with each other directly. Such mode may perform inefficiently with time-sensitive and location-sensitive tasks. For example, in the scenario of finding missing children, ideally, the most

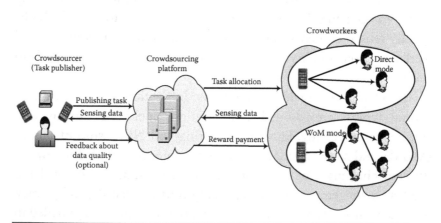

Figure 8.2 Illustration of MCS application scenarios.

effective action should be to inform people close to where the child was lost and immediately after the child was discovered lost to help locate the child.

8.2.2 Comparison among Typical MCS Applications

In this subsection, we briefly introduce and compare some typical MCS applications, which, respectively, belong to the categories of direct and WoM paradigms.

8.2.2.1 Direct Mode   Human sensor: The explosive growth in social network content suggests that the largest *sensor network* yet might be human. The concept of human sensor is mean *human-as-a-sensor* participates into the MCS system to collect individual sensing information and human observations, in addition to sensor measurements, from mobile devices owned by the public crowd.

8.2.2.1.1 Smart Transportation

8.2.2.1.1.1 TrafficInfo Reference 10 presented TrafficInfo, a prototype smartphone application to implement a participatory sensing-based live public transport information service. Modern cities continuously struggle with infrastructural problems especially when the population is massively growing. One affected area is public transportation. In default of offering convenient and reliable service the passengers tend to consider other transport alternatives. However, even a relatively simple functional enhancement, such as providing real-time timetable information, requires considerable investment and effort following traditional means, for example, deploying sensors and building a background communication and processing infrastructure. Using the power of crowd to gather the required data, sharing information and sending feedback are viable and cost-effective alternatives. In this demonstration, they present TrafficInfo; it visualizes the actual position of public transport vehicles with live updates on a map and gives support to crowdsourced data collection and passenger feedback.

In this case, the built-in sensors of the passengers' mobile devices, or the passengers themselves via reporting incidents, provide the required data for vehicle tracking and send instant route information to the service provider. The service provider then aggregates, cleans, analyzes the collected data, and derives and disseminates the real-time updates. For sensing, the built-in and ubiquitous sensors of the mobile phones are used either in participatory or opportunistic way depending on whether data collection happens with or without user involvement. The contribution of every traveler can be useful. Hence, passengers waiting for a trip can report the line number with a time-stamp of every arriving public transport vehicle at a stop during the waiting period. On the other hand, the on-board passengers can send actual position information of the moving vehicle periodically and report the event of arrival at/departure from a stop. Following the publish/subscribe communication model the passengers subscribe in TrafficInfo, according to their interest, to traffic information channels dedicated to different public transport lines or stops. Hence, they are informed about the live public transport situation, such as the actual vehicle positions, deviation from the static timetable, and crowded-ness information.

8.2.2.1.1.2 Sense2Health   Sense2Health [11] is an application that monitors personal exposure to environment pollution and assesses its health-related risks. The novelty of the application is that it requires little to no active involvement by users and unlike existing applications, it correlates the individual's well-being to their environment as opposed to their physical activity alone. Consequently, when health and environment data are acquired, this application enables users to better identify behavior changes toward enhancing their health by enhancing their environments. Furthermore, Sense2Health is an open platform for integrating existing domain-specific sensing applications (environmental and health monitoring) focused on decreasing required specialized development efforts. They present design of Sense2Health in addition to a proof-of-concept implementation for a noise-monitoring use case. Afterward, they assess its performance while integrating it with a dedicated open-source noise-sensing application. To assist users to track their well-being with respect to environmental pollution and understand its effect on their health, they provide an application that can leverage, altogether, biosensors and available domain-specific environment-monitoring mobile applications that track a phenomenon of interest, such as noise pollution.

8.2.2.1.1.3 EarPhone   Reference 12 presented the design, implementation, and performance evaluation of an end-to-end participatory urban noise mapping system called earphone. Earphone, for the first time, leverages compressive sensing to address the fundamental problem of recovering the noise map from an incomplete and random samples obtained by crowdsourcing data collection. It also addresses the challenge of collecting accurate noise pollution readings at a mobile device. It is used for monitoring environmental noise, especially roadside ambient noise. The key idea is to crowdsource the collection of environmental data in urban spaces to people who carry smartphones equipped with sensors and location-providing GPS receivers. The overall earphone architecture consists of a mobile phone component and a central server component. Noise levels are assessed on the mobile phones before being transmitted to the central server. The central server reconstructs the noise map based on the partial noise measurements. Reconstruction is done because the urban sensing framework cannot guarantee that noise measurements are available at all times and locations.

The system performance is evaluated in terms of noise-level measurement accuracy, resource (CPU, RAM, and energy) consumption, and noise-map generation. Extensive simulations and outdoor experiments have demonstrated that earphone is a feasible platform to assess noise pollution, incurring reasonable system resource consumption at mobile devices and providing high reconstruction accuracy of the noise map.

8.2.2.1.2 Image Processing   In imaging science, image processing is any form of signal processing for which the input is an image, such as a photograph or a video frame; the output of image processing may be either an image or a set of characteristics or parameters related to the image.

8.2.2.1.2.1 PetrolWatch   PetrolWatch is a system that automatically collects fuel prices using camera phones [13]. This system is implemented as a client-server program and has two principal modes of operation: fuel price collection and user query. The fuel price collection is completely automated and only requires the camera phone to be mounted on the dash board, with the camera lens pointing toward the road.

The system automatically triggers the mobile phones to photograph the road-side fuel price boards when they approach service stations. Sophisticated computer vision algorithms are used to scan these images and retrieve the fuel prices. To deal with a nonstructured environment and to reduce computer vision complexity, it relies on the geographic information system (GIS) database and GPS location to know the service station brand and uses the fact that each brand uses a specific color for its price board. The metadata (location coordinates, brand and time) are extracted and stored separately. The images and fuel brand data are passed on to the image-processing engine, which is implemented on the mobile phone. The first step detects the existence of a fuel price board. For each fuel brand, a tailored color thresholding is employed that can capture regions within the images, having a color scheme similar to the fuel brand price board. In certain situations, surrounding objects in the image may have colors resembling the board, for example, the blue sky may be similar to the mobile price board. In this case, postprocessing techniques are used to narrow the search. The price board dimensions are used to exclude some of the candidate regions selected by color thresholding. This is further refined by comparing the color histogram of all candidate regions with that of a sample price board image. The image is cropped to contain only the board, and is normalized to standard size and resolution. The color image is converted to binary, and connected component labeling is used to extract the individual numeral characters. A neural network algorithm is used to classify the digits. The extracted prices are uploaded to the server and stored in a database, linked to a GIS road network database populated with service station locations. The server updates fuel prices of the appropriate station if the current price has a newer timestamp. History is also maintained to analyze pricing trends.

8.2.2.1.2.2 CrowdSearch   CrowdSearch combines automated image search with real-time human validation of search results [14]. Automated image search uses a combination of local processing on mobile phones and remote processing on powerful servers. For a query image, this process generates a set of candidate search results that are packaged into tasks for validation by humans. Real-time validation uses the AMT, where tens of thousands of people are available to work on simple tasks for monetary rewards. Search results that have been validated are returned to the user.

The combination of automated search and human validation presents a complex set of tradeoffs involving delay, accuracy, monetary cost, and energy. From a crowdsourcing perspective, the key tradeoffs involve delay, accuracy, and cost. A single validation response is often insufficient since (1) humans have error and bias and (2) delay may be high if the individual person who has selected the task happens to be slow. While using multiple people for each validation task and aggregating the results can reduce error and delay, it incurs more monetary cost. From an automated image search perspective, the tradeoffs involve energy, delay, and accuracy. Exploiting local image search on mobile phones is efficient energywise but the resource constraints on the device limit accuracy and increase delay. In contrast, remote processing on powerful servers is fast and more accurate, but transmitting large raw images from a mobile phone consumes time and energy.

CrowdSearch addresses these challenges using three novel ideas. First, it develops accurate models of the delay-accuracy cost behavior of crowdsourcing users. Second, it uses the model to develop a predictive algorithm that determines which image search results need to be validated, when to validate the tasks, and how to price them. The algorithm dynamically makes these decisions based on the deadline requirements of image search queries as well as the behavior observed from recent validation results. Third, it describes methods for partitioning of automated image search between mobile phones and remote servers. This technique takes into account connectivity states of mobile phones (3G vs. Wi-Fi) as well as search accuracy requirements.

8.2.2.1.2.3 Shop Social   Shop Social is an experimental barcode-scanning mobile application that locates product information and relevant product video content based on barcodes the user scans [15]. Shop Social leverages the user's social graph along with product-related media (product descriptions, reviews, videos, photos, etc.) to help a shopper understand a particular product's potential for them, or for whomever they might be shopping. In an attempt to further encourage engagement, Shop Social also incorporates simple badging game mechanics. In addition to the native application running on user handsets, there is a set of nontrivial network services that feed content to the client applications and also serve as a communication conduit to enable users to interact together within the application.

The application supports a variety of features including product lookup by scanning or keying a product universal product code (UPC). The UPC is forwarded to the Google's App Engine (GAE) backend that returns the product description, product reviews, and a list of relevant video sand photos. Recently scanned products are retained temporarily in the scan history, and the user also has the ability to favorite the product by adding it to their *My Stuff* list. The user's social dashboard consists of a list of their Facebook friends, a gallery of badges that each friend has earned to-date, and a list of their favorite products. All of the functionality available to users when they scan a product directly is also available to users when they encounter products on their social dashboard. In addition to watching product videos, users are able to share the videos via Facebook, and/or rate videos with a thumbs-up or thumbs-down. The fact that a user watches a video and/or rates it is ultimately taken into account in future video relevancy scoring as described in the subsection 8.2.2.1.2.2.

8.2.2.1.2.4 SmartPhoto   A framework SmartPhoto is proposed to quantify the quality (utility) of crowdsourced photos [16]. Photos obtained via crowdsourcing can be used in many critical applications. Due to the limitations of communication bandwidth, storage, and processing capability, it is a challenge to transfer the huge amount of crowdsourced photos. To address this problem, Reference 16 proposed a resource-aware framework, called SmartPhoto, to optimize the selection of crowdsourced photos based on the accessible metadata of the smartphone including GPS location, phone orientation, and so on. SmartPhoto is based on the accessible geographical and geometrical information (called metadata) including the smartphone's orientation, position, and all related parameters of the built-in camera. From the metadata, they can infer where and how the photo is taken and then only can transmit the most useful photos. Three optimization problems regarding the tradeoffs between photo utility and resource constraints, namely the Max-Utility problem, the online Max-Utility problem, and the Min-Selection problem, are studied. Efficient algorithms are proposed and their performance bounds are theoretically proved. They have implemented SmartPhoto in a test-bed using Android-based smartphones and have proposed techniques to improve the accuracy of the collected metadata by reducing sensor reading errors and solving object occlusion issues. Results based on real implementations and extensive simulations demonstrate the effectiveness of the proposed algorithms.

8.2.2.1.3 Content Generation/Knowledge Sharing Mobile devices have significantly changed the way we seek information. People can now search large repositories like the web from anywhere. In addition, mobile question and answer (Q&A) services such as Naver Mobile Q&A, ChaCha, and Jisiklog make it possible to quickly and easily find information by asking questions via instant messaging or short message service (SMS) [17].

8.2.2.1.3.1 SmartTrace+ SmartTrace+ enables trajectory similarity functionalities without disclosing the user's trajectory [18]. One example of smartphone crowdsourcing is to ask a crowd of smartphone users to help identify mobility patterns or a given trajectory's popularity. Such a contribution can be utilized in large-scale urban and transit planning, transit rider information applications, shared-ride applications, social networking applications on smartphones, habitant monitoring, and so on.

Consider a transit authority that plans its bus routes and wants to know whether a specific route is taken by at least k users between 7:00 a.m. and 8:00 a.m. In such a scenario, the transit authority asks a crowd of users in a target area to participate with their local trace history through an open call. Users can opportunistically participate in the query's resolution without disclosing their traces to the authority for monetary benefit or for intellectual satisfaction. The SmartTrace+ project enables trace similarity search among smartphone users and optimizes queries with respect to response time and energy consumption. More importantly, SmartTrace+ is privacy-aware: it does not share user trajectories with the authority, but rather returns only matching scores.

8.2.2.1.3.2 VizWiz VizWiz aimed at enabling blind people to recruit remote-sighted workers to help them with visual problems in nearly real time [19]. Blind people use VizWiz on their existing camera phones. Users take a picture with their phone, speak a question, and then receive multiple spoken answers. Currently, answers are provided by workers on AMT. Prior work has demonstrated that such services need to work quickly, and so there have developed an approach (and accompanying implementation) called QuikTurkit that provides a layer of abstraction on top of Mechanical Turk to intelligently recruit multiple workers before they are needed. In a field deployment, users had to wait just over 2 minutes to get their first answer on average, but wait time decreased sharply when questions and photos were easy for workers to understand. Answers were returned at an average cost per question of only $0.07 for 3.3 answers. Given that many tools in this domain cost upward of $1000 (the equivalent of nearly 15,000 VizWiz uses), it is believed that nearly real-time human services can not only be more effective but also competitive with, or cheaper than, existing solutions [19]. Specifically, when set to maintain a steady pool of workers (at a cost of less than $5 per hour), VizWiz receives answers in less than 30 seconds.

As VizWiz uses real people to answer questions, the scope of questions it can help answer is quite large, and, as opposed to automatic approaches, users can phrase questions naturally assuming an intelligent system (or person) will answer their question. For instance, optical character recognition programs [20] require users to carefully center the text they want to read in the camera's view, and then return all of the text. In contrast, VizWiz users can simply ask what they really want to know, for instance, "How much is the cheeseburger?" As blind people cannot see the picture they are taking, pictures are often not framed well or have other problems. Real people can give guidance as to how to take a better picture. With real people answering questions, they can first target tools to what blind people want and then try to automate, rather than creating tools according to what can currently be done automatically.

8.2.2.1.3.3 Adverse Tracking   Adverse drug reactions (ADRs) have become a worldwide problem that draws the attention of people from all racial and ethnic groups [21]. The number of deaths caused by ADRs has greatly increased and led to many drug withdrawals in the last decades. Recent research findings indicate that most ADRs can be effectively prevented to some extent by using computer-aided information technologies. Though many spontaneous reporting systems (SRSs) have been built to enhance the pharmacovigilance, the ADR data are still very sparse because the large amount of reports obtained from consumers contains insufficient hints to identify a possible causal relationship between an adverse event and drug. Based on this motivation, Reference [21] developed adverse tracking, a SRS of ADRs through crowdsourcing. Their proposed system interacts with consumers through a Q&A interface and collects the ADR reports. The decision tree support vector machine based on the genetic algorithm is used in their system to automate the Q&A procedure.

Adverse tracking, a novel SRS, exploits a long-neglected opportunity—crowdsourcing-based interactive spontaneous reporting of ADRs from consumers for enriching the data source and reducing the fluctuations. The proposed system leverages the decision model and the system knowledge base to generate a series of multiple-choice questions to encourage the consumers reporting an adverse reaction event in a crowdsourcing manner. In an interactive Q&A procedure, ADRs corresponding to same symptoms are transformed into homogeneous data, since each multichoice question is considered as a dimension and its options take different values. Then a decision tree support machine based on the genetic algorithm can be employed to automatically complete a multiclass classification process such that the search space of appropriate disease–symptom pairs can be reduced. Their experimental results show that their system can strengthen the signal for improving the efficiency of SRS based on consumers' reports.

8.2.2.1.4 Natural Language Processing NLP is a field of computer science, artificial intelligence, and linguistics concerned with the interactions between computers and human (natural) languages. As such, NLP is related to the area of human–computer interaction. Many challenges in NLP involve natural language understanding, that is, enabling computers to derive meaning from human or natural language input, and others involve natural language generation.

There are many tasks in NLP, such as part-of-speech (POS) tagging. Given a sentence, determine the part of speech for each word. Many words, especially common ones, can serve as multiple parts of speech.

For example, *book* can be a noun (*the book on the table*) or verb (*to book a flight*); *set* can be a noun, verb, or adjective; and *out* can be any of at least five different parts of speech. Some languages have more such ambiguity than others. Languages with little inflectional morphology such as English are particularly prone to such ambiguity. Chinese is prone to such ambiguity because it is a tonal language during verbalization. Such inflection is not readily conveyed through the entities employed within the orthography to convey meaning. Information retrieval (IR) is concerned with storing, searching, and retrieving information. It is a separate field within computer science (closer to databases), but IR relies on some NLP methods (e.g., stemming). Some current research and applications seek to bridge the gap between IR and NLP [22].

8.2.2.1.4.1 CrowdDB Some queries cannot be answered by machines only [23]. Processing such queries requires human input for providing information that is missing from the database for performing computationally difficult functions and for matching, ranking, or aggregating results based on fuzzy criteria. CrowdDB uses human input via crowdsourcing to process queries that neither database systems nor search engines can adequately answer. It uses structured query language (SQL) both as a language for posing complex queries and as a way to model data. Although CrowdDB leverages many aspects of traditional database systems, there are also important differences. Conceptually, a major change is that the traditional closed-world assumption for query processing does not hold for human input. From an implementation perspective, human-oriented query operators are needed to solicit, integrate, and cleanse crowdsourced data. Recently, there have been problems that are impossible or too expensive to answer correctly using computers.

An application launches requests using CrowdSQL, a moderate extension of standard SQL. Application programmers can build their applications in the traditional way; the complexities of dealing with the crowd are encapsulated by CrowdDB. CrowdDB answers queries using data stored in local tables when possible and invokes the crowd otherwise. Results obtained from the crowd can be stored in the database for future use. CrowdDB incorporates the traditional query compilation, optimization, and execution components. These components are extended to cope with human-generated input as described in subsequent sections.

8.2.2.1.4.2 Part-of-Speech Tagging for Twitter   One of the most fundamental parts of the linguistic pipeline is POS tagging, a basic form of syntactic analysis that has countless applications in NLP [24]. Most POS taggers are trained from tree banks in the newswire domain, such as the WallStreet Journal corpus of the Penn Treebank. Tagging performance degrades on out-of-domain data, and Twitter poses additional challenges due to the conversational nature of the text, the lack of conventional orthography, and 140-character limit of each message (*tweet*).

Reference 24 addressed the problem of POS tagging for English data from the popular microblogging service Twitter. They develop a tag set, annotate data, develop features, and report tagging results nearing 90% accuracy. The data and tools have been made available to the research community with the goal of enabling richer text analysis of Twitter and related social media datasets.

8.2.2.2 WoM Mode

8.2.2.2.1 Red Balloon Challenge   The 2009 DARPA Red Balloon intentionally explored how the Internet and social networking can be used to solve a distributed, time-critical, and geolocation problem. A total of 10 red weather balloons were deployed at undisclosed locations across the continental United States, and $40,000 prize would be rewarded to the first team/individual to correctly identify the locations of all 10 balloons. A team from the Massachusetts Institute of Technology (MIT) won in less than 9 hours by exploiting the effect of WoM [25], which utilized the geometric reward mechanism (recursive incentive mechanism). The key idea is that a certain fraction a (e.g., 1/2 in MIT's scheme) of an individual's reward *bubbles up* to its parent (i.e., the specific participant who invites the individual to join the crowdsourcing campaign), a fraction a^2 bubbles up to its grandparents, and so on. This strategy combined the incentive of personal gain

Table 8.1 Taxonomy of Mobile-Crowdsourcing Applications/Systems

DIMENSIONS PARTICIPATION MODES	APPLICATION	TYPE OF SENSOR(S)	DATA WISDOM	INCENTIVE	INVOLVEMENT
Direct	TrafficInfo	Human, GPS	Collection	Service	Both
	Sense2Health	Human, ambient	Individual	Non	Opportunistic
	Earphone	Human	Collection	Ethical	Opportunistic
	CrowdSearch	Human	Collection	Money	Participatory
	Shop Social	Human, Camera	Collection	Ethical	Participatory
	SmartTrace+	GPS	Individual	Service	Opportunistic
	VizWiz	Camera	Individual	Money	Participatory
	Adverse tracking	Human	Individual	Service	Participatory
	CrowdDB	Human	Individual	Money	Participatory
	POS tagging	Human	Collection	Money	Participatory
WoM	Red Balloons	Human	Collection	Money	Participatory

with the power of social networks to efficiently connect people locating each balloon with the MIT team.

Table 8.1 summarizes and compares the typical MCS applications above. The *type of sensors* column shows which sensors are utilized in corresponding applications. The term of *data wisdom* can lie either in the individual or the collective contribution, which represents that the MCS system strives to benefit from each contribution in isolation or from an emerging property (i.e., by consolidating all crowdworkers' contributions to harvest the superadditive value). The *involvement* column denotes the specific crowdsourcing paradigm used: participatory or opportunistic. Incentive means the detail stimulus used to incentivize the crowd to participate and contribute sensing data or intelligent knowledge, including direct payment, altruism, enjoyment, and reputation.

8.3 Generic Framework

This section proposes a generic MCS framework, independent of specific applications which consists of multiple functional modules, and can accommodate multimodal data sources, and summarizes the various actions used to form the typical workflow of MCS applications [9].

8.3.1 A Generic Mobile-Crowdsourcing Framework

Figure 8.3 shows the comprehensive MCS framework and its modules. The basic functionality of each module is explained as follows:

- *Module 1*: Task management module characterizes the sensing specifications and use cases, including the types of participants, the required sampling rate for each type of sensors, and the requirements of data visualization and representation.
- *Module 2*: MCS front-end module provides crowdworkers (participants) with a cross-platform user interface (UI) for reporting crowdsourced data. The participants may be recruited in direct or WoM modes.
- *Module 3*: Crowdsourcer module should possess the following responsibilities: Publish the appropriate tasks to platform, by interacting with task management module; provide quality feedback about the sensed contents offered by crowdworkers;

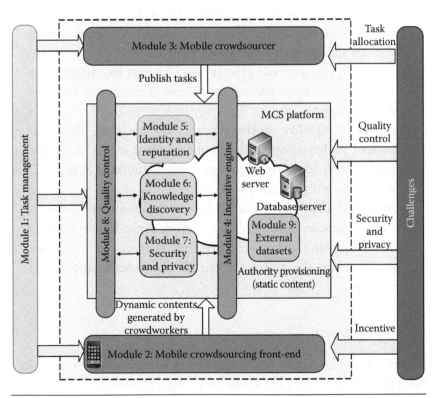

Figure 8.3   A generic mobile-crowdsourcing framework and its modules.

and pay the crowdsourcing platform and/or crowdworkers for using the service.

- *Module 4*: Incentive engine, generally, has two purposes: stimulate crowdworkers to actively participate and contribute high-quality data; and encourage crowdsourcers to provide truthful feedback about the quality of the crowdsourced data. Based on their behaviors, it may reward/punish crowdworkers and crowdsourcers with monetary, ethical, entertainment, and priority.

- *Module 5*: Identity and reputation management module will manage the identities of participants as well as crowdsourcers and will build reputation for them based on their past behaviors so as to enhance the quality of sensed data provided by crowd and trustworthiness of feedbacks by crowdsourcers.

- *Module 6*: Knowledge discovery module. In a MCS system, the submitted data may be unstructured, noisy, and falsified. In this regard, this module provides intelligent data-processing capability to extract and reconstruct useful information from the raw sensing data submitted by participants.

- *Module 7*: Security and privacy module aims to protect users' privacy, to increase privacy awareness of users, and to make app validation.

- *Module 8*: Quality control module. By analyzing the feedback of quality information about the crowdsourced data, this module will adjust the parameters of the modules 5, 6, and 7 to achieve a high-quality service.

- *Module 9*: External datasets. As some government authorities or agencies have provided open data malls, this module allows sensing data from external datasets to be incorporated into the system to enrich the functionalities and services.

8.3.2 *Typical Workflow of MCS Applications*

As described above, the MCS systems usually involve three stakeholders: crowdsourcers, crowdworkers, and crowdsourcing platform.

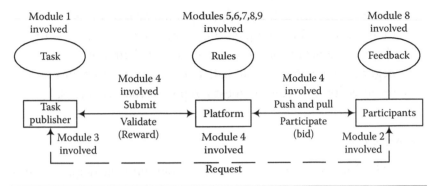

Figure 8.4   The typic workflow of the MCS applications.

Figure 8.4 illustrates the typical workflow of MCS applications, with the help of the aforementioned framework and modules.

- Between the crowdsourcer and the platform, three actions are usually distilled: submit, validate, and reward, in which submitting a task and its related requests exists at the early stage of crowdsourcing life cycle, while validating (evaluate the data feedback and select the satisfying ones) and rewarding (especially for the scenario of crowdsourcing contests) are the last two steps in the whole process.
- Between the crowd and the platform, three actions can be identified, that is, push and pull, participate, and bid. Specifically, push and pull signify the functionalities (e.g., personalized recommendation and customization) provided by the platform to attract, incent, and sustain the crowd. Participation happens when people have the intention to join and take some actions to respond to the tasks. Bidding is a submission state by the participants who have produced outcomes and joined in the competition (not necessarily for all types of crowdsourcing systems). Some studies indicate that sometimes there might be mass participation in a crowdsourcing project, however, only a small number of them work out a solution and submit for competition. Thus, it is feasible to differentiate these two actions.
- Besides the intermediation through MCS platform, the crowdsourcers and crowdworkers may have some direct

connections, through e-mail, telephone, or face-to-face communications. For example, the crowdworkers may inquire about some details of the task to support their works or may negotiate with the task publishers over the requirements and rewards. They can also request a reply for their concerns.

8.4 Key Challenges

Most MCS applications rely on users' voluntary participation and contribution. While participating in a crowdsourcing task, smartphone users not only consume their own resources, but also expose themselves to potential security and privacy threats, for instance, by sharing their sensed data with location tags. Furthermore, the MCS crowdsourcers hope to obtain high-quality contents from mobile crowdworkers who may be selfish, erroneous, or even malicious. Briefly, in MCS systems, four critical challenges arise: task management, incentive, security and privacy, as well as quality control.

8.4.1 Task Design and Allocation

Task design is the model under which the crowdsourcer describes her task, which consists of several components. Especially, the crowdsourcer might put a few criteria in place to ensure that only eligible people can do the task, or specify the evaluation and compensation policies. Three important factors that are related to crowdsourced task design are identified there: task definition, UI, and granularity [26].

- *Task definition*: The task definition is the information that the crowdsourcer gives potential crowdworkers, regarding the crowdsourced task. A main element is a short description of the task explaining its nature, time limitations, and so on. A second element is the qualification requirements for performing the task. These specify the eligibility criteria by which the crowdsourcer will evaluate crowdworkers' work.
- *User interface*: UI refers to the interface through which the crowdworkers access and contribute to the task. This can be a Web UI, an application programming interface (API), or any

other kind of UI. A user-friendly interface can attract more workers and increase the chance of a high-quality outcome. On one hand, a simple interface, such as one with nonverifiable questions, makes it easier for deceptive workers to exploit the system. On the other hand, an unnecessarily complicated interface will discourage honest workers and could lead to delays.

- *Granularity*: Basically, crowdsourced tasks can fall into two broad types: simple and complex. Simple tasks are the self-contained, appropriately short tasks that usually need little expertise to be solved, such as tagging or describing. Complex tasks usually need to be broken down into simpler subtasks. Solving a complex task might require more time, costs, and expertise, so fewer people will be interested or qualified to perform it. Crowdworkers solve the subtasks, and then their contributions are consolidated to build the final answer. A complex task's workflow defines how these simple subtasks are chained together to build the overall task, which could be iterative, parallel, or a combination.

Task allocation in MCS aims to assign a set of outsourced tasks to a specific set of mobile crowdworkers who can potentially finish these tasks accurately and efficiently. Some factors that may impact task allocation have been investigated in existing works, including temporal and geographic constraints, and social relationships. For example, the REAl-time schEduling for Crowd-based Tasks (REACT) crowdsourcing system was presented in Reference 27, which sought to address the issue of allocating tasks to crowdworkers to achieve successful completion of the tasks under real-time constraints, by combining crowdsourcing with scheduling techniques. Reference 28 explicitly took the mobility paths of mobile users into consideration for MCS applications. In addition, since crowdsourcing generally depends on human intelligence, the social attributes of mobile users (e.g., specialties and social activities) may highly impact the quality of crowdsourced tasks. Reference 29 presented a social-aware task allocation (SATA) scheme for direct mode-based MCS applications, which identified the well-suited mobile users for a specific crowdsourced task, explicitly based on matching degree between crowdworkers and task, including three factors: social attribute overlap degree, estimated task delay, and users' reputation.

Finally, note that, mobile users will only volunteer to contribute to MCS, if this process does not use up significant battery so as to prevent them from accessing their usual services, such as making calls and Internet access. Even though, most users charge their phones on a daily basis, it is still important that crowdsourced tasks do not incur significant energy costs for users. Basically, energy is consumed in all aspects of MCS applications ranging from sensing, processing and data transmission, and so on. In particular, some sensors such as GPS consume significantly more energy than others. As such, it is important for MCS applications to make use of these sensors in a conservative manner. For instance, Reference 30 proposed the adaptive scheme for obtaining phone location by switching between the accurate but energy-expensive GPS probing to energy-efficient but less accurate Wi-Fi/cellular localization. Similar approach can be employed for duty cycling other energy-hungry sensors [31].

8.4.2 Incentives

An incentive is a kind of stimulus or encouragement to stimulate one to take action, work harder, and so on. This issue is even more critical when the devices (e.g., mobile phones, wearable sensors) have very limited resources (e.g., energy and storage capacity) or the information revealed is highly sensitive. Deploying a mobile crowd sensing system on a wide scale requires a large number of participants. Participants may drop out of the collecting loop unless return on investment is greater than their expectations [32].

In general, in MCS, incentives could be classified into three categories. Probably, the most prominent incentive in today's crowdsourcing market is the financial incentive; another type of incentives is entertainment based. The crowdsourcer may provide a form of enjoyment or fun in the crowdsourced activity or may design a game around the crowdsourced activity; the third form of incentives is the social incentive. Some participants may take part in crowdsourcing activities to gain reputation or public recognition [33].

Economic incentives are the real money or any other commodity that the crowdworkers consider valuable. They are probably the most straightforward way to motivate participants. However, once money being involved, quality control becomes a major issue due to the anonymous and distributed nature of crowdworkers. Although the quantity of work performed by participants can be increased, the quality cannot, since crowdworkers may tend to cheat the system just to increase their overall earned payment.

The idea of taking entertaining as well as engaging elements from computer games and using them to incentivize participation in other contexts is increasingly studied in a variety of fields. However, in many situations, the tasks can be too boring or complicated to turn into any game that is actually enjoyable or fun to play. Other intrinsic incentive factors can include mental satisfaction gained from performing a crowdsourced activity, self-esteem, personal skill development, knowledge sharing through crowdsourcing (e.g., Wikipedia), and love of the community in which a crowdsourced task is being performed, and so on. However, like gamification, there is no consensus on how to quantitatively measure the effect of those elements on MCS.

Social psychological factor is another widely harnessed nonmonetary incentive mechanism to promote increased contributions to online systems. For example, social facilitation effect refers to the tendency of people to perform better on simple tasks while being watched by someone, rather than being alone, or when they are working alongside other people. Although social psychological incentives like historical reminders of past behavior or ranking of contributions can significantly increase repeated contributions, the required identity management and reputation measurement are extremely challenging, especially in dynamic MCS environment.

In brief, those three categories of incentives have corresponding advantages and disadvantages, thus, naturally, MCS systems' built-in incentives should exploit those methods in an integrated way to devise feasible and practical schemes to incentivize individuals to participate in and provide high-quality solutions to MCS systems.

8.4.3 Security and Privacy

Regardless of the type of crowdsourcing system and the method of interaction, there may be some level of privacy and security risk associated with the users. Some platforms facilitate anonymous contributions and may pose low risk, whereas others involve the collection and broadcast of various levels of personally identifiable information and may pose higher risk to contributors. Similarly, opportunistic systems may pose a higher security risk than participatory systems where users manually control data collection. This subsection will set out to identify potential privacy and security risks in MCS systems in general, and several popular MCS systems, in particular. It also reviews counter measures existing in the literature and suggests future research directions. Our hope is that this work can inspire and help designers and administrators of crowdsourcing systems to identify and mitigate privacy and security risks to their users. In addition, this work can be useful in helping potential users to identify and gauge the privacy and security risks they may face with various types of systems and to make more informed decisions about their usage.

8.4.3.1 General Privacy and Security Threats in Mobile Crowdsourcing Smartphones make an ideal platform for crowdsourcing systems not only because of their portability and easy access to the Internet, but also because they are equipped with environment sensors such as GPS, cameras, and accelerometers. Although these features may be considered highly desirable in this connected society, they may also expose users to new types of privacy and security threats. Yang Wang et al. [34] surveyed the common privacy and security threats discussed in research literature.

1. *Disclosing user identity*: Many MCS systems collect basic profile data from participants such as username, password, name, e-mail address, and phone number. In many cases, user profile information is visible to other registered users, making it hard to protect one's identity. Once a person's identity is revealed, he or she becomes vulnerable to various types of online attacks such as spams and phishing attacks.

2. *Disclosing user location and activity*: Smartphones running certain sensing applications have the ability to automatically update user location and movement from place to place, and to capture images, audio, and text from the surrounding environment. Although these capabilities may be desirable in some situations, such as the potential security risks of having the phones act as *miniature spies*, the ability to record

conversations and other potentially sensitive information such as user location can be viewed as an intrusion of user privacy and may act as a deterrent to potential contributors.

An example mobile sensing application CrowdSense@Place (CSP) was proposed and tested by Chon et al., further highlighting the issue of user privacy [35]. It functions by using data from images, texts, and spoken words which are opportunistically captured from smartphones as their owners travel to different places. This piece of data is used to help classify the type of place, such as hotels, hospitals, and parks. A privacy issue acknowledged by the developers is related to the exposure of too much detail on the activities of an individual and essentially eroding their ability to keep some of these details confidential. Although the users are generally aware that their activities are being monitored, there may be instances where they may visit a sensitive place which they may not want to share.

3. *Combining crowdsourced data with other user data*: It is difficult for ordinary people to know or reason the downstream use of their data. One of the potential risks is that businesses combine users' self-provided data with other user data retrieved from elsewhere. This combination of user data could yield new and sensitive insights about the users beyond their comfort level (e.g., this user is pregnant and lives in a certain area).

4. *Lack of user privacy awareness*: In many cases, users allow applications to access his or her data but are not aware of (1) how their data will be actually used and (2) the privacy risks associated with sharing their data. Many mobile applications do not make the process surrounding the collection, processing, and transmission of location-based data transparent to their users. The users are often left with no choice but to trust that their private data are being properly handled by the application and are not being redirected to unauthorized destinations such as advertisers or even the application developers. However, if users were more aware of those risks, they would be able to make more informed choices about which systems to contribute to and what type of information to share.

5. *Vulnerability of mobile devices*: A concern related to protecting user privacy is the greater vulnerability of small mobile devices to security breaches and attacks. According to Reference 36, there is a fundamental difference in the security of these small devices as compared to Internet security. These smaller devices have limitations related to computational speed and storage capacity, eliminating the viability of many traditional cryptographic techniques as protections. There is also the added threat of physical attacks such as resetting the device if it gets in the wrong hands.

6. *Relying on information that may be inaccurate*: In integrative crowdsourcing, individual submissions from contributors are aggregated into final outcome. Depending on the nature of the system, people may rely on information supplied by others to make critical decisions. For example, for systems where contributors report hazardous traffic conditions, or report emerging situations related to events such as natural disasters, human rights violations, or political unrest, there is a possibility of incorrect or inaccurate data being reported unintentionally or in some cases maliciously.

7. *Retaliation for reporting sensitive information*: Systems used to report sensitive information, such as incidents of violence or human rights abuses, though extremely valuable to responders are generally publicly viewable and can pose risks to victims and the people who report the incidents.

8.4.3.2 Countermeasures to Address Privacy and Security Threats One of the major concerns with automatic sensing is the possibility of compromising user privacy by recording and publicizing sensitive information (intentionally or unintentionally). We discuss two proposed approaches to ensuring sensitive data are not recorded by the system without express approval from the user in the following. The first of these methods requires manual intervention, whereas the second is automated.

1. *Allowing users to manually control which data are shared*: In this approach, users are given the option of manually controlling which of the captured data are eventually shared. This can

be achieved by providing a delay between the time when the data are sensed from the environment and the time when it is uploaded to the system. In the mobile-sensing application Crowd-Sense, Reference 35 proposed a 24-hour time lapse during which users can manually review and delete any data that are too sensitive to share. Users also have the ability to block the transmission of data in advance when anticipating activities of a sensitive nature. Although this method is useful, the developers acknowledge that on its own it is inadequate and may require the additional processing and filtering of data on the smartphone itself before being uploaded.

2. *Maintaining user anonymity*: Some applications try to protect user privacy by making the uploaded reports anonymous. However, there is still the possibility of using time and location information to deanonymize the data, revealing the owner of the device. Reference 37 proposed the AnonySense architecture as a means of protecting user privacy when reporting context-sensitive information, as it offers protection along multiple layers without manual intervention from the user. Some of the features of this architecture include: (a) a mechanism that allows delivery of tasks and submission of reports to be conducted using anonymous nodes; (b) tessellation that divides geographic areas into regions referred to as tiles. Reporting is carried out at the tile level, and each tile has a number of users large enough to protect the identity of individual users; and (c) k-anonymity—assuming that there are k users in each tile who submit reports to the system at regular time intervals, the reports can be indexed by tile and time interval leaving a high probability that users cannot be identified within a set of k users. A higher level of protection can be added by aggregating a second layer of reports before the data are submitted.

3. *Increasing privacy awareness of users*: Many users who contribute to crowdsourcing systems and other social networking sites may suffer from a lack of or inadequate knowledge of the security and privacy risks associated with those systems. In a study on user information privacy concerns in social networking sites, it is reported that awareness prompted users to learn

how to protect themselves and helped boost their confidence and ability in using privacy measures [38]. The measures presented below both focus on informing the users on how their data are handled by the various systems.

The functionality provided by TaintDroid acts as an extension to the Android platform and monitors how applications handle private user data accessed from their mobile phones. When personal data are downloaded, TaintDroid labels it and tracks its transmission over the network to its destination, or if it leaves the system. Users can therefore be supplied with real-time updates on the destination of their private data, facilitating the identification of misbehaving applications [39].

4. *App validation*: One approach to ensure that mobile applications adhere to certain privacy and security standards is to conduct some sort of inspection and validation before they get into the hands of consumers. Gilbert and Chun proposed the use of an automated security testing and validation system AppInspector, which can be used at the level of the centralized app markets like Apple App Store or Google Play [40]. According to them, testing and validation can potentially be done by the market provider or by a third-party entity. The proposed system has the ability to test security properties of apps by running *virtual* smartphones in the cloud, making use of information-flow and action-tracking components to detect and report malicious behavior as well as security and privacy violations. The researchers noted that the system is still in preliminary stages and there are potential challenges associated with the accurate characterization of the apps behavior from the logs generated.

8.4.4 Quality Control

MCS is increasingly being used as an effective tool to tackle problems requiring human sensing and intelligence. With the ever-growing crowdworker base that aims to complete microtasks on crowdsourcing platforms in exchange for financial gains, there is a need for stringent mechanisms to prevent exploitation of deployed tasks. Quality control

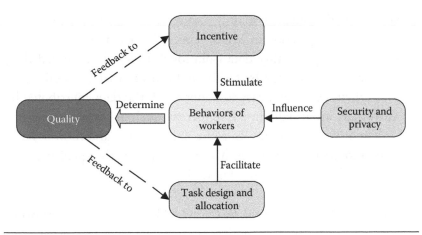

Figure 8.5 Relationship of quality with worker's behavior, incentive, task design, as well as security and privacy.

mechanisms need to accommodate a diverse pool of workers who may exhibit wide range of behaviors. Specifically, one major challenge with crowdsourcing is dealing with biases and errors in contributors' submitted work, whether malevolent or accidental.

As Figure 8.5 shows, basically, the quality of a MCS task's outcome is decided by crowdworkers' behaviors, which are affected by the following factors: incentive mechanism, task design and allocation, and security and privacy. Designing task with easy-to-user UI and proper granularity and allocating task to suitable workers are the premises for high-quality crowdsourced contents. The incentive mechanism stimulates workers to actively participate and put more effort on crowd-sourced tasks, which can improve quality greatly. According to the service quality, the platform can (re)regulate the rule of the incentive and task design. Security and privacy issues affect user's participation enthusiasm, thus influencing the quality of service.

Reference 41 has proposed multiple techniques to do quality control in MCS: Output agreement, input agreement, economic incentives, defensive task design, reputation systems, redundancy (combined with majority consensus), ground-truth seeding, statistical filtering, multilevel review, expert review, and contributor evaluation are also exploited in Reference 42. Furthermore, the MCS platform could explicitly explore the social interactions between contributors to improve the quality of contributions.

Note that, this chapter mainly focuses on framework and topics related to data acquisition in MCS, especially incentivizing individuals to provide high-quality data (solutions). It should be noted that, unlike the data generated by traditional sensors that are well structured, explicit, clean, and easy to understand, the data contributed by crowdworkers are usually in a free format, such as texts and images, and inherently noisy. Naturally, some data analysis and mining technologies should be utilized by MCS to improve the quality of the synthesized data, which is another topic.

8.4.4.1 The Malicious Behavior in the Crowd Malicious workers are workers deemed to have ulterior motives that deviate from the instructions and expectations as defined a priori by the microtask administrator. Reference 43 shows that varying types of malicious activity are prevalent in crowdsourced surveys and proposes measures to curtail such behavior. In Reference 43, the implicit behavioral patterns of malicious workers by the means of their responses are analyzed. Based on aspects such as (1) the eligibility of a worker to participate in a task; (2) whether responses from a worker conform to the preset rules; or (3) whether responses fully satisfy the requirements expected by the administrator, they determine the following types of behavioral patterns.

- *Ineligible workers*: Every microtask that is deployed on crowdsourcing platforms presents the workers in the crowd with a task description and a set of instructions that the workers must follow, for successful task completion. Those workers, who do not conform to the prior stated prerequisites, belong to this category. Such workers may or may not provide valid responses, and their responses cannot be used by the task administrator since they do not satisfy the prerequisites. For example, consider a prerequisite in a survey, "Please attempt this micro-task ONLY IF you have successfully completed 5 micro-tasks previously." Some workers responded to questions regarding their previous tasks with, "this is my first task," clearly violating the prerequisite.
- *Fast deceivers*: Malicious workers tend to exhibit a behavior that is strongly indicative of the intention to earn easy and quick money, by exploiting microtasks. In their attempt to

maximize their benefits in minimum time, such workers supply ill-fitting responses that may take advantage of a lack of response validators. These workers belong to the class of fast deceivers. For example, workers who copy–paste the same response for different questions.

- *Rule breakers*: Another kind of behavior prevalent among malicious workers is their lack of conformation to clear instructions with respect to each response. Data collected as a result of such behavior have little value. For instance, consider the question of "Please identify at least 5 keywords that represent this task." In response, some workers provided fewer keywords. In such cases, the resulting response may not be useful to the extent intended by the task administrator.

- *Smart deceivers*: Some eligible workers who are malicious try to deceive the task administrators by carefully conforming to the given rules. Such workers mask their real objective by simply not violating or triggering implicit validators. For example, consider the instruction, "Transcribe the words in the corresponding image and separate the words with commas." Here, workers who intentionally enter unrelated words, but conform to the instructions by separating the words with commas, may neutralize possible validators and achieve successful task completion. Although this type of workers behave to an extent like fast deceivers, the striking difference lies in the additional attempt of smart deceivers to hide their real goal and bypass any automatic validating mechanisms in place. Some workers provided irrelevant keywords such as "yes, no, please," "one, two, three," and so forth to represent their preferred task types. Some of these workers take special care to avoid triggering attention check.

8.4.4.2 Ensuring Trustworthiness of Data   In order to have a trustworthy system with accurate results, it is important to have a means of verifying and validating the data entered [34].

1. *Reputation system for devices*: A reputation system that assigns reputation scores to the various devices contributing to the system is the solution proposed by Huang et al. [44]. In this

system, reputation scores are assigned to devices by the server based on the perceived level of trust associated with the data uploaded over a period of time. The reputation system calculates reputation scores using the Gompertz function for computing devices. The system was tested in a noise-monitoring application using Apple iPhones, and based on experimental results the authors report three-fold improvement in comparison with the state-of-the-art Beta reputation scheme.

2. *Ratings for individual users (contributors)*: Reference 45 proposed a data mining algorithm to identify the highly *trustworthy results* from thousands of submissions. They experimented with a prototype of a photo-tagging site and reported that their algorithm yielded *substantial improvements* in accuracy. First, the algorithm assigned equal rating to all users, then used crowd consensus to compare results of the tagged photos. The next step was to calculate individual user accuracy, following which individuals were assigned new ratings based on their accuracy. Those with similar ratings were then placed into the same group. These steps were repeated until the number of users with rating changes was at a minimum.

3. *Report verification*: One strategy for dealing with accuracy of submissions is to get them verified by other members of the crowd. For example, the verification button in Ushahidi allows for collective feedback from the crowd. However, past use of the system shows that only a small number of people served as verifiers, limiting the effectiveness of this method [46].

8.5 Quality-Aware Incentive

There exist much works on participation incentive in MCS. Two system models are considered for economic-based incentive mechanisms in MCS: the platform/crowdsourcer-centric model (focusing on maximizing platform's profit) and the user-centric model (focusing on eliciting crowd's truthful value on tasks). However, those works have two weakpoints: First, they only investigate the utilities of partial roles in MCS, that is, crowdworkers' utilities and platform's profit. The behavior of the third components, crowdsourcer, is not examined at all; second, they only deal with the users' participation incentive, ignoring how to incentivize crowd to provide high-quality sensed contents [47].

Reference 48 proposed a novel class of incentive protocols in crowdsourcing applications based on social norms by integrating reputation mechanisms into the existing pricing schemes currently implemented on crowdsourcing websites (e.g., Yahoo! Answers, AMT). Especially, the problem of *freeriding* of crowd in ex-ante payment incentive was addressed, which means that if the crowdsourcer pays before the task starts, a worker always has the incentive to take the payment and provides no effort to solve the task. An incentive platform named TruCentive was presented by Hoh et al. [49], which addressed the problems of low user participation rate and data quality in mobile crowdsourced parking systems (i.e., smartphone users are exploited to collect real-time information about parking availability).

However, those works mainly related to the crowd side (incentivizing users to participate and provide high-quality contents). We argue that quality-aware incentive mechanisms should pertain to all MCS components: crowd, crowdsourcer, and crowdsourcing platform. Specifically, it is the crowdsourcers that publish tasks needed to be outsourced to the crowd, and the quality of contents provided by crowd should be naturally judged and reported by crowdsourcer. Furthermore, the report on content quality will significantly affect the reward that the crowd can obtain. Thus, there exist a social dilemma between crowd and crowdsourcers: the crowd should be stimulated to provide sensed contents with higher quality to get more reward that depends on the quality reports feedback by crowdsourcers; the crowdsourcers want to utilize high-quality contents provided by crowd, but pay crowd as little as possible, probably by false reporting the quality of sensed contents.

To address the above social dilemma, by appropriately integrating three popular incentive methods: reverse auction, reputation, and gamification, this chapter proposes a quality-aware incentive framework for MCS, QuaCentive, which can motivate crowd to provide high-quality sensed contents, to stimulate crowdsourcers to give truthful feedback about quality of sensed contents, and to make platform profitable.

8.5.1 Detailed Operation of QuaCentive Scheme

The whole process of QuaCentive scheme is described in Figure 8.6, which explicitly deals with two scenarios: good rating and bad rating on service (sensed data) reported by crowdsourcer. It mainly consists of nine steps:

1. A crowdsourcer *buys* a sensing service from the crowdsourcing platform, by depositing N points in platform. It could give a short description of the task such as explaining its nature, time limitations, and so on. Note that the payment from the crowdsourcer will be refunded, if and only if the crowdsourcer reports that this service is failed (bad rating on sensed data), and the report (bad rating) is successfully verified by platform through gamification. The deposit itself is only for fairness and voucher purpose.

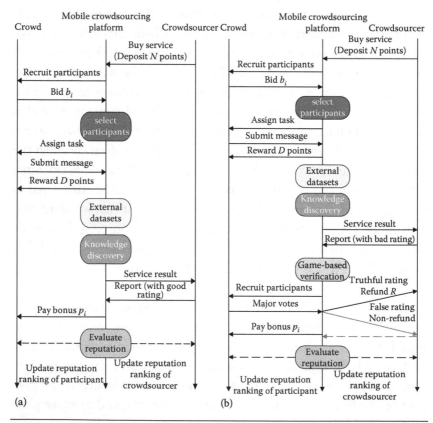

Figure 8.6 Detailed operation steps of QuaCentive. (a) Good rating by crowdsourcer and (b) bad rating by crowdsourcer.

2. Crowdsourcing platform distributes task details to the crowd to recruit participants and to receive the bids of the participants.

3. Platform selects participants based on their reputation rankings and bids, and then assigns task to the selected participants (contributors).

4. The contributors submit messages (sensed contents) to platform and get a static D reward.

5. Platform combines external datasets and/or knowledge discovery modules with the sensed data provided by participants and then provides the result to the crowdsourcer.

6. The crowdsourcer receives this result and sends back a report of whether the sensed contents led to a successful service.

7. If the crowdsourcer reports that the service is successful, then the contributor i gets p_i bonus.

8. Otherwise, the verification of crowdsourcer's bad rating (report of unsuccessful service) will be crowdsourced to crowd, in gamification way. If the result of major votes is positive (i.e., the crowdsourcer's bad rating is the ground-truth), then the crowdsourcer will receive R refund; otherwise, refund is not given.

9. Platform updates the reputation rankings for both contributors and crowdsourcers.

The key steps of QuaCentive scheme are described as follows in more detail.

8.5.1.1 Selecting Participants Based on Their Reputation Ranking and Bids (Step 3) A concept of reputation ranking was used to represent the trustworthiness (quality) of a participant. The less the value of a user's reputation ranking is, the higher his or her trustworthiness is. An analogical example is: in a school, if a student's ranking is No. 1, then he or she is considered as the best student in this school. This way of the representation of a participant's quality will help us choosing the participants with minimum valuation. Of course, the reputation rankings of participants and crowdsourcers should be computed according to their past behaviors and current performance when the service is completed.

Formally, mobile users who want to take part in a task present their bids (private information to them) in a sealed bid manner, and the wining participants are chosen through reverse auction, based on the measurement of bid multiplied by reputation ranking. We will formulate the process and properties of the selection method in the following subsection.

8.5.1.2 Rewarding Contributors (Step 7) To incentivize users to share their sensed contents, QuaCentive pays a contributor i two types of rewards: D points of static reward and a bonus of p_i points that is dynamically determined. The static reward is granted immediately after the sensed data are accepted by QuaCentive. This incentive is provided irrespective of whether the message is reported by crowdsourcers as good rating or bad rating. Such a participation incentive can be helpful to ensure a steady stream of available participants to make the crowdsourcing platform attractive to crowdsourcer. This constant reward D is expected to be a small amount. The bonus reward p_i is granted right after the service is reported as successful. The amount of bonus can be significantly larger than the constant reward D, since the sensed message may lead to a successful service needed by crowdsourcers. Furthermore, QuaCentive collects the statistics of successful service rate, infers each participant's reputation ranking, and accordingly increases the bonus to encourage contributors to offer high-quality data.

8.5.1.3 Verifying Crowdsourcers' Bad Ratings on the Quality of the Sensed Contents (Step 8) As described above, if the crowdsourcer provides positive confirmation on a sensing service, the contributor will receive a bonus of p_i. However, if the crowdsourcer reported that the quality of contents offered by crowd is bad, the crowdsourcer would have a chance to have R refund. Thus, without verification, crowdsourcer has the strong incentive to get refund by lying about the outcome of this task, the so-called *false reporting*. To make clear whether the failed service comes from the really useless contribution data or crowdsourcer's malicious behavior for paying less money to platform, QuaCentive framework distributes the primitive task, its associated sensed contents, and crowdsourcer's feedback, by crowdsourcing platform, and recruits the crowd to vote whether the task is successful or failed in gamification way. If the result of major votes is positive (i.e., truthful rating), the crowdsourcer will have R refund; otherwise, refund is not given.

8.5.1.4 Updating Reputation Rankings of Participants and Crowdsourcers In QuaCentive system, after completing each service and verification process, the platform rates each participant and crowdsourcer with a neutral, negative, and positive feedback. As a simple example, it could assign 0 point for each neutral feedback, −1 point for each negative feedback, and 1 point for each positive feedback, and the reputation of the participants could be computed as the sum of its points over a certain period. The users with high reputation are easy to be selected as winning participants to assume the task and to obtain more payment than participants with low reputation. For the crowdsourcers, the lower reputation will lead to more money consumption when buying the crowdsourced service from the platform.

8.5.2 Theoretical Analysis of QuaCentive

The properties of QuaCentive are thoroughly analyzed from the three players' viewpoints in MCS: crowdworkers, crowdsourcing platform, and crowdsourcer. Table 8.2 shows the notations and their meanings used in our analysis.

Table 8.2 Notations and Their Meanings in QuaCentive

NOTATION	MEANING
v_i	Valuation of a participant i in a reverse auction
b_i	Bid value of a participant i
c_i	Reputation ranking of a participant i
p_i	Bonus for a participant i for successful service
p	Probability that crowdsourcers report that a sensing service is successful (good rating)
q	Probability that the verification game returns the ground-truth
n	Number of participants in a sensing task
D	Static reward to any participant in MCS
X	Average bonus to a participant from crowdsourcing platform as a profitable enterprise
N	The number of points that a crowdsourcer deposits for buying a service
R	The refund that a crowdsourcer obtains if service is unsuccessful
C	The revenue that crowdsourcing platform receives from other ways (e.g., Ads)
m	Number of participants in the verification game

8.5.2.1 From Crowd Side: Individual Rationality, Incentive of Providing High-Quality Sensed Contents, and Truthful Bid Intuitively, the users who provide high-quality data and require low payment should be selected as participants to complete sensing tasks. Here we integrate reverse auction and reputation mechanisms to select appropriate participants. Specifically, assume there are n users in the auction, and user i takes part in the reverse auction by placing its bid in a sealed bid manner. The valuation of the participant i in the auction v_i, is defined as Equation (8.1):

$$v_i = b_i \cdot c_i \tag{8.1}$$

where b_i and c_i, respectively, represent the bid price and the reputation ranking of the participant i. Note that the less the value of c_i is, the higher the trustworthiness of the participant is. This way of the representation of participant's trustworthiness will help us choosing the participants with minimum valuation.

All users are sorted in an ascendant order according to the metric of valuation defined above. Then, in the sorted list, from top to bottom, bidders of the required number with the minimum valuations are selected as the winners to complete the task. Then, the payment should be given to each winner, which is an important step in QuaCentive to compensate each participant's cost and attract more users.

Assume that users $(i-1)$ and i are the two consecutive participants in the rank list, and user i is above $(i-1)$, the final payment p_i to participant i can be intentionally inferred through Equation 8.2:

$$p_i \cdot c_i = b_{i-1} \cdot c_{i-1} \tag{8.2}$$

That is, the payment p_i is

$$p_i = \frac{b_{i-1} \cdot c_{i-1}}{c_i} \tag{8.3}$$

Theorem 1: *The payment scheme in QuaCentive can incentivize users to actively participate in crowdsourced tasks (i.e., individual rationality), to provide high-quality sensed contents, and to bid truthful price.*

The schematic proof is as follows:

$$p_i \cdot c_i = b_{i-1} \cdot c_{i-1} > b_i \cdot c_i \tag{8.4}$$

where the last inequality term stems from the ranking rule of QuaCentive.

Obviously, $p_i \geq c_i$. It implies the total payment to winning participant i, $\left(D + p_i\right)$ is strictly larger than the bid price b_i, thus QuaCentive satisfies individual rationality, that is, users will actively participate in crowdsourced tasks.

Furthermore, Equation 8.3 illustrates the smaller the value of c_i is, the higher payment the participant i would obtain. Therefore, the winning participant i has incentive to provide high-quality sensed service and to build higher reputation (i.e., to be ranked at front positions), so that he or she could earn larger payment.

Intuitively, if a payment a bidder is offered in an auction is independent of the bidder's bid value, the auction is truthful. Formally, let $\vec{b}_{-i}$ denote the sequence of all users' bids except user i, that is, $\vec{b}_{-i} = \left(b_1, \ldots, b_{i-1}, b_{i+1}, \ldots, b_n\right)$. Then the deterministic bid-independent auction can be defined as follows:

The auction constructs a payment (price) schedule $p(\vec{b}_{-i})$. If $p(\vec{b}_{-i}) \geq b_i$, players i win at price $p_i = p(\vec{b}_{-i})$; otherwise player is rejected and $P_i = 0$.

According to the theorem in Reference 50, a deterministic auction is truthful if and only if it is equivalent to a deterministic bid-independent auction. Obviously, the payment mechanisms in QuaCentive are a deterministic bid-independent auction, thus, it can incentivize users to truthfully bid.

8.5.2.2 From Crowdsourcing Platform's Side: Condition of Being Profitable Under what conditions is QuaCentive a profitable business enterprise? We will analyze this question to narrow down the region of the values for average bonus X to participants.

We consider a single task, in which, there are n contributors submitting messages, the probability that a successful service is reported by a crowdsourcer is p, and the probability that the verification game can infer the ground-truth is q. For each task, the cost for the service platform is as follows:

1. Reward to the contributors is: $n \cdot (D + p \cdot X) + n \cdot (1 - p) \cdot q \cdot X$.
2. Refund to the crowdsourcer is: $(1 - p) \cdot (1 - q) \cdot R$, where $(1 - p) \cdot (1 - q)$ is the probability that the reported unsuccessful service by a crowdsourcer is confirmed by verification game.

3. When unsuccessful serviced reported, the cost to recruit m participants to vote $m \cdot (1 - p) \cdot D$ should be added. Here, we simply present the total payment $m \cdot D$ to all voters, since the game is easy to play, and the players joining the verification game is mainly for fun, small D points for each participant could be sufficient to incentivize users to play the game.

Usually, MCS platform typically has multiple sources of revenue as an example of targeted advertisements, so it may not rely solely on income from the hosted MCS service. Let us assume that platform has a peer-service revenue of C from these alternate sources. Our goal is to ensure the platform is profitable. In other words, we have the following constraint:

$$C + N \geq n \cdot (D + p \cdot X) + n \cdot (1 - p) \cdot q \cdot X \\ + (1 - p) \cdot (1 - q) \cdot R + m \cdot (1 - p) \cdot D \qquad (8.5)$$

That is:

$$X \leq \frac{(C + N) - n \cdot D - (1 - p) \cdot (1 - q) \cdot R - m \cdot (1 - p) \cdot D}{n \cdot p + n \cdot (1 - p) \cdot q} \qquad (8.6)$$

To ensure QuaCentive a profitable business enterprise, the average bonus to each participant should not be greater than the right term in Equation 8.6. So, once a crowdsourcer buys a service with N points, the platform will compute the possible average bonus to all participants immediately. If the participant's bid b_i is greater than the average bonus, she will be excluded from the candidates.

8.5.2.3 From Crowdsourcer's Side: Truthful Feedback about Quality of Sensed Contents   All players in MCS system are rational, and they may make a dishonest comment for self-interests. For crowdsourcers, they may use MCS to complete their tasks, but may give feedback that they have experienced a failed service, so as to receive refund from the platform. Furthermore, crowdsourcer's dishonesty has the negative consequence that the contributors who have provided high-quality sensed contents are not rewarded with bonus, which can discourage contributors' participation.

Table 8.3 The Gain for a Crowdsourcer *j* by Being Honest and Dishonest When Verification Game is Successful and Unsuccessful

VERIFICATION-GAME SIDE	TO BE HONEST	TO LIE
CROWDSOURCER SIDE		
Successful verification (with probability q)	$R + F(c_j - 1)$	$F(c_j - 1)$
Failed verification (with probability $(1 - q)$)	$-R + F(c_j - 1)$	$R + F(c_j - 1)$

We now present a game theoretical analysis how QuaCentive encourages a crowdsourcer, which gives truthful rating on crowd's job. With such a design, rational crowdsourcers would choose to be honest in our system.

The key idea behind our approach lies in the verification game and the effect of reputation on crowdsourcer's behavior choice. In the simplest way, the platform rates each contributor and crowdsourcer with a neutral (0), negative (−1), and positive feedback (+1). The reputation of the participants is computed as the sum of its points over a certain period. To simplify analysis, we assume $F(c_j - 1)$ denote the expected earnings of crowdsourcer *j* with reputation ranking $(c_j - 1)$. Note that in QuaCentive, the less reputation ranking is, the larger the player can obtain, thus, $F(c_j - 1) \geq F(c_j + 1)$.

Table 8.3 shows, under the condition that a crowdsourcer reports unsuccessful service (bad rating on the sensed contents), the gain for a crowdsourcer if he or she tells the truth versus lies, and when verification game can be successful (infer the ground-truth) versus unsuccessful (i.e., the crowdsourcer's lying is not found by the verification game).

Specifically, when a crowdsourcer *j* tells the ground-truth that the quality of sensed data is low, and the verification game can successfully confirm the ground-truth (with probability q), then the utility of the crowdsourcer *j* is the refund $R + F(c_j - 1)$; if unfortunately, the verification game draws the wrong conclusion (i.e., the crowdsourcer reports the ground-truth, however, the verification game finally determines that the crowdsourcer lies), then the utility is $-R + F(c_j + 1)$. The other terms in Table 8.2 can be explained similarly.

To ensure consumers maximize their gain by being honest, we need the following constraint to hold:

$$q \cdot (R + F(c_j - 1)) + (1 - q) \cdot (-R + F(c_j + 1)) \geq q \cdot F(c_j + 1)$$
$$+ (1 - q) \cdot (R + F(c_j - 1)) \tag{8.7}$$

That is:

$$(2 \cdot q - 1) \cdot (F(c_j - 1) - F(c_j + 1)) \geq 2 \cdot R - 3 \cdot q \cdot R \qquad (8.8)$$

Obviously, if $q \geq 2/3$, then inequality as shown in Equation 8.8 always holds. It implies that, if the probability that verification game can successfully infer the ground-truth (when crowdsourcers report unsuccessful service) is larger than 2/3, then, the best choice of crowdsourcer is to truthfully feedback the quality of sensed content.

8.5.2.4 Strategies to Deal to Whitewashing Users Note that, the reputation rankings are tightly associated with participants and crowdsourcers' identities. Since it is hard to distinguish between a legitimate newcomer and a whitewasher, this makes whitewashing a big problem in QuaCentive framework. Whitewashing means that when a player (crowdworker or crowdsourcer) has bad reputation in MCS system, then to avoid disincentives, he or she leaves the system and returns back with a new identity as a new comer to the system. Although the problem of whitewashing can be solved using permanent identities, it may take away the right of anonymity for users. In general, there exist two methodologies to alleviate the problem of whitewashing in QuaCentive framework.

- The first way is: The initial reputation for newcomers (including potential whitewashers) is adaptively adjusted according to the level of whitewashing in the current system. It means that if whitewashing level is low, the initial value will be kept high; whereas if whitewashing level increases, initial value will be decreased adaptively.
- Another way lies in that it is possible to counter the whitewashing by impose a penalty on all newcomers, including both legitimate newcomers and whitewashers, which is the so-called social cost.

8.6 Conclusion

Crowdsourcing is evolving as a distributed problem solving and business production model in recent years. In crowdsourcing paradigm, tasks are distributed to networked people to complete such that a crowdsourcer's production cost can be greatly reduced. On the other hand, the shift of desktop users to mobile platforms in the post-PC era along with the unique multisensing capabilities of modern mobile devices is expected to eventually unfold the full potential of crowdsourcing. As an emerging research and application area, MCS brings numerous issues to be addressed, such as quality control, task design, incentive, security and privacy, and so on. These challenges will bring unprecedented opportunities to academic researchers, industrial designers/developers, as well as policy makers. This chapter provides taxonomy for various MCS applications, in which, especially, two paradigms to mobilize rational individuals in MCS are identified: direct and WoM modes. Then, a generic MCS framework is presented, which consists of multiple functional modules, and the typical workflow of MCS applications is illustrated. Finally, this chapter introduces a quality-aware incentive framework for MCS, QuaCentive, which appropriately integrates three popular incentive methods: reverse auction, reputation, and gamification, which can motivate the crowd to provide high-quality sensed contents, can stimulate crowdsourcers to give truthful feedback about quality of sensed contents, and can make platform profitable. In brief, this chapter could facilitate the research, development, and deployment of MCS applications.

References

1. Michelucci, P. and J.L. Dickinson. The power of crowds. *Science*, 2016; 351(6268): 32–33.
2. Howe, J. The rise of crowdsourcing. *Wired Magazine*, 2006; 14(6): 1–4.
3. Hosseini, M., K. Phalp, J. Taylor, and R. Ali. The four pillars of crowdsourcing: A reference model. In: *IEEE Eighth International Conference on Research Challenges in Information Science (RCIS)*, Marrakesh, Morocco, May 28–30, 2014, pp. 1–12.
4. Christin, D., A. Reinhardt, S.S. Kanhere, and M. Hollick. A survey on privacy in mobile participatory sensing applications. *Journal of Systems and Software*, 2011; 84(11): 1928–1946.
5. Ganti, R.K., F. Ye, and H. Lei. Mobile crowdsensing: Current state and future challenges. *IEEE Communications Magazine*, 2011; 49(11): 32–39.
6. Khan, W.Z., Y. Xiang, M.Y. Aalsalem, and Q. Arshad. Mobile phone sensing systems: A survey. *IEEE Communications Surveys & Tutorials*, 2013; 15(1): 402–427.

7. Chatzimilioudis, G., A. Konstantinidis, C. Laoudias, and D. Zeinalipour-Yazti. Crowdsourcing with smartphones. *IEEE Internet Computing*, 2012; 16(5): 36–44.

8. Wu, F.J. and T. Luo. A generic participatory sensing framework for multi-modal datasets. In: *Proceedings of IEEE Ninth International Conference on Intelligent Sensors, Sensor Networks and Information Processing (ISSNIP)*, Singapore, April 21-24, 2014.

9. Wang, Y.F., X.Y. Jia, Q. Jin, and J. Ma. Mobile crowdsourcing: Framework, challenges, and solutions. *Concurrency and Computation: Practice and Experience*, 2017; 29(3).

10. Farkas, K., A.Z. Nagy, T. Tomas, and R. Szabo. Participatory sensing based real-time public transport information service. In: *Pervasive Computing and Communications Workshops (PERCOM Workshops)*, Budapest, Hungary, March 24–28, 2014.

11. Hachem, S., G. Mathioudakis, A. Pathak, V. Issarny, and R. Bhatia. Sense2health: A quantified self application for monitoring personal exposure to environmental pollution. In: *SENSORNETS*, Angers, France, February 11–13, 2015.

12. Rana, R.K., C.T. Chou, S.S. Kanhere, N. Bulusu, and W. Hu. Ear-phone: An end-to-end participatory urban noise mapping system. In: *The 9th ACM/IEEE International Conference on Information Processing in Sensor Networks*, Stockholm, Sweden, April 12–16, 2010, pp. 105–116.

13. Dong, Y.F., L. Blazeski, D. Sullivan, S.S. Kanhere, C.T. Chou, and N. Bulusu. Petrolwatch: Using mobile phones for sharing petrol prices. In: *Proceedings of the 7th Annual International Conference on Mobile Systems, Applications and Services*, Krakw, Poland, June 2009.

14. Yan, T., V. Kumar, and D. Ganesan. Crowdsearch: Exploiting crowds for accurate real-time image search on mobile phones. In: *The 8th Annual ACM International Conference on Mobile Systems, Applications, and Services*, San Francisco, CA, June 15–18, 2010, pp. 77–90.

15. Engelsma, J., F. Jumah, A. Montoya, J. Roth, V. Vasudevan, and G. Zavitz. Shop social: The adventures of a barcode scanning application in the wild. In: *The Fourth International Conference on Mobile Computing, Applications, and Services*, Springer, Seattle, WA, October 11–12, 2012, pp. 379–390.

16. Wang, Y., W. Hu, Y. Wu, and G. Cao. SmartPhoto: A resource-aware crowdsourcing approach for image sensing with smartphones. In: *Proceedings of the 15th ACM International Symposium on Mobile Ad hoc Networking and Computing*, Philadelphia, PA, August 11–14, 2014, pp. 113–122.

17. Lee, U., J. Kim, E. Yi, J. Sung, and M. Gerla. Analyzing crowd workers in mobile pay-for-answer q&a. In: *The SIGCHI Conference on Human Factors in Computing Systems*, Paris, France, April 27–May 2, 2013, pp. 533–542.

18. Chatzimilioudis, G., A. Konstantinidis, C. Laoudias, and D. Zeinalipour-Yazti. Crowdsourcing with smartphones. *IEEE Internet Computing*, 2012; 16(5): 36–44.

19. Bigham, J.P., C. Jayant, H. Ji, G. Little, A. Miller, R.C. Miller, and T. Yeh. VizWiz: Nearly real-time answers to visual questions. In: *The 23nd Annual ACM Symposium on User Interface Software and Technology*, New York, October 3–6, 2010, pp. 333–342.
20. kNFB reader. knfb Reading Technology, Inc., 2008. http://www. knfbreader.com/.
21. Chen, C., Y. Huang, Y. Liu, C. Liu, L. Meng, Y. Sun, and B. Jiao. Interactive crowdsourcing to spontaneous reporting of Adverse Drug Reactions. In: *IEEE International Conference on Communications (ICC)*, Sydney, Australia, June 10–14, 2014, pp. 4275–4280.
22. Natural language processing, Available online:https://en.wikipedia.org/ wiki/Natural_language_processing. Access date: April 18, 2017.
23. Franklin, M.J., D. Kossmann, T. Kraska, S. Ramesh, and R. Xin. CrowdDB: Answering queries with crowdsourcing. In: *Proceedings of the 2011 ACM SIGMOD International Conference on Management of data*, Greece, Athens, June 12–16, 2011, pp. 61–72.
24. Gimpel, K., N. Schneider, B. O'Connor, D. Das, D. Mills, J. Eisenstein, and N.A. Smith. Part-of-speech tagging for twitter: Annotation, features, and experiments. In: *Proceedings of the 49th Annual Meeting of the Association for Computational Linguistics: Human Language Technologies: Short Papers-Volume 2*, Stroudsburg, PA, 2011, pp. 42–47.
25. Tang, J.C., M. Cebrian, N.A. Giacobe, H.W. Kim, T. Kim, and D.B. Wickert. Reflecting on the DARPA red balloon challenge. *ACM Communications*, 2011; 54(4): 78–85.
26. Allahbakhsh, M., B. Benatallah, A. Ignjatovic, H.R. Motahari-Nezhad, E. Bertino, and S. Dustdar. Quality control in crowdsourcing systems: Issues and directions. *IEEE Internet Computing*, 2013; 17(2): 76–81.
27. Boutsis, I. and V. Kalogeraki. Crowdsourcing under real-time constraints. In: *The 27th IEEE International Parallel & Distributed Processing Symposium*, Boston, MA, May 20–24, 2013.
28. He, S., D.H. Shin, J. Zhang, and J. Chen. Toward optimal allocation of location dependent tasks in crowdsensing. In: *IEEE INFOCOM 2014- IEEE Conference on Computer Communications*, Toronto, Canada, April 27–May 2, 2014, pp. 745–753.
29. Ren, J., Y. Zhang, K. Zhang, and X.S. Shen. SACRM: Social aware crowdsourcing with reputation management in mobile sensing. *Computer Communications*, 2015; 65: 55–65.
30. Conti, M., C. Boldrini, S.S. Kanhere, E. Mingozzi, E. Pagani, P.M. Ruiz, and M. Younis. From MANET to people-centric networking: Milestones and open research challenges. *Computer Communications*, 2015; 71: 1–21.
31. Zhang, B., Y.F. Wang, Q. Jin, and J. Ma. Energy-efficient architecture and technologies for Device to Device (D2D) based proximity service. *China Communications*, 2015; 12(12): 32–42.
32. Guo, B., Z. Yu, X. Zhou, and D. Zhang. From participatory sensing to mobile crowd sensing. In: *Pervasive Computing and Communications Workshops (PERCOM Workshops)*, Budapest, Hungary, March 24–28, 2014.

33. Liu, Y., V. Lehdonvirta, T. Alexandrova, M. Liu, and T. Nakajima. Engaging social medias: Case mobile crowdsourcing. In *Proceedings of the First International Workshop on Social Media Engagement (SoME)*, Barcelona, Spain, February 17, 2011.
34. Wang, Y., Y. Huang, and C. Louis. Respecting user privacy in mobile crowdsourcing. *Science*, 2013; 2(2): 50.
35. Chon, Y., N.D. Lane, F. Li, H. Cha, and F. Zhao. Automatically characterizing places with opportunistic crowdsensing using smartphones. In: *The 14th International Conference on Ubiquitous Computing*, Pittsburgh, PA, September 5–8, 2012, pp. 481–490.
36. Chatzigiannakis, I., G. Mylonas, and A. Vitaletti. Urban pervasive applications: Challenges, scenarios and case studies. *Computer Science Review*, 2011; 5(1): 103–118.
37. Land, M.B. Peer producing human rights. *Alberta Law Review*, 2009; 46(4): 1115.
38. Mohamed, N. and I.H. Ahmad. Information privacy concerns, antecedents and privacy measure use in social networking sites: Evidence from Malaysia. *Computers in Human Behavior*, 2012; 28(6): 2366–2375.
39. Enck, W., P. Gilbert, S. Han, V. Tendulkar, B.G. Chun, L.P. Cox, and A.N. Sheth. TaintDroid: An information-flow tracking system for realtime privacy monitoring on smartphones. *ACM Transactions on Computer Systems (TOCS)*, 2014; 32(2): 5.
40. Gilbert, P., B.G. Chun, L.P. Cox, and J. Jung. Vision: Automated security validation of mobile apps at app markets. In: *Proceedings of the Second ACM International Workshop on Mobile Cloud Computing and Services*, Bethesda, MD, June 28–July 01, 2011.
41. Quinn, A.J. and B.B. Bederson. Human computation: A survey and taxonomy of a growing field. In: *Proceedings of the SIGCHI Conference on Human Factors in Computing Systems*, Vancouver, BC, Canada, May 07–12, 2011.
42. Allahbakhsh, M., B. Benatallah, A. Ignjatovic, H.R. Motahari-Nezhad, E. Bertino, and S. Dustdar. Quality control in crowdsourcing systems. *IEEE Internet Computing*, 2013; 17(2): 76–81.
43. Gadiraju, U., R. Kawase, S. Dietze, and G. Demartini. Understanding malicious behavior in crowdsourcing platforms: The case of online surveys. In: *Proceedings of the 33rd Annual ACM Conference on Human Factors in Computing Systems*, Seoul, South Korea, April 18–23, 2015, pp. 1631–1640.
44. Huang, K.L., S.S. Kanhere, and W. Hu. Preserving privacy in participatory sensing systems. *Computer Communications*, 2010; 33(11): 1266–1280.
45. Zhai, Z., P. Sempolinski, D. Thain, G. Madey, D. Wei, and A. Kareem. Expert-citizen engineering: "Crowdsourcing" skilled citizens. In: *The IEEE Ninth International Conference Dependable, Autonomic and Secure Computing (DASC)*, Sydney, Australia, December 12–14, 2011, pp. 879–886.

46. Gao, H., G. Barbier, R. Goolsby, and D. Zeng. *Harnessing the crowd-sourcing power of social media for disaster relief.* Tempe, AZ: Arizona State University, 2011.
47. Wang, Y.F., X.Y. Jia, Q. Jin, and J. Ma. QuaCentive: A quality-aware incentive mechanism in mobile crowdsourced sensing (MCS). *The Journal of Supercomputing*, 2016; 72: 2924–2941.
48. Zhang, Y. and M. van der Schaar. Reputation-based incentive protocols in crowdsourcing applications. In: *The 31st Annual IEEE International Conference on Computer Communications*, Orlando, FL, March 25–30, 2012, pp. 2140–2148.
49. Hoh, B., T. Yan, D. Ganesan, K. Tracton, T. Iwuchukwu, and J.S. Lee. TruCentive: A game-theoretic incentive platform for trustworthy mobile crowdsourcing parking services. In: *15th International IEEE Conference on Intelligent Transportation Systems*, Anchorage, AK, September 16–19, 2012, pp. 160–166.
50. Goldberg, A.V., J.D. Hartline, A.R. Karlin, M. Saks, and A. Wright. Competitive auctions. *Games and Economic Behavior*, 2006; 55(2): 242–269.

9

ENERGY-EFFICIENT TECHNOLOGIES IN DEVICE-TO-DEVICE BASED PROXIMITY SERVICE

9.1 Introduction

Recently, proximity services (ProSe) have become a promising mobile industry that enables to create numerous interesting mobile services. Basically, ProSe can be composed of two main groups of use cases: public safety communications and discovery mode (commercial applications) [1]. Proximity applications allow users in local area to share experiences about specific places in real time, which can give them more fun and can increase users' engagement. It is estimated that this emerging ProSe market has grown to $1.9 billion in revenues by 2016 [2].

Existing technologies used to serve the proximity awareness can be broadly divided into over-the-top (OTT) and device-to-device (D2D) (peer-to-peer [P2P]) solutions. In the OTT model, a server usually located in the cloud receives periodic location updates from user's mobile devices (e.g., using global positioning system, GPS), and then the server determines proximal relationships based on users' locations and interests. The constant location updates not only quickly drain the battery (due to the relatively large power consumption of GPS usage and cellular connections), but also cause serious privacy problem. Moreover, OTT approaches may incur undesired network overheads and latency for discovery and communication.

Today, modern mobile phones have the capability to detect users' proximity and offer means to communicate and share data in ad hoc way, with the people in the proximity, which nurtures the so-called D2D-based ProSe. Different from OTT, D2D schemes forego centralized processing instead of autonomously determining proximal values at the device level by transmitting and monitoring for relevant attributes. This mode offers the crucial privacy benefit.

Although, D2D-based ProSe schemes are relatively energy-efficient approaches (due to the property of short-range communications), energy is still one of the critical bottlenecks for ProSe on smartphones. Currently, most smartphones are powered by lithium-ion batteries, and at the moment the typical way to create more powerful batteries is to make them bigger. However, this does not match well with the evolution of mobile terminals, which tend to have less room available for the battery in order to accommodate additional components and technologies. Typically, the amount of energy in a battery is growing only 5% annually [3]. Thus, the constrained battery power of mobile devices makes a serious impact on users' experience.

Various energy management techniques in modern mobile handsets were thoroughly summarized by Vallina-Rodriguez and Crowcroft [4]. However, this work only investigated the energy efficient technologies for stand-alone smartphones and centralized architecture. There exist no comprehensive energy-efficient frameworks and technologies specially designed for the features of D2D-based ProSe, which is the main motivation of this chapter. Specifically, the contribution of this chapter lies in that: by incorporating the special features of ProSe, the chapter presents the energy-efficient architecture and technologies for ProSe from the following four aspects: underlying networking technologies, localization schemes (outdoor and indoor), leveraging application and architecture features, and utilizing users' context. Moreover, for each aspect, we summarize the related challenges and point out the potential solutions.

The remainder of this chapter is organized as follows: Section 9.2 presents the framework of energy-efficient architecture for D2D-based ProSe, which is composed of design principle, detailed technologies, as well as power model and measurement. Energy-efficient underlying network technologies are summarized in Section 9.3. Section 9.4 provides energy-efficient localization methodologies and schemes, for outdoor and indoor environments. Section 9.5 provides several schemes, which save energy by incorporating special features of applications and architecture in ProSe. Section 9.6 gives the context-aware energy-efficient schemes, including three kinds of contexts: physical, usage, and social contexts. Section 9.7 summarizes power modeling and energy measurements on smartphones. Finally, we briefly conclude this chapter.

9.2 Architecture of Energy-Efficient Proximity Services

Before delving into the detailed energy-efficient schemes in ProSe, it is imperative to first provide the special features of ProSe which are listed as follows:

- D2D communication technologies underlie ProSe. Basically, there exist various D2D communication technologies in unlicensed radio bands (e.g., Bluetooth, Wi-Fi Direct) and cellular spectrum (e.g., long-term evolution [LTE] Direct). Especially, peer and service discoveries are major issues in D2D communications.
- Location information is central to ProSe. Not only are contents tagged with locations, but ProSe uses geoproximity as the primary filter in determining who are discoverable. Therefore, it is necessary to thoroughly investigate the energy-efficient localization technologies in outdoor and indoor environments.
- Nowadays, smartphones are equipped with the rich collection of sensors (GPS, accelerometer, camera, etc.) and multiple ratio interfaces (Bluetooth, Wi-Fi, etc.) that can be fully exploited to enable novel applications that sense the individual's environment and provide a unique view on the surrounding world.

Sustainability and energy reduction have emerged as important issues in the social, political, and technical agendas in recent decades. On the one hand, individuals express interests in adopting more sustainable products and behaviors; but on the other hand, they do not wish to do so at the expense of their comforts. Hence it is important that energy reduction schemes in ProSe should be context aware, and should take users' behaviors and preferences into account.

To incorporate the special features of ProSe, as shown in Figure 9.1 (the middle block), energy-efficient technologies and architecture should be designed from the following four aspects (arranged from bottom to up): underlying networking technologies (including

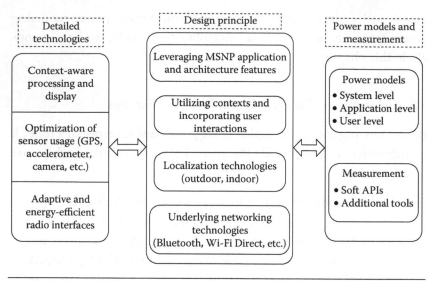

Figure 9.1   Architecture of energy-efficient ProSe.

Bluetooth, Wi-Fi Direct, LTE Direct, etc.), localization schemes (outdoor and indoor), leveraging application and architecture features, and utilizing users' context.

Basically, the power consumption of a smartphone mainly arises from: (1) wireless radio interfaces, for example, Wi-Fi, Bluetooth and Cellular, and so on; (2) various sensors including GPS, accelerometer, camera, and so on; and (3) processor and touch/display screen. Therefore, as shown at the left block in Figure 9.1, energy-efficient ProSe should design adaptive radio interfaces (e.g., dynamically adjust the idle and scan frequency of wireless interfaces), should optimize sensor usage (e.g., properly change sampling frequency of sensors), and should utilize context-aware processing and display usage (e.g., adaptively adjust the CPU mode and the brightness of the screen according to users' contexts).

Power/energy models can predict energy consumption through usage data provided by mobile operating systems and can play an important role in design energy-efficient mechanisms. Especially, energy measurement is one of the essential aspects in power models (shown at the right block in Figure 9.1).

It should be explicitly noted that, from the engineering point of view, energy-efficient ProSe can be interpreted as the reduction of the energy utilized to accomplish ProSe (e.g., networking, localization and content sharing), while maintaining the required performance level (e.g., data transmission rate, delay, localization accuracy).

9.3 Energy-Efficient D2D Networking for Proximity Services

9.3.1 Energy-Efficient Device-to-Device Communications Technologies

The most widely deployed direct communication technology, Bluetooth, was designed for interconnecting personal devices, such as mobile phones and headsets. The most favorable feature of Bluetooth is that it can stay operational the whole day without depleting the power consumption profile significantly. However, Bluetooth itself cannot be utilized for the D2D file transfers or video streaming, as it does not have the high data transmission rate. In addition, Bluetooth was not designed for building opportunistic-based ProSe applications. This is reflected by characteristics such as relatively short radio range, secure pairing, and lack of broadcast support in Bluetooth stack of smartphones.

Wi-Fi Direct [5] initially called Wi-Fi P2P is a Wi-Fi standard that enables devices to connect easily with each other without requiring a WLAN access point (AP), at typical Wi-Fi communications speeds. In Wi-Fi Direct, battery-constrained devices have to act as the role of P2P group owner (GO, i.e., soft AP), and therefore energy efficiency of soft AP is extremely challenging. To solve the issue, Wi-Fi Direct defines two power-saving mechanisms: opportunistic power save protocol and notice of absence protocol. However, neither are enough to meet power efficiency and reliability requirements for various services. Therefore, an enhanced power management for Wi-Fi Direct was proposed in [6], which exploited the periodic property of data transmission in applications and dynamically adjusted the duty cycle of Wi-Fi Direct devices. The fair role switching between soft AP and client is proposed to further balance the energy consumption, by estimating the residual contact duration [7].

In summary, basically, there exist two methodologies to design energy efficient Wi-Fi-based direct networking technologies, and they should be appropriately incorporated into the existing power-saving mechanisms in Wi-Fi-based networking technologies in smooth and unobstructive way.

- According to application properties, Wi-Fi interface of a soft AP can be dynamically set as sleep to save power. However, putting a soft AP into sleep imposes two challenges: First, without a careful design, it may cause packet loss; second, it may introduce increasing network latency and impair user experience if the extra latency is perceivable by users.
- Considering the fact that the role of soft AP of a connection in ProSe consumes more energy than a *client*, role switching could increase fairness and could substantially extend the battery lifetime of mobile devices.

LTE Direct is a new emerging infrastructure-assisted D2D communication paradigm using licensed spectrum. It is reported by Qualcomm that LTE Direct can discover as many as 7200 devices in 0.64 s in comparison to 369 terminals found in using Wi-Fi Direct that took 82–119 s. The simulation results showed that LTE Direct can achieve improved energy-efficient communication by 5053 (Mb/J) [8].

However, unfortunately, given many technical challenges and disjoint opinions of 3GPP member companies, commercial *products* would not be expected for the near future years, thus the immediate attention of industrial players is on D2D in the unlicensed band.

9.3.2 Energy-Efficient Peer and Service Discovery for Proximity Services

Peer and service discovery are the fundamental issues in ProSe. That is, before two devices can directly communicate with one another, they must first know (discover) that they are near to each other.

In general, two methodologies for peer and service discovery can be identified: infrastructure-assisted mode and direct mode, which are explained as follows:

- The design goal of infrastructure-assisted mode is to make peer and service discoveries and pairing procedures faster, more efficient in terms of energy consumption and more user friendly.
- In direct mode, mobile devices have to continuously transmit synchronization signals without knowing the physical location of the intended mobile devices. Hence, each mobile device

needs to frequently search for synchronization signal, which is more energy and time consuming than that in network-assisted mode, but it gives more control to individuals and alleviates the problem of privacy breach existing in infrastructure mode.

For direct peer discovery, combining Wi-Fi Direct and Bluetooth could provide an interesting solution. For instance, CQuest [9] designed an efficient long-range neighbor discovery mechanism, based on the clustering of individuals and the radio heterogeneity of mobile devices.

Specifically, due to the mobility of individuals in real life, temporary clusters are always formed around different locations like gym, shopping mall, and so on. A low-power short-range radio, such as Bluetooth, is sufficient for the devices inside such a cluster to discover each other and build small communication groups (refer to the Figure 9.2), but neighbors in proximity who can be only reachable via the long-range high-power radio remain undiscovered. To discover these neighbors, CQuest enables cooperative scanning over the set of high-power radios (e.g., Wi-Fi). That is, given this overlap in dense environment (the area of common high-power neighbors), not all mobile devices need to discover high-power neighbors. Instead, a subset of devices can perform high-power discovery and disseminate the resulting information through low power ratio to nondiscovery devices.

However, the weak point of this hybrid system lies in that: both short-range radios in unlicensed spectrum have to simultaneously work together, which brings interference to each other.

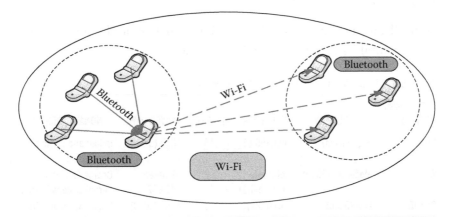

Figure 9.2   Illustration of peer discovery through collaboration between Bluetooth and Wi-Fi.

9.4 Energy-Efficient Localization Technologies

Smart mobile devices provide various positioning services via GPS, Wi-Fi-based positioning system (WPS), or cell-ID positioning. Basically, the following several metrics can be used to characterize the performance of various localization systems [10].

- *Accuracy*: The veracity of a position of being determined (measured as deviation in meters)
- *Precision*: Degree of reproducibility (measured as percentage of fixes within a specific accuracy)
- *Time to first fix* (*TTFF*): The used time until location can be determined initially (in seconds)
- *Power*: Average energy consumption (usually in milliwatt [mW])
- *Nonavailability*: Situations where positioning is limited

The characteristics of those three popular positioning methods are summarized in Table 9.1.

Note that accuracy and precision are only coarse values to exemplify the magnitude of the attributes of today's smartphones. Especially the accuracy of Wi-Fi and cell-ID positioning can differ in areas with a low density of APs or cell towers. The energy consumption is described as the amount of energy that is needed to obtain one position sample. Obviously, among those techniques, there exist a trade-off between the localization accuracy and energy consumption.

9.4.1 Technical Design Principle

Basically, in energy-efficient LBS design principles and solutions, two fundamental issues should be taken into account.

Table 9.1 Characteristics of Various Positioning Methods

POSITIONING METHOD	ACCURACY	PRECISION	TTFF (s)	POWER (mW)	NONAVAILABILITY
GPS	High (10 m)	High (95%)	15	High (6.616)	Indoor and canyon, and so on
WPS	Moderate (50 m)	Relatively high (90%)	3	Moderate (2.852)	Coarse AP (rural area) or density area
Cell-ID positioning	Low (5 km)	Low (65%)	3	Low (1.013)	Region without cells (rural area)

First, what are the main power consumption components in localization? The main culprit is the energy-hungry sensors. According to [11], accelerometers consume only a tiny amount of energy (0.0488 mW at 10 Hz), and gyroscope's energy consumption rate is also small (2.14 mW at 10 Hz), compared to the widely used localization technique, GPS (150 mW at 1 Hz). Those data suggest that, intuitively, under the condition of satisfying the performance constraints, it is better to use energy-efficient sensor for localization, instead of high energy-consuming sensors.

Second, what is the main reason of energy inefficiency? Generally, low utilization of energy-consuming components (if no other substitutable techniques are suitable) will definitely lead to the inefficiency of energy usage. Therefore, intuitively, increasing the utilization rate of underlying localization component through collaborative methodology would improve the energy efficiency in smartphone-based LBS applications.

Inspired by the aforementioned consideration, the Subsection 9.4.1 explicitly proposes three technical design principles in system level: substitution, adaption, and collaboration (as shown in Figure 9.3), by comprehensively summarizing various existing sporadic energy-efficient LBS schemes, which could guide to design new energy-efficient LBS applications. It should be explicitly noted that the proposed design principle should comply with special environments that LBS applications work for. Moreover, context-assisting methodologies are complementary to the aforementioned three design principles. For instance, when a LBS application can effectively detect the users move from outdoor to indoor, then power-hungry GPS that cannot be used indoors should be switched off immediately.

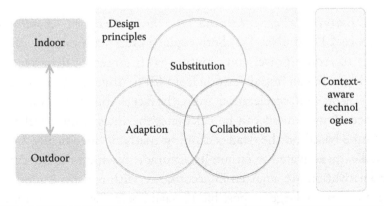

Figure 9.3 Methodologies of energy-efficient localization technologies.

- *Substitution*: If performance metrics permit, low energy-consuming localization methods should be used as substitute for power-hungry localization technique (e.g., GPS), such as proper integration of GPS, WPS, and cell-ID positioning and any other localization techniques to save energy, depending on users' circumstance.
- *Adaption*: Adaptively change localization factors (e.g., lowering sampling rates), depending upon users' real-time contexts and the localization accuracy required.
- *Collaboration*: Cooperative localization with the surrounding neighbors or other techniques to reduce energy consumption in localization.

As Figure 9.3 shows, some typical energy-efficient indoor and outdoor localization schemes will be elaborated in Section 9.4.2, which actually adopt/follow one or more of these three principles. Several detailed context-assisting schemes are introduced in Section 9.4.3, which are complementary to/lay the underlying foundation for these three principles.

9.4.2 Energy-Efficient Localization Schemes

As embodiment of the three technical design principles, several typical energy-efficient LBS schemes are summarized and compared, including two common environments: outdoor and indoor. The design principle of these schemes often does not work alone, and appropriately integrated principles could make the effect more impressive.

9.4.2.1 Energy-Efficient Localization for Outdoor Environment

Reference 12 and eNav [11] both replace GPS with energy-efficient sensors to reduce power consumption with target localization accuracy. By getting an initial position from GPS, direction from geomagnetic sensor, and acceleration from the accelerometer, velocity and distance of movement can be calculated by integrating acceleration over time based on the dead-reckoning method. Specifically, GPS is periodically activated to ensure its accuracy in which the intervals of GPS activation are adjusted in accordance with real-time conditions and are kept as large as possible to save energy within the limit of providing target accuracy [12]. The actual GPS sampling is returned only when an accurate estimate is needed, that is, the vehicle is close to the next navigation waypoint [11].

GloCal [13] leverages various sensors to infer user walking characteristics and further to depict the entire user trajectory instead of GPS. Specifically, GloCal uses accelerometer to identify user walking steps, and acceleration feature is further investigated to determine the accurate stride length of a specific user. The walking displacement is then derived by multiplying the step counts with the stride length. It can at last reduce 30% of global positioning errors of GPS with only negligible extra energy consumption.

Proximity alert [14] utilizes distances to the point-of-interests (POIs) and user's transportation mode to dynamically decide on location providers and location-sensing interval. SensTrack [15] smartly selects the location-sensing methods when Wi-Fi or GPS is available, and it periodically collects data from the corresponding sensor to detect a turning point or estimate current speed and the distance from the last recorded location to reduce the GPS sampling rate.

Sharing the position information with proxy devices using the short-range communication (SRC) such as Bluetooth or ZigBee would consume less energy than GPS. Along this direction, Reference 16 proposes two methods: (1) a distributed location-sharing (DLS) protocol, which is operated by proxy device sensing and waiting time to turn off GPS, and (2) an optimal mean waiting time decision (OWD) algorithm, which makes DLS protocol to achieve an optimal GPS on/off time portion of each device using only past GPS on/off statistics. The scheme achieves remarkable power saving compared to no collaboration.

Incremental scanning [17] reduces the energy consumption of Wi-Fi localization by scanning the available channels one-by-one and terminating early once a sufficient subset of APs has been discovered. This work identifies three mechanisms that help achieve this goal. The results of the user-study show that the three mechanisms underlying the incremental scanning policy only trade-off a bit of accuracy for a serious reduction in energy consumption.

Following the principle of substitution, assisted GPS (A-GPS) attains location information through cellular network instead of GPS itself, which can save energy of mobile terminals. Similarly, a scheme utilizing the collaboration principle proposes an energy-efficient GPS tethering mechanism, called GPS-Tether (GTR), which conserves battery power of a smartphone (GTR client) by opportunistically retrieving GPS coordinates from its neighboring smartphone (GTR server) rather than activating its own GPS interface [18]. This is similar to A-GPS, but the help comes from the proximity and without incurring the complexity corresponding to the dedicated network protocols and the worldwide deployment of aforementioned location servers. GTR can extend smartphones battery usage as much as 19% compared to A-GPS.

We compare the aforementioned schemes in Table 9.2. From this table, we can observe that substitution and adaption have prevalent usage in outdoor localization to reduce power consumption. Collaboration can reuse the existed sensing/localization results by sharing those results with neighboring smartphones using energy-efficient short-range communications technologies, thus eliminating some unnecessary usage of sensors. Most of these schemes are stand-alone to localize, without following the principle of collaboration. Collaboratively using dead reckoning and adaption to real-time conditions would greatly further help save localization energy of smartphones.

9.4.2.2 Energy-Efficient Localization for Indoor Environments   Different from outdoor localization, the distinguished features of localization for indoor environments lie in that: first, GPS does not work for indoor; second, indoor environment is more complicated, but, fortunately, several short-range (direct) communications techniques can be utilized to facilitate the energy-efficient localization.

References 19 and 20 utilize a low-power ZigBee interface to substitute for the energy-hungry Wi-Fi interface. In detail, in ZiFind [19], mobile user detects the unique interference signatures induced by the Wi-Fi infrastructure and uses them as fingerprints to estimate the current location. ZIL [20] uses ZigBee interfaces on the clients to capture beacon frames broadcasted by Wi-Fi APs, without the mappers designed in ZiFind. Furthermore, a novel fingerprint-matching algorithm and three variants of the K-nearest neighbor (KNN) algorithm are designed in ZIL to align a pair of fingerprints effectively and to improve the localization accuracy, respectively. ZIL outperforms ZiFind in terms of energy consumption while keeping the competitive localization accuracy.

Instead of Wi-Fi signal fingerprinting, sensors or site-specific input, practical indoor localization system (PILS) [21] utilizes the freely available cellular signal information by using the detailed statistical

Table 9.2 Comparisons of the Typical Energy-Efficient Localization Schemes in Outdoor Environment

SCHEMES	FEATURES	PERFORMANCE	DESIGN PRINCIPLES	TO BE IMPROVED
ALMGES [12]	Periodically activates GPS; adjusts GPS intervals; utilizes dead reckoning	Reduces power consumption with target accuracy	Substitution and adaption	Experiments only on an Android-based smartphones
eNav [11]	Dead reckoning and adaptive GPS sampling	Around 80% energy savings	Substitution and adaption	Less rigorous usability study and no crowdsourcing server consideration
GloCal [13]	Multiply step counts with stride length	Reduces 30% of global positioning errors	Substitution	Future work of assisting GPS positioning in unmanned vehicles
Proximity alert [14]	Dynamically decides on location providers and location-sensing interval	Increases battery lifetime by 75.71%	Adaption	Without user's mobility and transportation modes
SensTrack [15]	Smartly selects sensing methods and reduces sampling rate	Significantly reduces the usage of GPS and still achieves a high-tracking accuracy	Adaption and substitution	Without exploiting the resilient accelerometer data processing, tracking for multiple mobility patterns, and joint optimization of energy and accuracy
DLS+OWD [16]	Collaborates with proxy devices for localization	65.5% power saving	Collaboration and substitution	Localization accuracy is not specific
Incremental scanning [17]	Scans Wi-Fi just on a few selected channels	Reduces the energy consumption between 20.64% and 57.79%	Substitution and adaption	Without other collaborative localization methods
GPS tethering [18]	Opportunistically retrieves GPS coordinates from its neighboring smartphone	Extends smartphones battery usage as much as 19% compared to A-GPS	Collaboration	Without investigating how to incentivize rational smartphone users to help each other

properties of physical layer information of 4G LTE networks for both connected and neighbor BSs to statistically map the indoor locations. PILS runs in the background and reads cellular signals based on a scheduling policy and hence, it consumes minimal energy overhead. The results show promise for improvements in current indoor localization systems using cellular signals.

Instead of using signature entries from APs, motion-assisted device tracking algorithm (MADT) [22] utilizes the fundamental rules of received signal strength indication (RSSI) and environmental factors such as distance and direction, so as to guide a user with a device receiving Bluetooth signal from the target to gradually approach them. Extensive experiments show that the algorithm is efficient in terms of localization accuracy, searching time, and energy consumption.

Besides the previous substitution methods, GreenLoc [23] leverages mobile sensors, short-range wireless technologies, and group mobility patterns to save energy and is designed to be easily integrated with any Wi-Fi-based indoor localization system. Mobile clients connect to the GreenLoc server and send different sensed data. The server adaptively delegates a node representative to act as a designated cluster head. The scheme shows extremely promising reductions in energy.

References 24 and 25 follow the principle of collaboration and substitution. LearnLoc [24] combines three regression-based machine-learning techniques with smart Wi-Fi fingerprinting and supplemental sensors to improve the accuracy of indoor localization to create a low-cost, infrastructure-less indoor navigation solution. LearnLoc framework using the KNN variant provides a superior output across several diverse indoor environments, with the best localization accuracy and a competitive energy consumption overhead when compared to the other alternatives. Indoor sparse filter [25] is updated using a localization method combining pedestrian dead reckoning (PDR) and real-time learning. Landmarks such as intersections and elevators signal fingerprints such as Wi-Fi, and magnetic signatures are used to trigger the learning procedure. Field experimental results show that the proposed method can perform localization with much higher efficiency, high reliability, and merely slight loss of accuracy.

The typical indoor energy-efficient localization techniques are compared in Table 9.3. We can know that the most existing methods are under the guidance of substitution. Furthermore, unlike outdoor

Table 9.3 Comparisons of Typical Energy-Efficient Localization Schemes in Indoor Environment

SCHEMES	FEATURES	PERFORMANCE	DESIGN PRINCIPLES	TO BE IMPROVED
ZiFind [19]	Exploits the cross-technology interference in the unlicensed 2.4 GHz frequency spectrum	Energy consumption is only 32% of the Wi-Fi-based method	Substitution	Introducing mappers inevitably complicates the system implementation and increases the cost
ZIL [20]	Uses low-power 802.15.4 ZigBee interfaces and a novel fingerprint matching algorithm	Saving about 68% energy and 87% localization accuracy	Substitution	Further considers cooperating with other indoor localizations
PILS [21]	Utilizes cellular signal information to map the indoor locations	Accuracy up to 91% with an average localization error of less than 2.3 m	Substitution	Multilevel ranking algorithm to use a variety of factors
MADT [22]	Leverage Bluetooth RSSI and user motions	The battery level drops from 51% to 50% during the localization	Substitution	Experiment scale and range are not large enough
GreenLoc [23]	Short-range wireless technologies, sensors, and group mobility patterns	60% energy reduction in localization accuracy averaging 2 m	Adaption and collaboration	Privacy issues and valuation of the system without a large-scale test bed
LearnLoc [24]	Wi-Fi fingerprinting and inertial sensors with machine learning techniques	Approximately 1–3 m accuracy	Collaboration and substitution	Needs to perform more aggressive trade-offs between accuracy and energy
Indoor sparse filter [25]	Incorporates PDR, fingerprinting, and real-time learning	Energy consumption rate improved by 45%	Collaboration and substitution	Only uses simulations to evaluate the phone battery usage

localizations, the schemes for indoor seldom follow the adaption principle. We think that the underlying reason for this phenomenon partially lies in the following fact: indoor environments are extremely complicated (and even intractable) than outdoor, and the substitution principle is relatively easy to be utilized in this circumstance. However, appropriately integrating all those design principles could greatly improve the energy efficiency for indoor localization.

9.4.3 Complementary Context–Assisting Energy-Efficient Schemes

Although a great deal of schemes following the three technical principles can decrease power usage, many environmental factors also can be helpful or harmful to energy consumption, so when confronted with specific environments, becoming context-aware will be conducive to energy efficiency. Here, we summarize several typical complementary schemes: indoor–outdoor detection, battery interaction, and image-based localization. Timely recognition and distinguishability of the indoor and outdoor environments apparently can save energy for localization methods. Battery interaction can exploit the users' behavior to make energy efficient. In addition, if reasonably used, pictures taken by cameras can play an important role in localization.

9.4.3.1 Detection: Indoor or Outdoor Indoor and outdoor are very different, thus knowing whether the user is indoor or outdoor is useful in invoking the appropriate location-tracking solution. To save power, the phone itself may be able to turn on or off certain sensors depending on the specific environment, indoor or outdoor. Moreover, the detection method of its own should be energy efficient. Specifically, a semisupervised learning method called cotraining is presented in [26] to detect if a device is indoor or outdoor, which is able to automatically learn characteristics of new environments and devices, and provides a detection accuracy exceeding 90% even in unfamiliar circumstances. Implementation of the indoor–outdoor detection service is lightweight in energy use—it can sleep when not in use and does not need to track the device state continuously. It is shown to outperform existing indoor–outdoor detection techniques that rely on static algorithms or GPS, in terms of both accuracy and energy efficiency.

BlueDetect [27] is based on the emerging low-power iBeacon technology in which only a few small-sized, low-cost, and battery-powered Bluetooth low energy (BLE) beacons are required by BlueDetect, which are placed at landmarks, such as the boundary of covered corridors and entrances/exits of buildings, with a sparse density in intermediate regions between indoor and outdoor environments (classified as a semioutdoor environment). The GPS module is turned on for LBS only in the outdoor environment. When it comes from outdoors to semioutdoors, the decrease in the mean GPS signals is utilized as a trigger to turn off GPS and turn on Bluetooth, and the iBeacon mode of BlueDetect is responsible for providing LBS within semioutdoor environments. Transitions between semioutdoor and indoor environments are achieved seamlessly by comparing the signals of two BLE beacons placed on both sides of the entrance of the building.

9.4.3.2 Battery Interactions   In order to manage power usage properly, people should constantly interact with battery in mobile handsets. As we know the battery real-time state, the next charging opportunity, and other contexts related to energy consumption, we have more probability to maintain power levels of mobile phones to support essential mobile applications.

Some surveys about human–battery interaction show that mobile phone users do not fully understand and appreciate their actual need from mobile battery charging [28]. Automatic logging is an efficient way to obtain continuously battery information. These human–battery interaction data can highlight how we can improve users' experience with their battery life and educate them about the limited power that his or her devices have. Although the aforementioned technique cannot help mobile handset truly save energy, it manages to consume the remaining power in the most reasonable way, in a sense, in an energy-efficient way.

9.4.3.3 Image-Based Localization   Positioning systems can exploit signatures hidden in a user's environment to identify a location. Images are often used to locate a place by identifying landmarks. In detail, WhereAmI [29] uses texts appearing together in an outdoor image, which is taken by smartphone cameras as a unique signature for localization. The texts from the user-generated image can either be determined automatically using optical character recognition (OCR), or the user can aid the OCR process by identifying the text. The process of matching is triggered by uploading only the texts from the phone, thereby reducing the bandwidth requirement compared to uploading an image. Compared to GPS, WhereAmI has lower response time, less energy consumption, and better position accuracy in urban areas.

9.5 Application and Architecture Features

It is expected that user involvement in ProSe will open undiscovered usages and interactions. For instance, people with similar interests can leverage geographical proximity and interpersonal affinities for personal benefits in ProSe. Therefore, incorporating the special application and networking architecture can greatly improve the energy efficiency in ProSe.

9.5.1 Collaborative Downloading

Given a scenario, for example, in a same classroom, it is traditional method that students use smartphones to download the same teaching documents, respectively. A novel approach presented by Bojic et al. [30] is that mobile users, who are interested in the same content, collaborate and download the desired material together. The main idea of the collaborative approach is to identify a set of mobile users who are physically at a small distance to each other (e.g., 10 m) and who are interested in the same documents. Each of those mobile users downloads only a part of the required content via the conventional mobile network and shares them with other mobile users within the newly formed mobile ad hoc network via short-range communication technology (e.g., Bluetooth or Wi-Fi Direct). The collaborative approach is enabled by filtering mobile users with their physical location and clustering those mobile users into groups of similar interests.

9.5.2 Store–Carry–Forward Mechanism

Most of the existing content dissemination schemes in ProSe rely on the movement of individuals and his or her encounter opportunities to relay data to the destination, that is, the so-called store-carry-forward (SCF) mode. If there is no connection available at a particular time, a mobile device can store and carry the data until it encounters other nodes. When the node has such a forwarding opportunity, all encountered nodes could be the candidates to relay the data. Thus, relaying selection and forwarding decision need to be made by the very current node based on certain forwarding criteria. Various studies have been conducted in literature to collect and analyze real mobility traces. The findings of those studies are that: (1) human being's mobility in real life is not random, but exhibits repetition to a certain extent and (2) the mobility of humans is affected by his or her social relationships. Thus, in ProSe, it is a promising research direction to exploit how users' social behaviors influence the message forwarding in an energy-efficient way. Specifically, a SCF-like cellular architecture is proposed in Reference 31, in which, when the application data are not delay sensitive, an individual can first transmit the data to a proper mobile relay, which carries the message close to the base station (BS), and then the mobile relay retransmits the data to the BS. Numerical simulations show that, for delay insensitive services, a factor of more than 30 in energy savings can be obtained by SCF, in comparison with direct transmission.

9.6 Context Awareness for Energy-Efficient ProSe

Context is any information that can be used to characterize the situation of an entity. In ProSe, contexts could include information about people, objects, and surroundings, and collecting such data can be done from a number of sources. Specifically, the following contexts should be taken into account in energy-efficient ProSe.

- *Physical context*: The significant development of hardware, such as multiple CPUs and sensors, makes smartphones more powerful; however, the power consumption of those function-rich mobile terminals increases rapidly. Therefore, for instance, dynamically adjusting the working frequency of CPUs and sensors can reduce energy consumption.
- *Usage context*: The context in which mobile devices are used is significantly different from a stationary compute at home or in offices. Specifically, smartphones are equipped with a

touch screen; thus, adaptive changing of the display screen according to the usage habits can greatly reduce the energy consumption.

- *Social context*: In ProSe, content exchanges are only possible within close physical proximity. Therefore, users' proximity and mobility are typical social contexts that could be exploited to design energy-efficient ProSe applications.

9.6.1 Physical Context

In most ProSe applications, on one hand, the sensors embedded in mobile devices have to be sampled often in order to capture users' behaviors, which, however, may lead to faster depletion of the battery; on the other hand, if the sensors are sampled at a slower rate, then it may not be possible to accurately capture the user's behaviors. To meet the challenges posed by phone sensing, three adaptive schemes were designed by Rachuri [32]. First, an adaptive sampling framework was designed in which sensors sample data by intentionally considering the user's context to conserve energy, while providing the required accuracy to the applications. Second, to further reduce the energy consumption of capturing data, a specific framework is proposed, which exploits the sensors in buildings and dynamically distributes the sensing tasks between the local phone and the sensors in buildings. Third, to efficiently process the data, a computation offloading scheme is provided, which determines whether to locally compute the classification tasks on the mobile device or remotely in the cloud by considering various dimensions such as latency, energy and data traffic, and so on.

9.6.2 Usage Context

Displays of smartphones often account for a significant amount of the total energy consumption, which makes them one of the primary targets for energy optimization. Organic light-emitting diodes display whose energy consumption is directly related to the color and brightness of the pixels, present opportunities for energy saving, by, for example, modifying the colors and intensity of pixels in regions that are less important to viewers.

Specifically, when a user interacts with a smartphone via the touch screen, the screen areas are covered by the user's fingers and even some of the neighboring areas could be safely dimmed. Based on the aforementioned idea, FingerShadow was proposed by Chen et al. [33], which conducted local dimming for the screen areas covered by fingers of an individual to save power efficiently, without compromising the user visual experience.

9.6.3 Social Context

Usually, in proximity alert framework, once a device is within the radius of a region, the alert of entering would be fired, and once the device is out of the radius, the alert of exit would be fired. Unfortunately, the aforementioned proximity alert service always depends on frequent checking of physical location (at the level of seconds) by using energy-consuming sensing of GPS, Wi-Fi, GSM, and so on; however, this is often unnecessary. Especially, when the user is far away from the PoI, it is not necessary to request location with high frequency, as the user could not pass through the location in such a short time. Rather than fixing the sensing interval, changing it dynamically based on the distance to the PoI and the speed of the user would lead to significant energy saving. An energy-efficient proximity alert service was implemented for Android [14] in which this adaptive method is adopted to keep track of distances to the PoIs and to detect the specific context such as transportation mode (idle, walking or driving, etc.) with the low energy-consuming accelerometer, to dynamically make decision on localization sensing interval and the proper location provider strategy.

9.7 Energy Consumption Models and Measurement

In literature, there exist several attempts to define power models for mobile devices, which can predict energy consumption under the statistic power-usage data provided by mobile operation systems (OS) and can play an important role in design energy-efficient mechanisms. Based on the abstraction level used to gather data, the taxonomy of power modeling of smartphones was provided by Ferroni et al. [34], as shown in Figure 9.4.

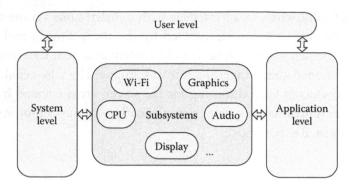

Figure 9.4 Energy consumption models on smartphones.

- *System level*: As the primary goal, the development of a model is able to compute the whole system power consumption at a given time. This category is further divided by considering the way in which power measurements are performed, that is, using external systems, internal hardware/software components, or via internal application programming interface (API) of the operating system.
- *Application level*: The power consumption is modeled per single application or per applications that have specific workload characteristics, for example an intensive network usage.
- *User level*: It includes all the works taking into account the user behavior and the impact of different usage patterns also in overall power consumption.

Standing on the side of the whole parts of the communication, it can be classified on system level, which is the development of a model able to compute the whole system power consumption at a given time. Sometimes from the perspective of system level, we have to consider device-specific condition [35]. Being able to generate a device-specific scalable power consumption model is crucial for understanding, designing, and implementing better mobile application software. Of course, the consumption measurement and modeling system must have a low footprint on its own while it collects system metrics; otherwise, it will affect the validity of the measurements. A system that significantly alters the power consumption of the device can cause changes that will invalidate the accuracy of the measurements and of the produced consumption model. A proper energy-accounting infrastructure will assist both application developers and smartphone users to extend the battery life of their devices and to make informed decisions about where to spend the remaining device power, potentially in real time.

Earlier research in power analysis for smartphone devices has produced power models with differing features and focuses. These research efforts can be differed in the following key aspects:

- Measurements and data collection of each device subsystem's power consumption can be obtained by leveraging existing system hooks in the Android kernel wake locks driver. This enables us to acquire both accurate measurements and at the same time reduces the power and computational requirements of model generation on the phone. Furthermore, analysis can be performed at different timescales because the measurements happen cumulatively and in real time.
- Operation states should be taken into account for individual subsystems in order to construct the model. However, the calculations are independent of the time interval used in creating the model and do not depend on assigning fixed power usage ratings to the operation states considered.
- All the primary device subsystems under different conditions that provide an in-depth understanding of how the power is consumed should be conducted on a smartphone.

On mobile systems, the battery discharge is reported by the kernel as a global value. There is no existing commercially available method to divide the measurements into per-subsystem or per-application readings on any platform. Therefore, a statistical model that breaks the power consumption has been created into subsystems.

Power used = CPU + display + graphics + GPS + audio + MIC + ... + Wi-Fi. (also in Figure 9.4.)

To accomplish this, subsystem usage measurements are collected from the operating system. In the sample implementation, the usage information is obtained from the power suspend driver, which was implemented by Google for the Android operating system. In addition, the CPU utilization, screen brightness, and pixel color strength on the screen are recorded by the researcher, and other such parameters, which are not controlled by a subsystem or device driver, but are controlled by the operating system itself.

Each subsystem is dependent on various factors that determine their power consumption. This includes the amount of time they are kept alive, but more importantly, the state of the hardware and other environmental factors. Various experiments were performed in order to tailor the aimed regression model to fit each of the subsystems.

In display subsystem, the effect of the display brightness level and pixel color on the rate of current discharge was studied. The results of the experiments concluded that red, green, and blue pixels each have a different effect on the rate of current discharge. Blue pixels cause a higher rate of current discharge than green pixels, and green pixels cause a higher rate of current discharge than red pixels. The discovery is that the effect of the remaining colors can be determined by performing calculations on their RGB components.

The effects of CPU utilization and CPU frequency on the rate of current discharge in CPU subsystem were also studied. The summary was concluded: the effect of the percentage CPU utilization on the rate of current discharge can be accurately represented linearly. In addition, the CPU frequency has a large effect on the rate of current discharge and the higher the CPU frequency, the greater the rate of current discharge.

The power consumption by the Wi-Fi subsystem during transmission depends on the rate of packets sent or received. Experiments were conducted to determine the exact relationship between these factors and to create equations to fit the findings into the aimed model. Generalizing the readings can be arrived at one single equation to govern all states of transmission.

The audio subsystem wake lock records the amount of time that the device was active. The conclusion is that the recorded wake lock uptime is the same value at all three volume levels.

Other subsystems can also be characterized using the same method as described for display, Wi-Fi, and audio.

If referred properly, the series of discoveries can be efficient enough to be used frequently in order to keep the subsystem constants current and adaptable according to environmental changes. It could help developers make better decisions for power efficiency, through knowing the variation in power consumption pertaining to the pixel strength of the display and the exact behavior of the Wi-Fi subsystem under various modes.

9.7.1 Specific Measurements to Record Energy Consumption on the Smartphone

Some energy consumption models are introduced in Section 9.7, nevertheless energy measurements are one of the essential aspects of it. Basically, there are three specific ways to record the total energy consumption on the smartphone [36]. One is to write an app to log the battery level and export the log to computer for analysis. It is widely used in IOS energy analysis. Another way is using battery measurement tool such as Monsoon, which provides the battery supply and directly records the power consumed by the phone with proper granularity, which is concluded that all samples are recorded at a frequency of 5 kHz with current scale accuracy of ±1%. The third way to measure energy consumption is to use a model introduced in OS to check the battery each application is taking. However, the numbers are normalized, and it does not provide the detailed power measurement. The total measurement category is further divided by considering the way in which power measurements are performed [34] (Table 9.4).

Table 9.4 Categories of Energy Measurements

MEASUREMENTS METHODS	WORK	CLASS	FEATURE	PREDICTED ERROR
System-level power model without laboratory measurements	[37]	APIs	Relies only on data provided by the OS and easily adapts to new hardware	2.62%
Context-aware system able to predict battery lifetime	[38]	APIs	Sum of single components	10%
Utilization-based and nonutilization-based power behavior	[39]	APIs	A fine-grained energy estimation, obtained by tracing applications system calls	3.5%~20.3% and 0.2%~3.6%
Self-modeling approach	[40]	ACPI	Collects data traces and uses iterative techniques	12% (1 Hz)
Unsupervised method	[41]	Energy Profiler tool	Based on genetic algorithms	95th percentile error of 0.313
Specifically designed for online generation of power models	[42]	Custom measurement tool	Modified to create power models from data coming from sensors within each device	4.1%

Many works are conducted by using only system APIs. Reference 37 presents a methodology to build a system-level power model without laboratory measurements. This approach can be adapted to any device for that it relies only on data provided by the OS. It has possibility to adapt the model to new hardware. In reality, when a new hardware component is installed into the mobile device, it is possible to add regression variables describing the activity levels of the new component, to define new test cases to stress it, and finally, to fit the new data with a regression model. They reported a median error of 2.62% in the power estimation in real mobile Internet services. Moreover, the power model proposed is independent from usage scenarios and can be used for runtime power estimation with reasonable accuracy.

A context-aware system that is able to accurately predict the battery lifetime is proposed in Reference 38. The energy consumed by each system component is considered dependent on its operational state and on the amount of time spent in the state. As a consequence, the system power consumption is modeled as the sum of single components. The proposed model was able to predict the remaining battery lifetime with a relative error of 10%.

Another remarkable contribution in Reference 39 proposes a new power modeling approach, which is able to capture both utilization-based and nonutilization-based power behavior. This methodology is based on a fine-grained energy estimation, obtained by tracing applications system calls. Its error for the whole application, with 1 s granularity, varies between 0.2% and 3.6%, compared to the error 3.5%–20.3% given by utilization-based model. Several works have been specifically designed to exploit interfaces already available on specific devices or OSs, such as advanced configuration and power interface (ACPI).

Reference 40 proposed an interesting system called Sesame. It is a self-modeling approach to build high-rate mobile system models without any need for external measurements. Sesame first collects data traces; then, the measurement model is built using two iterative techniques: model molding and predictor transformation. The model is composed by a set of submodels, each corresponding to a different system configuration in the model generation.

In Reference 41, an unsupervised method to measure device energy consumption is proposed, based on genetic algorithms. This solution relies on online measurements through system APIs to generate on-demand models. Data retrieved through the available APIs include the battery voltage, current, and other information. The model predicts power consumption with a 95th percentile error of 0.313. This error increases when the application is released *into the wild* with more frequent and overlapping feature usage.

The last remarkable contribution in the energy measurement is by Zhang et al. [42], which is specifically designed for online generation of power models. With the help of external tools, they started their study analyzing each component separately. The power consumption of major components affects the system independently. Different devices have power models pretty far from each other. In order to measure different devices, they create power models starting from data coming from sensors within each device. The power measurement model built with PowerBooter is accurate to within 4.1% of measured values for 10-second intervals.

In brief, the existed energy measurement methods have been divided into different kinds according to real conditions. Proper energy models along with fine measurement methods contribute to the accurate metric of mobile handsets so as to design energy-saving technologies for proximity service.

9.7.2 Trade-Off between Energy Efficiency and Performance

As far as ProSe (especially for location based service [LBS]) schemes concern, besides considering the energy consumption, performance of localization is the most crucial factor that should be emphasized. Specifically, localization performance has different metrics such as accuracy, fairness, latency, stability, robustness, and so on. Although localization energy can be saved through the proposed technical principles or context assistance, it is meaningless if localization schemes only pursue for low energy consumption without consideration of performance. Naturally, trade-off between energy efficiency and localization performance should be investigated.

Localization accuracy (commonly measured as the error range between the estimated position and the real position) is one of the popular and significant metrics of localization performance. Intuitively, energy and accuracy is a pair of contradiction, as high localization accuracy would consume much energy, vice versa. Therefore, almost all schemes take energy-accuracy trade-off into consideration, when designing energy-efficient solutions in localization and tracking.

In literature, various energy-efficient schemes of localization and tracking utilize different metrics to quantitatively measure the trade-off between energy and accuracy. This section explicitly classifies those works into two categories: One is based on the weighted combination of localization accuracy and energy consumption; another is based on the ratio of the variation of energy to accuracy.

Most localization schemes just qualitatively illustrate the energy-accuracy trade-off (e.g., graphically show the scheme can maximize accuracy and meanwhile, minimize energy simultaneously), but lack the quantitative metric to characterize the trade-off. One technique to solve such joint optimization problems is to consider a combined objective function, such as a linearly weighted sum of energy and the reciprocal of accuracy, and to optimize the combined objective:

$$C(t) = E(t) + \frac{\lambda}{A(t)} \tag{9.1}$$

where:

$C(t)$ is the objective function to be minimized

$E(t)$ represents energy used

$A(t)$ represents the location accuracy achieved

λ is a scalar weight that reflects the relative importance of energy and accuracy for a given scenario

According to the objective function, various localization schemes can compute different values of $C(t)$ and can figure out what is the best trade-off of energy and accuracy. It should be noted that References 43 and 44 belong to this category.

Another measurement type is to keep one of the dimensions (energy or accuracy) unchanged, and to compare another to illustrate the energy efficiency of various localization schemes. References 13 and 14 belong to this category.

Quantitatively in this type, we can set a more general metric energy to accuracy ratio (EAR) as

$$\text{EAR} = \left| \frac{E_2 - E_1}{A_2 - A_1} \right| \qquad (9.2)$$

where:

E_i $(i = 1,2)$ represents energy consumption of the compared schemes

A_i $(i = 1,2)$ represents the corresponding localization accuracy

Intuitively, for any localization schemes, the larger EAR is, the poorer energy efficiency is.

9.8 Conclusion

Smartphones have become a commonplace during the past years. Applications and services using Internet technologies previously available only for devices with limited mobility, such as laptops, are now readily accessible on devices fitting our pockets. Unfortunately, the battery performance and energy efficiency of small-scale mobile devices have not progressed at a desired pace, resulting in users having to frequently recharge their devices and to pay considerable attention to battery consumption.

Considering the special features of ProSe, this chapter provides the energy-efficient architecture in which corresponding energy-efficient technologies are investigated from five aspects: underlying networking technologies, localization schemes, leveraging application and architecture features, utilizing users' context, and energy consumption models and measurement. Furthermore, besides exploring the specific schemes pertaining to each aspect, this chapter offers a perspective for research and applications and could provide some guidelines for developers to design energy-efficient ProSe.

References

1. Wang, Y.F., J. Tang, Q. Jin, and J.H. Ma. BWMesh: A multi-hop connectivity framework on android for proximity service. In: *Proceedings of the 12th IEEE International Conference on Ubiquitous Intelligence and Computing (UIC)*, Beijing, China, October 10–14, 2015, pp. 278–283.

2. Crocker, P.B. Proximity based mobile social networking: Applications and technology, making new connections in the physical world. Smith's Point Analytics, November 2011.
3. LTE Direct: The case for Device-to-Device proximate discovery. Qualcomm Technologies, Sandiego, CA, February 18, 2013.
4. Vallina-Rodriguez, N. and J. Crowcroft. Energy management techniques in modern mobile handsets. *IEEE Communications Surveys & Tutorials*, 2013; 15(1): 179–198.
5. Camps-Mur, D., A. Garcia-Saavedra, and P. Serrano. Device to device communications with WiFi Direct: Overview and experimentation. *IEEE Wireless Communications*, 2013; 20(3): 96–104.
6. Lim, K.W., W.S. Jung, H. Kim, J. Han, and Y.B. Ko. Enhanced power management for Wi-Fi Direct. In: *Proceedings of the IEEE Wireless Communications and Networking Conference (WCNC)*, Shanghai, China, April 7–10, 2013, pp. 123–128.
7. Trifunovic, S., A. Picu, T. Hossmann, and K.A. Hummel. Slicing the battery pie: Fair and efficient energy usage in device-to-device communication via role switching. In: *Proceedings of the 8th ACM MobiCom Workshop on Challenged Networks*, New York, 2013, pp. 31–36.
8. Mumtaz, S., H. Lundqvist, K.M.S. Huq, J. Rodriguez, and A. Radwan. Smart direct-LTE communication: An energy saving perspective. *Elsevier Ad Hoc Networks*, 2014; 13: 296–311.
9. Kravet, R.H. Enabling social interactions off the grid. *IEEE PERVASIVE Computing*, 2012; 11(2): 8–11.
10. Bareth, U. and A. Kupper. Energy-efficient position tracking in proactive location-based services for smartphone environments. In: *Proceedings of the IEEE 35th Annual Computer Software and Applications Conference*, Munich, Germany, July 18–22, 2011, pp. 516–521.
11. Radu, V., P. Katsikouli, R. Sarkar, and M.K. Marina. A semi-supervised learning approach for robust indoor-outdoor detection with smartphones. In: *Proceedings of the 12th ACM Conference on Embedded Network Sensor Systems*, New York, 2014, pp. 280–294.
12. Lee, Y., J. Lee, D.S. Kim, and H. Choo. Energy-efficient adaptive localization middleware based on GPS and sensors for smart mobiles. In: *Proceedings of the IEEE Fourth International Conference on Consumer Electronics (ICCE)*, Berlin, Germany, 2014, pp. 126–130.
13. Wu, C., Z. Yang, Y. Xu, Y. Zhao, and Y. Liu. Human mobility enhances global positioning accuracy for mobile phone localization. *IEEE Transactions on Parallel and Distributed Systems*, 2015; 26(1): 131–141.
14. Bulut, M.F. and M. Demirbas. Energy efficient proximity alert on android. In: *Proceedings of the IEEE International Conference on Pervasive Computing and Communications Workshops*, San Diego, CA, March, 18–22, 2013, pp. 18–22.
15. Zhang, L., J. Liu, H. Jiang, and Y. Guan. SensTrack: Energy-efficient location tracking with smartphone sensors. *IEEE Sensors Journal*, 2013; 13(10): 3775–3784.

16. Kwak, J., J. Kim, and S. Chong. Energy-optimal collaborative GPS localization with short range communication. In: *Proceedings of the 11th International Symposium on Modeling & Optimization in Mobile, Ad Hoc & Wireless Networks,* Tsukuba Science City, Japan, May 13–17, 2013, pp. 256–263.

17. Brouwers, N., M. Zuniga, and K. Langendoen. Incremental Wi-Fi scanning for energy-efficient localization. In: *Proceedings of the IEEE International Conference on Pervasive Computing and Communications,* Budapest, Hungary, March 24–28, 2014, pp. 156–162.

18. Qi, Y., C. Yu, Y.J. Suh, and S.Y. Jang. GPS tethering for energy conservation. In: *Proceedings of the IEEE Wireless Communications and Networking Conference (WCNC),* New Orleans, LA, March, 9–12, 2015, pp. 1320–1325.

19. Gao, Y., J. Niu, R. Zhou, and G. Xing. ZiFind: Exploiting cross-technology interference signatures for energy-efficient indoor localization. In: *Proceedings IEEE INFOCOM,* Turin, Italy, April 14–19, 2013, pp. 2940–2948.

20. Niu, J., B. Wang, L. Shu, T.Q. Duong, and Y. Chen. ZIL: An energy-efficient indoor localization system using ZigBee radio to detect WiFi fingerprints. *IEEE Journal on Selected Areas in Communications,* 2015; 33(7): 1431–1442.

21. Poosamani, N. and I. Rhee. Towards a practical indoor location matching system using 4G LTE PHY layer information. In: *Proceedings of the IEEE International Conference on Pervasive Computing and Communication Workshops,* St. Louis, MI, March 23–27, 2015, pp. 284–287.

22. Abdellatif, M., A. Mtibaa, K.A. Harras, and M. Youssef. GreenLoc: An energy efficient architecture for WiFi-based indoor localization on mobile phones. In: *Proceedings of the IEEE International Conference on Communications,* Budapest, Hungary, June 9–13, 2013, pp. 4425–4430.

23. Gu, Y. and F. Ren. Energy-efficient indoor localization of smart hand-held devices using Bluetooth. *IEEE Access,* 2015; (3):1450–1461.

24. Pasricha, S., V. Ugave, C.W. Anderson, and Q. Han. LearnLoc: A framework for smart indoor localization with embedded mobile devices. In: *Proceedings of the 10th International Conference on Hardware/Software Codesign and System Synthesis (CODES),* Amsterdam, the Netherlands, October 4–9, 2015, pp. 37–44.

25. Dang, C. and K. Sezaki. A sparse particle filter for indoor localization using mobile phones. *IEICE Communications Express,* 2014; 3(4): 144–149.

26. Radu, V., P. Katsikouli, R. Sarkar, and M.K. Marina. A semi-supervised learning approach for robust indoor-outdoor detection with smart-phones. In: *Proceedings of the 12th ACM Conference on Embedded Network Sensor Systems,* New York, 2014, pp. 280–294.

27. Zou, H., H. Jiang, Y. Luo, J. Zhu, X. Lu, and L. Xie. BlueDetect: An iBeacon-enabled scheme for accurate and energy-efficient Indoor-outdoor detection and seamless location-based service. *Sensors,* 2016; 16(2): 268.

28. Vallina-Rodriguez, N. and J. Crowcroft. Energy management techniques in modern mobile handsets. *IEEE Communications Surveys & Tutorials*, 2013; 15(1): 179–198.

29. Vo, Q.D., D. Coelho, K. Mueller, and P. De. WhereAmI: Energy efficient positioning using partial textual signatures. In: *Proceedings of the IEEE International Conference on Mobile Services (MS)*, New York, June 27–July 2, 2015, pp. 9–16.

30. Bojic, I., V. Podobnik, and A. Petric. Swarm-oriented mobile services: Step towards green communication. *Expert Systems with Applications*, 2012; 39(9): 7874–7886.

31. Coll-Perales, B., J. Gozalvez, and V. Friderikos. Store, carry and forward for energy efficiency in multi-hop cellular networks with mobile relays. In: *Proceedings of IFIP Wireless Days (WD)*, Valencia, Spain, November 13–15, 2013, pp. 1–6.

32. Rachuri, K.K. Smartphones based social sensing: Adaptive sampling, sensing and computation offloading. University of Cambridge Computer Laboratory St. John's College, Cambridge, England, 2012.

33. Chen, X., K.W. Nixon, H. Zhou, Y. Liu, and Y. Chen. FingerShadow: An OLED power optimization based on smartphone touch interaction. In: *Proceedings of the 6th Workshop on Power-Aware Computing and Systems (HotPower)*, Broomfield, CO, October 5, 2014.

34. Ferroni, M., A. Cazzola, F. Trovò, D. Sciuto, and M.D. Santambrogio. On power and energy consumption modeling for smart mobile devices. In: *Proceedings of the 12th IEEE International Conference on Embedded and Ubiquitous Computing (EUC)*, Milano Italy, August, 26–28, 2014, pp. 273–280.

35. Murmuria, R., J. Medsger, A. Stavrou, and J.M. Voas. Mobile application and device power usage measurements. 2012 *IEEE Sixth International Conference on Software Security and Reliability (SERE)*, Gaithersburg, MD, June 20–22, 2012, pp. 147–156.

36. Liu, S. Potential of opportunistic relaying performance study across wireless interfaces on smartphones. Ph.D dissertation, University of Notre Dame, 2014.

37. Xiao, Y., R. Bhaumik, Z. Yang, M. Siekkinen, P. Savolainen, and A. Yla-Jaaski. A system-level model for runtime power estimation on mobile devices. In: *Proceedings of the 2010 IEEE/ACM International Conference on Green Computing and Communications & International Conference on Cyber, Physical and Social Computing (CPSCom)*, Hangzhou, China, December 18–20, 2010, pp. 27–34.

38. Zhao, X., Y. Guo, Q. Feng, and X. Chen. A system context-aware approach for battery lifetime prediction in smart phones. In: *Proceedings of the 2011 ACM Symposium on Applied Computing (SAC)*, New York, 2011, pp. 641–646.

39. Kang, J.M., S.S. Seo, and J.W.K. Hong. Usage pattern analysis of smartphones. In: *Network Operations and Management Symposium (APNOMS)*, Taipei, September 21–23, 2011, pp. 1–8.

40. Dong, M. and L. Zhong. Self-constructive high-rate system energy modeling for battery-powered mobile systems. In: *Proceedings of the 9th International Conference on Mobile Systems, Applications, and Services (MobiSys)*, New York, 2011, pp. 335–348.
41. Kjærgaard, M.B. and H. Blunck. Unsupervised power profiling for mobile devices. In: *Proceedings of the 8th International Conference on Mobile and Ubiquitous Systems: Computing, Networking and Services (Mobiquitous)*, Copenhagen, Denmark, December 6–9, 2011, pp. 138–149.
42. Zhang, L., B. Tiwana, Z. Qian, Z. Wang, R.P. Dick, Z.M. Mao, and L. Yang. Accurate online power estimation and automatic battery behavior based power model generation for smartphones. *IEEE/ACM/IFIP International Conference on Hardware/Software Codesign and System Synthesis (CODES+ISSS)*, New York, October 24–29, 2010, pp. 105–114.
43. Jurdak, R., P. Corke, A. Cotillon, D. Dharman, C. Crossman, and G. Salagnac. Energy-efficient localization: GPS duty cycling with radio ranging. *ACM Transactions on Sensor Networks (TOSN)*, 2013; 9(2): 23.
44. Neishaboori, A. and K. Harras. Energy saving strategies in WiFi indoor localization. In: *Proceedings of the 16th ACM International Conference on Modeling, Analysis & Simulation of Wireless and Mobile Systems (MSWiM)*, New York, 2013, pp. 399–404.

PART III

APPLICATIONS OF PROXIMITY SERVICE

10
VEHICULAR SOCIAL NETWORKS

10.1 Introduction

As a promising application of device-to-device (D2D)-based proximity service, vehicular social networks (VSNs) have received more and more attentions from academia and governments recent years, VSNs intentionally exploit social characteristics and human behaviors, to share and disseminate data among vehicles and drivers, in a context of temporal and spatial proximity on the roads.

In a sense, VSNs can be regarded as a significant extension to the traditional vehicular ad hoc networks (VANETs), which not only include traditional vehicle-to-vehicle (V2V) and vehicle-to-roadside infrastructure (V2I) communication protocols, but more importantly accommodate and exploit the human factors, for example, human mobility, selfishness, and user preferences affecting vehicular connectivity. In other words, VSNs are formed when vehicles (individuals) *socialize* and share common interests [1]. VSNs are emerging as a technology to integrate capabilities of new generation wireless networking to vehicles and human to efficiently provide ubiquitous connectivity either on the road to mobile users, or connected to the outside world through other networks at home or work place. It is believed that it will play a vital function in intelligent transportation systems (ITS), which would be a realization in the near future. Different from the existing mobile social networks (MSNs), where the participants are mainly human beings who interact with one another using smartphones and mobile devices, the participants of VSNs are heterogeneous, and include vehicles, devices onboard vehicles, as well as drivers and passengers. Thus, three types of relationships are found in VSNs: (1) between humans, (2) between humans and machines, and (3) between machines and machines. Also, due to the features of

Table 10.1 Comparison between Traditional MSNs and VSNs

	NORMAL MSNs	VSNs
Types of social relationships	Between human and human	• Between human and human • Between human and machine • Between machine and machine
Network architecture	Internet and opportunistic networks	Internet and VANETs
Dynamic rate of network topology	Random; depends on the specific scenarios	Usually very high due to the movement speed of vehicles
Energy constrains	High constrains and sensitive for most mobile devices	It may not be sensitive, as mobile devices could be charged in vehicles
Real-time requirements	Normal; limited time-latency is acceptable	High, especially for safety applications

vehicles and the special application environments, the VSN systems have unique characteristics that distinguish them from normal MSN systems, as summarized in Table 10.1 [2].

It is shown that knowledge of the social interactions of nodes can help to improve the performance of mobile systems [3]. Therefore, it is anticipated that VSN applications can be widely used in many field. The most three common types of applications over VSNs are: (1) safety improvements: applications that improve the safety of people on the roads by notifying the occupants of vehicles about any dangerous situation in their neighborhoods; (2) traffic management: applications that provide users with up-to-date traffic information enabling them to improve route selection, traffic efficiency, and driving behavior; (3) infotainment: applications that enable the dissemination, streaming, downloading, or sharing of location-dependent information such as advertisement, and multimedia files such as audio and video over the VSN [4].

VSN systems are built on top of VANETs that provide connectivity between users and devices participating in the VSN as well as the Internet at large. Although cellular networks can provide such connectivity, the cost may be too high, and the latency may be too large. Instead, VANETs may be established to connect the users and devices onboard vehicles that are physically close to each other [5,6]. For example, drivers or passengers may carry smartphones with various forms of wireless communication capabilities (e.g., cellular interfaces, Wi-Fi Direct), or vehicles may be equipped with embedded computing and networking systems, which support the use of dedicated short-range communications (DSRC) for vehicle-to-pedestrian (V2P) [7], V2V, and vehicle-to-roadside (V2R) communications, with deployed roadside infrastructures [2].

This chapter is organized as follows: Section 10.2 overviews the basic concepts, architectures, and characteristics of VSNs. Various VSNs applications are presented in Section 10.3. Section 10.4 discusses the three main challenges in VSNs: mobility models, routing protocols, and security issues. Finally, we conclude this chapter and point out some future directions.

10.2 Overview on VSNs

10.2.1 Main Components

As shown in Figure 10.1, the main physical components of VSNs include application unit (AU), onboard unit (OBU), and roadside unit (RSU). Usually RSU hosts an application that provides services, and OBU is a peer device that uses the services. The application may reside in the RSU or in the OBU, and the device that hosts the application is called provider, whereas the device using the application is described as the user. Each vehicle is equipped with an OBU and a set of sensors to collect and process the information, then send it to other vehicles or RSUs. Vehicle also carries one or more AUs that use the applications

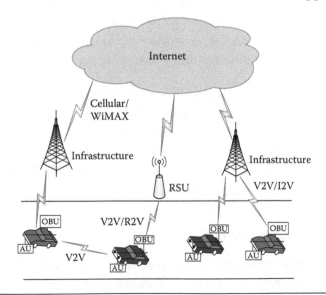

Figure 10.1 Main physical components in VSNs.

provided by the provider using OBU connection capabilities. The RSU can also connect to the Internet or other servers, which allows AU to connect to the Internet [8].

10.2.1.1 Onboard Unit An OBU is a wave device usually mounted onboard a vehicle, used for exchanging information with RSUs or other OBUs. It includes a network device for short-range wireless communication based on IEEE 802.11p radio technology. It may additionally include another network device for nonsafety applications based on other radio technologies such as IEEE 802.11a/b/g/n. The OBU connects to the RSU or OBUs through a wireless link based on the IEEE 802.11p radio frequency channel. It also provides a communication service to the AU and forward data on behalf of other OBUs on the network. The main functions of the OBU are wireless radio access, ad hoc and geographical routing, network congestion control, reliable message transfer, data security, and IP mobility.

10.2.1.2 Application Unit The AU is the device equipped within the vehicle that utilizes applications provided by the provider with the help of OBU. The AU can be a dedicated device for safety applications or a normal device such as a personal digital assistant (PDA) to run the Internet. AU can be connected to OBU through a wired or wireless connection or can be resided in OBU as a single physical unit. The distinction between the AU and the OBU is logical. The AU communicates with the network solely via OBU, which takes the responsibility for all mobility and networking functions.

10.2.1.3 Roadside Unit The RSU is a wave device usually fixed along the roadside or in dedicated locations (junctions, parking spaces, etc.). The RSU is equipped with one network device for a DSRC based on IEEE 802.11p radio technology, and it can also be equipped with other network devices so as to be used for the purpose of communication within the infrastructural network.

10.2.2 Network Architectures

Communication types can be categorized into four types [9,10]. The fundamental functions of each communication type are shown in Figure 10.2.

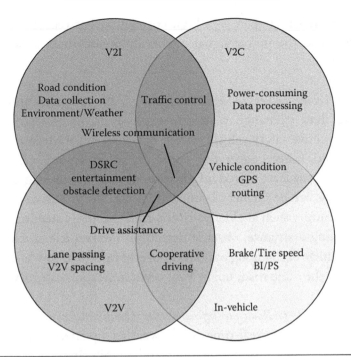

Figure 10.2 Network architectures in VAENTs.

10.2.2.1 In-Vehicle Communications It is a necessary and important component in VSN research. In-vehicle communication system can detect a vehicle's performance and a driver's fatigue and drowsiness, which is critical for driver and public safety. This is becoming a major stream of research in the area of intelligent vehicle systems. Intelligent in-vehicle systems, mainly, onboard equipments (OBE) collect information from the driver or vehicle, analyze, and classify the data collectively to predict or detect driver fatigue.

Machine learning techniques are extensively used for such data classification. This platform collects standard vehicle information such as the speed, pressure on the brake or gas pedal, steering wheel rotation, and global positioning system (GPS) routing.

In addition to standard vehicle information, driver behavioral information (BI) such as facial expression (e.g., blink rate, yawning, eyebrow raise, chin drop, and head movements) can be collected and analyzed. Even physiological signals (PS) such as heart-rate variability and electroencephalogram (EEG) signal behavior can be sampled to determine the drowsiness (nonalert) level of the driver. Researchers have reported that there is a high correlation between the level of alertness and the power signal in the alpha and theta band of the EEG signal.

10.2.2.2 Vehicle-to-Vehicle Communications V2V communications can provide data exchange platform, expand driver assistance, and facilitate active safety vehicle system development. Driver assistance is provided using cooperative communication among vehicles to adaptively broadcast shared information or warning messages for the driver. This can be further customized for specific groups of people in the community such as elderly drivers. Lane keeping, steering control and parking assistance, obstacle detection, intervehicle spacing, and driver/vehicle exchanging optional or useful information while traveling along the same road, fall in V2V communication category.

10.2.2.3 Vehicle-to-Roadside Infrastructure Communications V2I communications enable real-time traffic/weather updates for the driver and provide environmental sensing and monitoring. This platform ultimately enables driver safety by providing the right information at the right time, such as speed limit, and weather condition information collected using various roadside sensors. This platform is capable of automatically informing the driver of hazardous road conditions. Sensed data of road surface and spacing can be transmitted to vehicles over intervehicle communication using the 5.9 GHz DSRC. For example, wet conditions of the road surface can be detected based on processing the polarized light from the road surface with a vision system. Anticollision detection systems based on vehicle and obstacle spacing using adaptive cruise control are another V2R communication application.

V2I communications enable real-time weather/traffic updates for the driver, which ultimately makes the transportation systems more informative. As one option, researchers in academia and industry have shown that the V2I transmitter can actually be placed on the vehicle's tire.

10.2.2.4 Vehicle-to-Cloud Communications It means that vehicles may communicate via wireless broadband mechanisms such as 3G/4G. This type of communication will specifically be useful for active driver assistance and vehicle tracking in network fleet management. In particular, the smartphones/gadgets could be used as a gateway in this platform to send/receive data to and from a central monitoring data centers connected to the broadband cloud. Vehicle-to-cloud (V2C) networks can provide useful information in two ways:

- Outgoing data that may include vehicle-centric information (e.g., speed, global positioning, routing, device functionality, and performance), driver-centric information such as driver's specific behavior (e.g., drowsiness and length of continuous driving), audio/video, and others. All of this data may be optionally forwarded to a central monitoring server for further analysis and storage.
- Incoming data that may include receiving data from a central office for various communications with driver or vehicle system.

10.2.3 Characteristics

VSN has its own unique characteristics compared with other wireless ad hoc networks [8,11].

10.2.3.1 Predictable Mobility
In VSNs, vehicles are usually constrained by prebuilt highway, roads, and street, by the requirement to obey road signs, traffic lights and to respond to other moving vehicles, so given the speed and the street map, the future position of the vehicle can be predicated.

10.2.3.2 Highly Dynamic Topology and Frequently Disconnected Network
Due to high speed of movement of vehicles and driver behaviors affected by various situations on the roads, VSN network topology and link connection change frequently. The lifetime of the link between vehicles is affected by the radio communication range and the direction of the vehicles; thus increasing the radio communication range leads to an increase in the lifetime of the link. The lifetime of the link between vehicles moving in opposite directions is very short compared with case in the same direction. The rapid changes in link connectivity cause the

effective network diameter to be small, while many paths are disconnected before they can be utilized.

10.2.3.3 Various Communications Environments VSNs usually operated in two typical communications environments: highway traffic scenarios and urban. The former is relatively simple and straightforward (e.g., constrained one-dimensional movement), whereas in urban the condition becomes much more complex. The streets in urban are often separated by buildings, trees, and other obstacles. Therefore, there is not always a direct line of communications in the direction of intended data communication.

10.2.3.4 Hard Delay Constraints In some VSN's applications, the network does not require high data rates but has hard delay constraints. For example, in an automatic highway scene, when brake event happens, the message should be transferred and arrived in a certain time to avoid car crash. In this kind of applications, instead of average delay, the maximum delay will be crucial.

10.2.3.5 Sufficient Energy, Storage, and High Computational Ability Because the nodes are vehicles, they have ample energy and can be equipped with a sufficient number of sensors and computational resources (such as processors, a large memory capacity, advanced antenna technology, and GPS). These resources increase the computational capacity of vehicles.

10.3 Applications of VSNs

VSN applications are typically classified in active road safety applications, traffic efficiency and management applications, and convenience and comfort/infotainment applications [8,12].

10.3.1 Active Road Safety Applications

This kind of application aims to avoid the risk of car accidents and to make safe driving by distributing information about hazards and obstacles. The basic idea is to broaden the drivers' range of perception, allowing them to response much quicker.

10.3.1.1 Intersection Collision Avoidance Improving intersection collision avoidance systems will lead to the avoidance of many road accidents. This system is based on I2V or V2I communication. The sensors at infrastructure gather process and analyze the information from the vehicles moving close to the intersection. Based on the analysis of data, if there is a probability of an accident or a hazardous situation, warning message will be sent to the vehicles to warn them about the accident, so that they can take appropriate action to avoid it. There are many applications that belong to intersection collision avoidance systems, all of them relying on I2V or V2I communication.

- *Warning about violating traffic signal*: It is designed to send a warning message to vehicles to warn drivers about a dangerous situation, if the driver does not stop. The messages sent depend on several factors: traffic status, timing, the vehicle's speed, the vehicle's position, and the road surface.
- *Warning information assistance*: This application includes left turn assistant, stop sign movement assistant, and intersection collision avoidance assistant. It aims to warn drivers about hazardous situations that may occur when their vehicles pass by an intersection. Data information collected from road sensors was sent to the vehicles, so that effective measures can be taken to avoid the accident.
- *Warning about blind merge detection*: It aims to prevent a collision at the merge point where the visibility is poor. The system collects and processes the data at the intersection and if an unsafe situation was detected, it will generate a warning message to vehicles; at the same time, it will warn the remaining vehicles on the road.
- *Pedestrian crossing information designated intersection*: The main goal of this application is to warn drivers when a pedestrian goes across the road. Information about pedestrians and the vehicles was collected by sensors installed nearby and was sent to the system. Meanwhile the system analyzes and processes the collected data. If somebody has pressed the located walk button and if there is a possibility of collision to occur, the systems will send a warning message to the vehicles.

10.3.1.2 Sign Extension   The main goal of this application is to alert inattentive drivers to signs placed on the roadsides in order to prevent accidents. Most of the sign extension applications rely on I2V communication and with a communication range of 100–500 m.

- *In-vehicle signage*: It relies on the RSUs fixed in a specific area, for example, in a school zone, hospital zone or animal passing area to send an alert message to vehicles approaching the area.

- *Curve speed warning*: It relies on the RSUs fixed before the curve to disseminate a message to approaching vehicles alerting them about the location of the curve and about the limited speed to pass through the curve safely.

- *Low parking structure and bridge warning*: It is designed to alert the driver neglecting the minimum height of the park they are trying to enter or the height of the bridge they are trying to pass under, by sending a warning message to the vehicle via an RSU installed close to the parking facility, bridge, then the OBU can determine whether it is safe to enter or pass the structure.

- *In-vehicle amber alert*: It depends on I2V communication and sends amber warning messages to vehicles. The messages are disseminated when the police confirm that there is a vehicle involved in the crime, and it is issued to all vehicles in the area, except for the suspect vehicle.

- *Postcrash warning*: It aims to prevent potential accidents. A vehicle that is disabled because of foggy weather or an accident will send warning messages to other vehicles that are traveling in the same direction or the opposite direction by V2V or V2I communications to inform them about the status information and its location, heading, and direction.

10.3.1.3 Information from Other Vehicles   This type of applications relies on V2V communication or I2V communication or both to perform applications functions, and the warning message may be triggered based on event-driven or periodic messages.

- *Cooperative forward collision warning*: It accomplishes the goals to assist a vehicle in avoiding being involved in an accident with the vehicle traveling ahead of it. The system uses V2V communication with a multihop technique to send warning messages about the situation to a driver. These messages include information of position, direction, velocity, and acceleration, each vehicle processes the information after receiving it and decides on the danger level then forward it to other vehicles.

- *Vehicle-based road condition warning*: It is based on V2V communication; vehicle collects sufficient information about the road status via its sensors, processes it to determine the road situation in order to initiate a warning to the driver, and sends a warning message to other vehicles.

- *Blind spot warning*: It alerts the driver if he or she decides to change lane and if there is a vehicle in the blind spot. It uses V2V communication to send warning messages to other vehicles on the road.

- *Highway merge assistant*: It prevents accidents from occurring on the condition that a vehicle is attempting to merge on the highway. If the vehicles are moving on a ramp or if there are other vehicles in the vehicle's blind spot, then the system should initialize a warning message and should send it to other vehicles informing them about the speed, position, and direction of the vehicle to prevent the accident.

- *Highway/rail collision warning*: It aims to prevent vehicles from being involved in an accident with train by using I2V communication at intersections. The warning message may come from the train directly in order to warn drivers to take corrective action.

- *Cooperative vehicle–highway automation system*: This system controls the velocity and position of vehicles and makes them travel on the highway as a platoon relying on V2V and V2I communication. The system collects vehicular information and merges it with the map data in order to control the vehicle's movements and enhance the traffic flow on the highway.

10.3.2 Traffic Efficiency and Management Applications

This category focuses on optimizing flows of vehicles by reducing travel time, avoiding traffic jam situations and makes the best choice when an unexpected event happened (such as choosing an appropriate speed to avoid waiting in the crossroad or changing the route when an accident occurred ahead of the road) so that it improves the efficiency. Applications similar to enhanced route guidance/navigation, traffic light optimal scheduling, and lane merging assistance are intended to optimize routes [8].

- *Approaching emergency vehicle warning*: It is designed to satisfy the requirements to provide a clear road allowing emergency vehicles to reach their destinations without waiting in traffic. It was accomplished by disseminating alert messages to vehicles relying on one-way V2V communication. The message includes information about the emergency vehicle's velocity, direction, position lane information, and driven route.
- *Emergency vehicle signal preemption*: Available infrastructures at each intersection support emergency vehicles passing through crossroad at a high-priority V2I communication. The device will turn traffic lights to green when a emergency vehicle approaches the traffic signals, which will minimize the driving time and will reduce the possibility of an accident to occur involving it.
- *Wrong-way driver warning*: It is designed to alert a vehicle whether it is traveling in the correct way. By using V2V communication, a vehicle traveling the wrong way can alert the other vehicles around it via warning messages to prevent it happen again.
- *Proper speed to go across the intersection*: This application is designed to send a useful message to vehicles to help the drivers arrange a proper speed to escape the waiting time in the traffic light on the intersection. The message that is sent depends on several factors such as traffic status, timing, the vehicle's speed, the current distance between the vehicle and the stop sign, and the road surface.
- *Change the route to avoid the accident*: The main goal of this application is to remind drivers about any accidents or

collisions that have happened in front of the road. Relying on V2V communication or the V2I communication, the drivers who have received the message can change their route in case of being stopped by the accidents.

10.3.3 Comfort and Infotainment Applications

Concerning on comfort and infotainment applications, he or she have very different communication requirements, from no special real-time requirements of traveler information support applications, to guaranteed QoS needs for multimedia and interactive entertainment applications.

The aim of comfort and infotainment applications is to offer convenience and comfort to drivers or passengers. Various traveler information applications belong to this category such as a platform for peer-to-peer file transfer and gaming on the road, a real-time parking navigation system to inform drivers of any available parking space, and digital billboards through vehicular networks for advertisement (e.g., local information regarding restaurants, hotels and in general, point of interest).

Because Internet access can be provided through V2I communications, business activities can be performed as usual in a vehicular environment, realizing the notion of mobile office. On the road, media streaming between vehicles also can be available, making long travel more pleasant and an acquired goal, the moving movie to embed human–vehicle interfaces, such as color reconfigurable head-up and head-down displays, and large touch screen active matrix liquid crystal displays (LCDs), for high-quality video-streaming services.

Table 10.2 gives a summary of the various applications of VSNs.

10.4 Challenges of VSNs

10.4.1 Mobility Model

Evaluating large-scale VSN solutions is a great challenge, because real experiments are not feasible and there are no large-scale test beds available. Thus the most used evaluation technique is simulation, which requires a mobility model to provide relatively accurate results. Mobility model as one of the key components in VSN's simulation

Table 10.2 Summary of Various VSNs Applications

Applications of VSNs	Active road safety application	**Intersection collision avoidance** Warning about violating traffic signal Warning information assistance Warning about blind merge detection Pedestrian crossing information designated intersection **Sign extension** In-vehicle signage Curve speed warning Low parking structure and bridge warning In-vehicle amber alert Postcrash warning **Information from other vehicles** Cooperative forward collision warning Vehicle-based road condition warning Blind spot warning Highway merge assistant Highway/rail collision warning Cooperative vehicle-highway automation system
	Traffic efficiency and manage application	Approaching emergency vehicle warning Emergency vehicle signal preemption Wrong way drive warning Proper speed to go through the intersection Change the route to avoid the accident

	Convenience and comfort/infotainment	A platform for peer-to-peer file transfer
		Gaming on the road
		A real-time parking navigation system
		Digital billboards (local information regarding
		  restaurants, hotels, and in general, point of
		  interest)
		Mobile office
		On the road media streaming (moving movie for
		  high-quality video-streaming services)

provides the position of nodes in the topology at any instant of time, which mainly affects network throughput and connectivity. In other words, mobility models represent real-world scenarios for VSN and play a vital role in the performance evaluation of routing protocols. In addition, more accurate and correct results would be acquired by mobility model when there is more similarity to the realistic vehicular mobility. So in this part, we present several realistic vehicular mobility traces first. Then, some typical vehicular mobility models were introduced.

10.4.1.1 Available Vehicular Mobility Trace

- *Microscopic vehicular mobility trace of Europarc roundabout, Creteil, France* [13]: The vehicular mobility dataset is mainly based on the real data collected by the General Departmental Council of Val-de-Marne (94) in France.

  Different simulations tools and models are brought together to characterize and synthesize this trace:
  - The street layout of Creteil roundabout area is obtained from the OpenStreetMap (OSM) database
  - The traffic demand information on the traffic flow is derived from car counting and camera video analysis
  - The traffic assignment of the vehicular flows is performed by Gawron's dynamic user assignment algorithm, included in the SUMO—simulation of urban mobility—simulator
  - The traffic light mechanism is derived from sequence adaptation manually documented in the regional transportation system

  The resulting synthetic trace includes a roundabout with 6 entrances/exits, 2 or 3 lane roads, 1 bus road, 4 changing-lane spots, and 15 traffic lights. It comprises around 10,000 trips, over rush hour periods of 2 hours in the morning (7–9 a.m.) and 2 hours in the evening (5–7 p.m.).

  The trace is available at: http://vehicular-mobility-trace. github.io/.

- *Vehicular mobility trace of the city of Cologne, Germany* [14]: This vehicular mobility dataset is mainly based on the data made available by the TAPASCologne project. TAPASCologne is an initiative by the Institute of Transportation Systems at the German Aerospace Center (ITS-DLR), aimed at reproducing, with the highest level of realism possible, car traffic in the greater urban area of the city of Cologne, in Germany.

  To that end, different state-of-art data sources and simulation tools are brought together to cover all of the specific aspects required for a proper characterization of vehicular traffic:
  - The street layout of the Cologne urban area is obtained from the OSM database.

- The microscopic mobility of vehicles is simulated with the SUMO software.
- The traffic demand information on the macroscopic traffic flows across the Cologne urban area (i.e., the O/D matrix) is derived through the travel and activity patterns simulation (TAPAS) methodology.
- The traffic assignment of the vehicular flows described by the TAPASCologne O/D matrix over the road topology is performed by means of Gawron's dynamic user assignment algorithm.

The resulting synthetic trace of the car traffic in the city of Cologne covers a region of 400 km² for a period of 24 hours, comprising more than 700 individual car trips. It provides a 2-hour mobility trace (6–8 a.m.) in different formats.

The trace is available at: http://kolntrace.project.citi-lab. fr/#availability. More information about the dataset can be found in Reference 15.

10.4.1.2 Typical Mobility Models   VSNs mobility model can be classified as macroscopic and microscopic. The macroscopic mobility models describe gross quantities of interest similar to vehicular density or mean velocity and treat vehicular traffic according to the fluid dynamics, whereas the microscopic description treats each vehicle as a unique entity. Then, we will introduce mobility models in these two aspects.

10.4.1.2.1 Microscopic Mobility Models

- *Random waypoint (RWP) mobility*: The most popular random mobility model is the RWP, largely used for simulating ad hoc networks and available in many simulators like ns2, GloMoSim, Qualnet, and so on. According to RWP model, a mobile node stays in a location and waits for a certain amount of time known as pause time, before it selects a new random location and travels toward it with a chosen speed, uniformly distributed in a predefined interval. Initially mobile nodes are distributed randomly and uniformly in the simulation area. However, mobile nodes do not distribute themselves uniformly while moving.

- *Freeway mobility model and Manhattan mobility model*: The model was introduced by Kumar [16] and Akhtar et al. [17] to generate the movement pattern of mobile nodes on streets. The movement pattern is defined by maps and includes many freeways in the simulation area. Each road is composed of many lanes, and no urban routes are considered. The model does not take road intersections into considerations. Initially, the mobile nodes are randomly placed in the simulation area. This model is similar to RWP model. The speed of each vehicle is given by

$$V(t+1) = V(t) + \text{random}() \times A(t) \qquad (10.1)$$

where:
$V(t)$ is the vehicle's speed at any time instant t
random() function returns a random value in the interval
$[-1, 1]$
$A(t)$ is the acceleration of a vehicle

A safety gap should be maintained between any two subsequent vehicles moving in a lane. If the distance between two subsequent vehicles is less than the required minimal distance, the second will reduce its acceleration $A(t)$ and let the former moves away. Lane changing is not allowed in this model thus the vehicle have limited movement. Vehicles move in a given lane until they reach the simulation area limit. Obviously the scenario is unrealistic.

Unlike the freeway model, which simulates highway environment, this model simulates an urban environment. The simulation area contains vertical and horizontal roads made up of multiple lanes and allows the motion in both directions (north–south and east–west). Initially, vehicles are placed on the roads randomly. Then, vehicles start moving according to history-based speeds as in freeway mobility model. When a vehicle reaches crossroad, it randomly selects a direction (turning left, turning right, or may be going straightforward) to follow. In contrast to the freeway model, a vehicle can change lane when it passes through a crossroad and as there is no control mechanism defined, vehicles will continue their movements with no stopping [16,18].

- *Stop sign model (SSM) and traffic sign model (TSM)*: SSM [16] is the first mobility model that incorporates a traffic control mechanism. At each road intersection, a stop signal is installed to control vehicles movement and to slow down and make a pause accordingly. The model is based on the TIGER/ Lines database real maps. As each road is only having a single lane, no overtaking mechanism for vehicles is given and similar to all previous models, a vehicle should keep a speed to maintain the safety distance. If large number of vehicles arrive the road intersection at the same time, then vehicles should make a queue and wait for its front vehicles to travel the road intersection.

  Considering the weakness in SSM, the authors in Reference 16 proposed TSM. In TSM, stop signs are replaced with traffic lights. A vehicle stops at a road intersection only when it encounters a red signal. When a vehicle reaches the road intersection, the traffic light may be randomly turned red with probability p, and therefore the probability of traffic light turning green is $1 - p$. The traffic light turns red for a random time period and forces the vehicles to stop. As soon as the traffic light turns green, vehicles traverse the road intersection one after another until the queue is empty.

- *Integrated mobility model (IMM)*: IMM was proposed by Alam et al. [19], which integrated the Manhattan mobility model, freeway mobility model, SSM, TSM, and some other characteristics.

In IMM, the vehicles move on a straight road as in freeway mobility model with high speed and enters in the city area where nodes spread in a Manhattan mobility model fashion. During their mobility, the vehicles stop at certain signs, reduce their speed, and wait for some seconds on traffic lights at intersections. There are some stationary vehicles on the roadside that communicate with the vehicles moving on the road. It also maintained a safe intervehicle distance of 1.5 m in IMM. Due to these characteristics, the ratio of old connections breaking and establishing of new ones is high. IMM is a representation of both rural and urban scenario. Although IMM provides a more detailed scenario for the simulation of VSNs by representing both the rural and urban area, it should include more realistic parameters and simulate for more congested traffic scenarios.

10.4.1.2.2 Macroscopic Mobility Model

- *A trace-based O-D macroscopic model*: Reference 20 proposed a realistic macroscopic origin-destination (O-D) vehicular mobility model. The model takes several aspects such as departures and arrivals in space, departures in time, total distance traveled, and total travel time into consideration, and it adopts a trace-based approach to produce the model.

  The proposed O-D mobility model was built considering the publicly available mobility trace from the city of Cologne, Germany. The trace used is a collection of raw data describing each vehicle's position at a given time with 1 second granularity. Before the statistical analyses, it had to transform the raw data into measures representing the macroscopic aspects that comprise an O-D mobility model.

  It assumes that each considered characteristic (i.e., departures and arrivals, travel distance, and total travel time) is represented by a random variable X, and its sample data $X_1, X_2, \ldots, X_n$ are obtained from the preprocessed trace. Steps of the statistical analysis are given as follows:

  Step 1: Hypothesizing families of distributions—The first step is the analysis of the data's statistical details, such as the histogram and summary. After this analysis, it is possible to formulate a hypothesis about the most suitable probability distribution functions $f(X, \theta)$ to describe the data. Based on the sample data statistical analysis, the probability distributions could be uniform, geometric, negative binomial, zero differentiate geometric, and zero differentiate negative binomial.

  Step 2: Estimation of distribution parameters—The maximum likelihood estimator (MLE) technique is adopted to estimate the θ parameter values.

  Step 3: Determine the representativeness of the fitted distributions—Finally, all candidate probability functions are evaluated in terms of their ability to represent the sample data.

To provide a generic model that may be adopted in different scenarios, it is important to evaluate how the specific characteristics of Cologne are related to the obtained results. In other words, the results obtained in subsection 10.4.1.2.2 are specific to the city of Cologne and need to be adapted and parameterized to be applied to other scenarios.

To this end, the demographic density (habitants/area) of each of the nine regions of Cologne was used in a linear regression model to infer how these values are related to the mean of departures and arrivals of each region. It concluded that there is an approximately linear relationship between the demographic density of a region and its mean number of departures and arrivals per area.

After computing the mean number of departures and arrivals relative to the demographic density of the regions, it is possible to apply regression techniques to estimate the parameters of the distributions selected in methodology part relatively to the density, creating a generic model that can be adopted by different scenarios.

- *A Markov jump process mobility model for urban scenario*: Reference 21 explores the use of Markov jump process to model the macroscopic level vehicular mobility. The proposed simple model can accurately describe the vehicular mobility; moreover, it can predict various measures of network-level performance, such as the vehicular distribution, and vehicular-level performance, such as average sojourn time in each area and the number of sojourned areas in the networks.

  - *Modeling motivation*: Consider a vehicle moving in the roads of a city. It will travel along a road and come across an intersection. It may wait at the traffic light for some time and choose the direction at the intersection, and then travels to another road to drive on. In the downtown of a big city, the roads are usually very crowded, and the intersections are very dense, which lead to very long waiting time at intersections and relatively short deriving time along roads. Therefore, intersection is an important factor in modeling the urban vehicular mobility. Thus, viewing from the sky above the city, you can observe a crowd of vehicles waiting at the area of each intersection, and streams of traffic moving from one area to another area. In other words, in order to describe the vehicular distribution, it needs to pay particular attention to the areas around intersections and to understand the vehicular behaviors of transition from one area to another from the system viewpoint.

Thus, if it divides the whole urban city into different areas each including at least one intersection, it can model the vehicles moving from one area to another adjacent area and therefore can model the vehicular traffic transiting from one area to another.

- *Model description*: Using the method of area partitioning described in the step of modeling motivation section, the whole vehicular system can be partitioned into many areas, the number of which are denoted by N. Vehicles move into the system, transit from one area to another, and finally move out the system. It uses a Markov jump process to model this system, which is illustrated in Figure 10.3. The Markov jump process includes N states to represent the N partitioned areas in the system. The states are numbered by the set $N = \{1, 2, ..., N\}$, and the corresponding areas are denoted by the set $A = \{A_1, A_2, ..., A_n\}$, which is also used

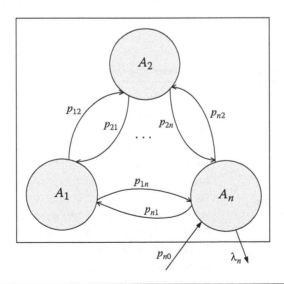

Figure 10.3 Markov jump process-based mobility model.

to represent the set of all N states. A vehicle entering the system and moving from one area to another is modeled by the entrance into the Markov jump process and the transition from one state to another. It now specifies the parameters of this Markov jump process that describes the dynamic behaviors of this vehicular mobility system.

From the viewpoint of the Markov jump process, vehicles enter the system with certain rate, stay in a state for some time, and then transfer to other state with the events of turn on and turn off. In this vehicular mobility system with multiareas, the vehicles' dynamic behaviors occur on two different timescales. Vehicles may enter and depart the system, or appear and disappear from the system due to turn on/off, which happens in a long timescale. To help understand Markov jump process model, some key parameters are summarized in Table 10.3.

To recap, it models the vehicular mobility system partitioned into N areas as a Markov jump process of N states. In this open system with M vehicles, vehicles join and leave the system freely. The exogenous arrival rate for state A_n is λ_n, and the effective arrival rate for state A_n is γ_n, which takes into account the exogenous arrival rate λ_n and all the transitions from other areas A_m $(m \neq n)$. After staying in the state A_n for the time period of $1/\mu_n$, a vehicle switches to

Table 10.3   Parameters of Markov Jump Process Model

NOTATIONS	MEANING
A	The set of areas of the set of states
N	The number of states or areas in the system
λ_n	The exogenous arrival rate for state n
γ_n	The effective arrival rate for state n
$1/\mu_n$	The expected vehicle residence time at area n
$\rho_n = \gamma_n/\mu_n$	The load of area or state n
$\rho_{nm}(m \neq n)$	Probability of vehicles moving from area n to area m
ρ_{n0}	Probability of vehicles leaving the system from area n
M	The number of vehicles in the system
W_n	The number of vehicles in area n

another state A_m with probability p_{nm}, or it may leave the process with probability p_{n0}. The load of state A_n is denoted as $\rho_n = \gamma_n/\mu_n$.

For the aforementioned Markov jump process used to model the vehicular mobility system, if the exogenous arrival to each state follows a Poisson process, it becomes a Markov jump process with superimposed Poisson arrival.

10.4.2 Routing Protocols

The routing of data in a VSN is a challenging task due to the high dynamics, unstable connection environments, and transfer direction limit (real road planning) involved. What is more, it has been discussed that radio obstacles, as found in communication process, have a significant negative impact on the performance of routing. Therefore, a suitable routing protocol for VSNs is an important issue, especially in regard to intervehicle communication applications.

10.4.2.1 Classification of VSNs Routing Mechanisms Routing in VSNs can be classified under transmission strategies and routing information. The routing information mainly focuses on topology-based and graphic-based routing, whereas the transmission strategies has a significant impact in protocol design and network performance (in case of network overhead, delay, and packet loss) [22]. The routing classification of VSNs can refer to Figure 10.4.

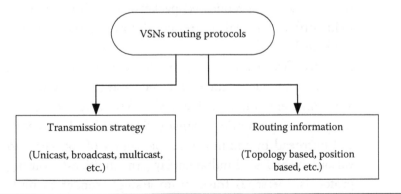

Figure 10.4 Classification of routing protocols of VSNs.

- *Transmission strategies*

  According to transmission strategies, routing can be classified under unicast, broadcast, multicast, and so on. Multicast can be further partitioned into geocast and cluster-based routing protocols.

  - *Unicast*: Unicast routing refers to information delivery from a single source to a single destination using the wireless multihop scheme; where the intermediate nodes are used to forward data from the source to the destination, or by using the store and forward scheme. It is the most widely used class in the general ad hoc networks. This scheme required the source vehicle to hold its data for a time and then forward it. There are many unicast routing protocols proposed for VSNs; most of the topology-based routing protocols belong to a unicast class such as vehicle-assisted data delivery (VADD) in VANETs, ad hoc on-demand distance vector (AODV), and dynamic source routing protocol (DSR).

  - *Broadcast*: Broadcasting routing enables packets to flood into the network to all available nodes inside the broadcast domain. Broadcasting routing is widely used in VSNs; it is mainly used in the route discovery process; some protocols (such as AODV) allow nodes to rebroadcast the received packets. This routing scheme allows packets to deliver via many nodes, which may achieve a reliable packet transmission; however, it could consume the network bandwidth by sending replicated packets, so each node needs to identify which packet is replica (it has received it before) to discard.

  - *Multicast*: Multicast routing protocols are the most active research area due to their efficiency and mobility within a dynamic environment like VSNs. Multicasting reduces the power consumption, transmission overhead, and control overhead by sending multiple copies of messages to various vehicles simultaneously. In multicast routing protocols, messages travel from a single sender to multiple destinations or toward a group of interested nodes.

Multicast routing protocols are further classified into geo-cast and cluster-based routing [23].

- *Geocast-based routing protocol*: Geocast routing protocols use location information for route establishment. Therefore, this approach aims to deliver messages from a single source to multiple destinations within a specified geographical location, which is called a zone of relevance (ZOR) and the area next to ZOR is a zone of forwarding (ZOF). In ZOF, messages are directed toward specified nodes rather than flooding of packets to all nodes of the network. Therefore, ZOF strategy reduces the control overhead and network traffic congestion during message dissemination. Multicast-based geocast routing protocols are useful for achieving safety and convenience during driving. These routing protocols also provide scalability within VSNs.

- *Cluster-based routing protocol*: Cluster-based routing behaves efficiently in multicast communication. All nodes of the network are arranged in virtual groups, called as clusters. Each cluster has a cluster head, which is elected upon several parameters like mobility, position, behavior, node degree, nodes ID, and so forth. The remaining nodes of a cluster are called cluster members. Cluster head is responsible for communicating with cluster members and other cluster heads. Therefore, the network overhead is reduced by dividing the network load into two phases. In the first phase, the cluster head communicates with cluster members and in the second phase, cluster head communicates with other cluster heads. Hence, the network load is directly proportional to the number of clusters in the network.

• *Routing information*: According to routing information, routing can be divided into topology-based and position-based routing protocols. In topology-based routing, each node should be aware of the network layout, also should be able to forward packets using information about available nodes and links in the network. In contrast, position-based routing should be aware of the nodes locations in the packet forwarding.

- *Topology-based routing protocols*: Topology-based routing protocols, usually traditional mobile ad hoc network (MANET) routing protocols, utilize link's information stored in the routing table as a basis to forward packets from source node to destination, which are commonly divided into three categories (based on underlying architecture): proactive (periodic), reactive (on-demand), and hybrid [24].

  - *Proactive protocols*: Proactive protocols allow a network node to use the routing table to store routes information for all other nodes, each entry in the table contains the next hop node used in the path to the destination, regardless of whether the route is actually needed or not. The table must be updated frequently to reflect the network topology changes, and it should be broadcasted periodically to the neighbors. This scheme may cause more overhead especially in the high mobility network. However, routes to destinations will always be available when needed [25].

  - *Reactive routing protocols*: Reactive routing protocols (also called on-demand) reduce the network overhead; by maintaining routes only when needed, the source node starts a route discovery process; if it needs a non-existing route to a destination, it does this process by flooding the network by a route request message. After the message reaches the destination node (or to the node, which has a route to the destination), this node will send a route reply message back to the source node using unicast communication. Reactive routing protocols are applicable to the large size of the mobile ad hoc networks, which have high mobility and frequent topology changes.

  - *Hybrid protocol*: Hybrid protocol is a mixture of both proactive and reactive protocols; it aims to minimize the proactive routing protocol control overhead and reduce the delay of the route discovery process within on-demand routing protocols. Usually the hybrid protocol divides the network to many zones to provide more reliability for route discovery and maintenance processes. Each node divides the network into two regions: inside and outside regions; it uses a proactive routing mechanism to maintain routes to inside region nodes and uses a route discovery mechanism to reach the outside region nodes [24].

- *Position-based routing protocol*: Position or geographic routing protocol is based on the positional information in routing process; where the source sends a packet to the destination using its geographic position rather than using the network address. This protocol required that each node has the ability to decide its own location and the locations of its neighbors through the geographic position system (GPS) assistance. The node identifies its neighbor that is located inside the node's radio range. When the source needs to send a packet, it usually stores the position of the destination in the packet header, which will help in forwarding the packet to the destination without the need to route discovery, route maintenance, or even awareness of the network topology [24,25]. Thus the position routing protocols are considered to be more stable and suitable for VSNs with a high mobility environment, compared to topology-based routing protocols.

10.4.2.2 Several Typical Routing Protocols In VSNs, a data package would go through multihop before it arrives at the destination. However, the links in VSNs get broken easily, which result in the package lost in the process of dissemination, so the proper routing protocol is necessary to different scenarios. In this part, we introduce some typical routing protocols. Table 10.4 gives a summary of these routing protocols.

- *Beaconing-based opportunistic service discovery protocol* (*OSDP*): Opportunities to instantaneously advertise and disseminate services arise with the technological advances of the vehicular networks. Points-of-interest (POI) to drivers and passengers such as shops and gas stations will broadcast advertisements with local services and facilities to vehicles nearby by exploiting the RSU infrastructure. The coverage limitation of RSU can prevent users (drivers or passengers) who are genuinely interested in services from finding them even with a short distance from the POI. To improve the service discovery, Reference 26 proposed an opportunistic service discovery protocol, a protocol based on opportunistic vehicle contact, and a store and forward technique to discover and advertise services in vehicular networks. OSDP is a light-weighted protocol for location-aware applications, which supports the distributed services fully independent from Internet access. OSDP is intended for nonsafety services as an alternative for users. Therefore, users sometimes may experience nonsatisfactory or failed requests.

OSDP handles the types of services shown in Figure 10.5. Roadside services periodically broadcast messages by beacons with their local information. Vehicles around the *RSU* can hear these beacons and can receive and store the information. A query occurs when a user wants to find a service. The vehicle broadcasts a query to the neighbors. Vehicle received the query checks its cache to determine whether related service exists. If at least one service is found, then the related vehicle will send a probe response to the inquiring one.

- *Data dissemination protocol in extreme traffic conditions*: Depending on the time of day or the geographical location in a city, the network topologies can change dramatically. Hence, protocols proposed to operate under these environments should be able to adapt themselves to the traffic condition. A light-weighted broadcast data dissemination protocol (U-HyDi) is proposed by Maia et al. [27], which is fully distributed and suited for urban scenarios not relying on any infrastructure or map information. By using solely one-hop neighbor information, U-HyDi can seamless operate under intermittently connected networks by applying store-carry-forward (*SCF*) techniques to deliver messages even when no end-to-end path exists.

  Moreover, under well-connected networks, U-HyDi employs a combination of sender-based and receiver-based broadcast suppression (BS) techniques to avoid excessive contention at

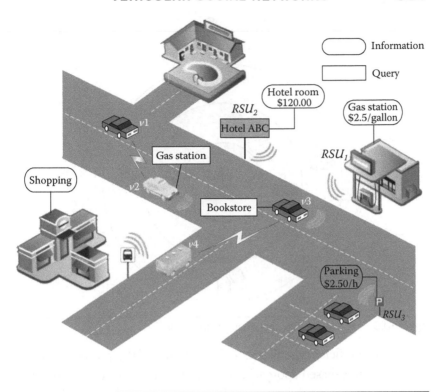

Figure 10.5 Opportunistic service discovery and information dissemination scenarios in VSNs. (e.g., the user of vehicle $v2$ wants to find a service through broadcasting a query to its neighbors. All vehicles that received this query will check their caches to determine if a related service exists. $v1$ happens to cache a related information when it is passed by gas station; then, he or she sends the information to $v2$.)

the link layer. A vehicle under U-HyDi can be in one of two regimes: the BS regime and the SCF regime.

- *BS regime (U-HyDi)*: Initially, when a vehicle receives a new data message, it determines whether it is a boundary vehicle or not, using its own GPS information or the information received from its neighbors. A boundary vehicle is a node that is on the boundary of a connected component such as vehicles B and C in Figure 10.6.

    As these vehicles do not have any other neighbor to whom they can further propagate the message, as soon as they receive a new message, they enter SCF regime. Alternatively, if a vehicle is not a boundary vehicle, it will enter the BS regime.

    U-HyDi employs a hybrid approach relying on both sender-based and receiver-based BS mechanisms to increase

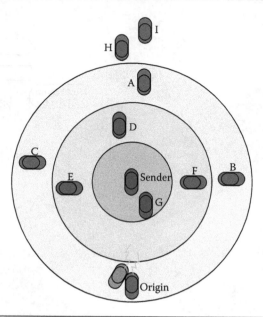

Figure 10.6 Example of the BS mechanism used by U-HyDi.

the efficiency, efficacy, and reliability of the protocol. In a sender-based approach, vehicles choose a priori next hop to rebroadcast the message. In a receiver-based approach, the rebroadcast decision is taken by the receivers after processing the received message. Here, the adopted sender-based approach is to select the farthest vehicle both from the sender and the origin, for these nodes have a greater probability of covering more an *uncovered area*. For instances in Figure 10.6, the sender chooses vehicle A to his next hop node (assuming vehicle A is the farthest vehicle that can connect to the sender in theory). Then the sender addresses the message to A to identify it as the next forward vehicle and rebroadcasts the message. Finally, A receives the message and realizes that it should continue this process.

However, suppose that these nodes do not receive the broadcast from the sender, it may be due to the outdated information at the sender. So, Guilherme and Azzedine combine the sender-based approach with a timer-based suppression mechanism, which is a receiver-based strategy. Under this scheme, whenever vehicle j receives a message for the first time from vehicle i, and j is not the chosen next hop, then vehicle j schedules and waits $T_{S_{ij}}$ to broadcast the message. $T_{S_{ij}}$ can be calculated by Equations 10.2 and 10.3:

$$T_{S_{ij}} = (S_{ij} \times t) + t \qquad (10.2)$$

$$S_{ij} = \left[N_s \times \left(1 - \left[\frac{min(D_{ij}, R)}{R} \right] \right) \right] \qquad (10.3)$$

where:

    t is the estimated one-hop delay, which includes both medium access control (MAC) and propagation delays

    N_s is the number of available time slot

    D_{ij} is the distance between nodes i and j

    R is the estimated transmission range

- *The SCF regime*: As described earlier, vehicles identified as boundary vehicles enter into the SCF regime until they reach new uninformed neighbors. U-HyDi uses message acknowledgment piggybacked in periodic beacon messages to identify uniformed neighbors. This acknowledgment appears in an additional field (stores the ID of the last message received by a vehicle) of periodic beacon messages. Hence, when a vehicle under the SCF regime receives a beacon from an uninformed neighbor, it schedules a rebroadcast using Equation 10.4. Notice that Equations 10.4 and 10.5 are very similar to Equations 10.2 and 10.3, with two main differences. First, Equation 10.4 does not have the additional t, which is required in Equation 10.2 to give time for the vehicle chosen by the sender-based approach an opportunity to rebroadcast the vehicles that are under the receiver-based approach. Second, Equation 10.5 implies that vehicles closer to the uniformed neighbor have a higher priority to rebroadcast, where in Equation 10.3, vehicles farther from the sender have the higher priority.

$$T_{S_{ij}} = S_{ij} \times t \qquad (10.4)$$

$$S_{ij} = \left[N_s \times \frac{min(D_{ij}, R)}{R} \right] \qquad (10.5)$$

For instance, in Figure 10.6, after receiving the message from the Sender, vehicles B and C enter into the SCF regime, assuming that they are boundary vehicles. When they are in communication range of a new neighbor and receive a beacon with the ID of an old message, they schedule to rebroadcast the message, thus resuming the dissemination process.

• *Socially inspired data dissemination method*: People have routines and their mobility patterns vary during the day, which have a direct impact on vehicular mobility. Therefore, protocols and applications design needs to adapt to these routines in order to provide better services. Reference 28 proposed a data dissemination solution that takes people's social metrics (clustering coefficient and node degrees) into account to select the best vehicles to rebroadcast data message. They focus on these two metrics because they provide the possibility to be aware of vehicles density in a region and consequently, to adjust the dissemination in an efficient way. The main goal of the proposal is to guarantee message delivery to all vehicles inside the region of interest (ROI) independently of the road traffic condition.

This process is similar to Reference 27 also including two regimes BS and SCF, with the relay node's choice and the waiting time's calculation being the difference. In this chapter, periodically beacon was broadcasted by every vehicle; these beacons contain some context information about the vehicle, for instance, the position and IDs of the data message, which have been received and are being carried by the vehicle. When a vehicle receives a beacon b from a neighbor s, it verifiers whether there is a data message that has not been acknowledged by s in b. For that, the vehicle looks into its list of received messages and compares their IDs with the IDs contained in b. If the vehicle finds any message m that has not been acknowledged, then it arranges a waiting delay t to rebroadcast m. The waiting time concerned those two social metrics and will be calculated later.

The clustering coefficient for a vehicle v is the number of connections between neighbors of v divided by the total number of possible connections between neighbors of v. Therefore, to accurately calculate the clustering coefficient for vehicle v, it is necessary to know the two-hop neighborhood knowledge of v. Here, Felipe and Guilherme use beacons information to estimate the clustering coefficient. As they know, for lower densities, the clustering coefficient is also low, but the variability is high. On the other hand, when the density is high, the value for the estimated clustering coefficient is also high, but the variability is low. For this purpose, the greater variability, the better. Otherwise, it will risk assigning the same or similar waiting delay to all vehicles. Therefore, for the first proposal, they give a higher priority to rebroadcast for vehicles that have a low-estimated clustering coefficient. In other words, the lower the estimated clustering coefficient, the lower the waiting delays. And it calculates the waiting delay by the equation:

$$t_{cc} = T_{max} \times \text{estimated } CC \qquad (10.6)$$

where the estimated CC ranges in the interval $[0, 1]$

To the node degree, as they has known when vehicle is low, the degree and its variability is also low. However, when the density increases, both the degree and its variability increase. Therefore, in the proposal based on the node degree, it uses an opposite approach. That is, the higher the degree of a vehicle at a given neighborhood, the higher its priority to rebroadcast the message, and it calculates the waiting delay by the equation:

$$t_{degree} = T_{max} \times \left(1 - \left(\frac{degree}{maxDegree} \right) \right) \qquad (10.7)$$

where:

   degree denote the degree of the vehicle, which is calculat-
      ing the waiting time

   maxDegree is the max one among all the degree of the
      neighbors and the degree of the vehicle

They also propose a joint solution using both the estimated clustering coefficient and the node degree. The idea is that, assuming that a single metric may not be adequate for all traffic density scenarios, a combination of the two may produce better results. The waiting delay can be calculated using this joint approach, defined by the equation: $t = \alpha \cdot t_{cc} + \beta \cdot t_{degree}$. In this work, to balance the equation delay, they assume that $\beta = 0.5$.

- *Integrity-oriented content transmission*: Content file transmissions (such as image, music, and video clips) in VSNs over the volatile and spotty V2V channels are susceptible to frequent interruptions and failures, which would result in the significant waste of precious vehicular bandwidth. On addressing this issue, Tom H. Luan et al. [29] proposed a strategy, which targets on provisioning the integrity-oriented intervehicle content transmissions.

The work toward the integrity-oriented content transmission was divided into three folds. First, supposing the initial distance and mobility statistics of vehicles were given, they develop an analytical framework to evaluate the data volume that can be transmitted upon the short-lived and spotty V2V connection. Second, given the content file size, they are able to evaluate the likelihood of successful content transmissions. Third, an admission control scheme was given based on the analysis to filter those suspicious contents.

To prove the accuracy of the analytical model and effectiveness of the proposed admission control scheme, they carry a serial of simulations. In the simulated scenario, with the proposed admission control scheme applied, it is observed that about 30% of the network bandwidth can be saved for effective content transmissions (Table 10.4).

10.4.3 Security Issues in Vehicular Social Network

Security has always remained the most significant concern in VSNs, because it is mandatory to assure public and transportation safety. This section summarizes several VSNs security related issues, including security requirement, major security threats and security solutions.

10.4.3.1 Security Requirement   For widespread deployment of secure VSNs, security solution designers should meet some basic and significant requirements. In this section, we primarily focus on security issues and requirements for safety-related applications. A security requirement was related to several aspects, as shown in Figure 10.7 [30,31].

Table 10.4 Summaries of Various VSN Routing Protocols

ROUTING PROTOCOLS	PROBLEMS OF SOLVING	SCENARIOS	MAIN TECHNOLOGIES	CATEGORY
OSDP	Solve coverage limitation of RSR and improve service discovery ability of users	Urban	SCF	Broadcast
U-HyDi	Operation on both intermittently connected topologies and well-connected topologies	Urban	BS SCF sender-based and receiver-based approach	Broadcast/ position based
Socially inspired data dissemination method	Routing design should take people's social characteristics into consideration for providing better services	Urban	BS SCF clustering coefficient and node degrees	Broadcast/ position based
Integrity-oriented content transmission	Solve waste of bandwidth because of invalid dissemination	Highway	Mathematical reasoning	

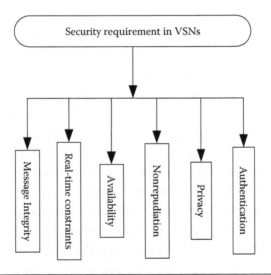

Figure 10.7 Security requirements in VSNs.

- *Authentication*: It is a majority and mandatory requirement in VSNs as it ensures that the messages are sent by the actual nodes and hence attacks done by the greedy drivers or the other adversaries can be reduced to a greater extent. Authentication can be divided into two categories: Entity authentication and ID authentication. Entity authentication ensures that the recently received message is fresh and live. It ascertains that a message is sent and received in a reasonably small time frame. ID authentication ensures that a message is trustable by correctly identifying the sender of the message.

- *Privacy*: It means to ensure that the information is not being leaked to the unauthorized people who are not allowed to access this information. For authorization, authentication is required, whereas for authentication purposes, if some authority publicly discloses identities of vehicles, then it may be dangerous for drivers for their travel activities may be monitored. So, in such type of networks, anonymity will always be critical. But in liability-related cases, maintaining the anonymity is not possible because some specified authorities such as traffic police needed to be allowed to trace the user identities for legal investigation.

- *Nonrepudiation*: It assures that it will be impossible for an entity or a driver to deny having sent or received messages in the network. It is needed for the sender in V2V warnings. In this way, if a vehicle sends some data, which is malicious, there is a proof for the liability purposes.

- *Availability*: In safety applications like postcrash warning, the wireless channel has to be available so that approaching vehicles can receive the warning messages. If the radio channel goes out, then the warning cannot be broadcasted, and the application itself becomes useless. Hence availability should be also supported by alternative means.

- *Real-time constraints*: Since, VSN is specially designed for safety applications that are dependent on strict time guarantees. Adversaries may try to delay the messages; that delay of seconds means that the message may become meaningless and it may lead to devastating results.

- *Message integrity*: The integrity of exchanged data in a system is to ensure that these data have not been altered in transit. Integrity mechanisms help therefore to protect information against modification, deletion or addition attacks. In the case of VSNs this category targets mainly V2V communications compared to V2I communication because of their fragility.

10.4.3.2 Major Security Threats   There are varieties of possible attacks in a vehicular network [30–32].

10.4.3.2.1 Attacks on Authenticity

- *Sybil attack*: In such attack, attacker uses multiple identities to send multiple messages; different identities are used at the same time. In this way, the attacker creates an illusion that messages are being sent from different nodes. The basic objective of this attack is to mislead other vehicles. Moreover, this attack can also disrupt routing protocols, leading to distributed DoS and unfair distribution of resources.
- *Node impersonation attack*: In node impersonation attack, an attacker can change its identity and acts like a real originator of the message. An attacker receives the message from the originator of the message and changes the contents of the message for his or her benefits. After that, an attacker sends this message to the other vehicles.
- *Message suppression*: In this attack, attackers can selectively cast away the packet from the network, which may contain significant information required by the receiver. An attacker also can remove the congestion alerts it receives in order to prevent the nodes to select an alternative path to the destination and force them to wait in traffic.

10.4.3.2.2 Attacks on Nonrepudiation
- *Loss of events traceability*: Despite its importance, we have not seen any document that addresses this attack that we find quite feasible in a VSNs environment. In fact, this no repudiation attacks consist of taking action, allowing subsequently an attacker to deny having made one or more actions. This kind of attack is essentially based on the erasure of action traces and creating confusion for the audit entity. Some attacks can serve as preliminary to nonrepudiation attack such as Sybil attack and duplication of keys and certificates.

10.4.3.2.3 Attacks on Availability

- *Black hole attack*: The black hole attack is a conventional attack against the availability in ad hoc networks, it exists also for VSNs. In black hole attack, the malicious node receives packets from the network, but it refuses to participate in the operations of routing data. This disrupts the routing tables and prevents the arrival of vital data to recipients mainly because the malicious node always declares being part of the network and able to participate, which is not the case practically. A Black hole node can e.g. redirect the traffic that receives to a specific node which does not exist in fact and this causes data loss.

- *Malware*: Given the existence of software components to operate the OBU and RSU, the possibility of infiltration of malware (malicious software) is possible in the network during the software update of VSNs units. The effect of a malware is similar to the effect of viruses and worms in an ordinary computer network, except that in a VSNs network, disruption of normal functionality is always followed by serious consequences.

- *Spamming*: The spam messages such as advertisements are of no use for users. However, in a VSN network, which is a mobile radio environment, this type of attack aims to consume bandwidth and causes voluntary collisions. Given the lack of a centralized management of the transmission medium, this makes the control of such attacks more difficult.

- *Denial of Service (DoS) attack*: DoS attack can be done by the network insiders or outsiders. An insider attacker may jam the channel after transmitting dummy messages and thus, stops the network connection. An outsider attacker can launch a DoS attack by repeatedly disseminating forged messages with invalid signatures to consume the bandwidth or other resources of a targeted vehicle. The impact of this attack is that VSNs losses its ability to provide services to the legitimate vehicles.

10.4.3.2.4 Attacks on Confidentiality

- *Eavesdropping*: It is the most important attack over the VSNs network against confidentiality. To perform this attack, attackers can be a vehicle or in a false RSU, they show that they are the part of the network. Their goal is to illegally get access to confidential data.

• *Traffic analysis attack*: In a VSN, the traffic analysis attack is a passive serious threat against confidentiality and privacy of the users. The attackers analyze the collected information after a phase of listening to the network; it tries to extract the useful information for its own purposes.

10.4.3.2.5 Attacks on Integrity and Data Trust

• *Masquerading attack*: The attacker actively pretends to be another vehicle by using false identities and can be motivated by malicious or rational objectives. Message fabrication, alteration, and replay can also be used toward masquerading. For example, assume that an attacker tries to act as an emergency vehicle to defraud other vehicles to slow down and yield.

• *Replay attack*: This is a classic attack. It consists in replaying (broadcast) a message already sent to take the benefit of the message at the moment of its submission. Therefore, the attacker injects it again in the network packets previously received. This attack can be used e.g. to replay beacons frames, so the attacker can manipulate the location and the nodes routing tables. Unlike other attacks, replay attack can be performed by non-legitimate users.

• *Message fabrication/Alteration*: As its name implies, this attack is against integrity; it consists of modifying, deleting, constructing, or altering existing data. It can occur by modifying a specific part of the message to be sent. In this attack, the attacker can also delete a part of the message, can alter, or can make new messages, which help him or her achieving its intended purpose of the attack.

10.4.3.3 Security Solutions Similar to any other network security mechanisms, security solutions can be based on prevention (proactive) and detection (reactive) techniques in VSNs. Considering the criticality nature of VSNs application, preventive security mechanisms are important than the reactive ones. Hence, most of the existing security mechanisms in VSNs aim to prevent security attacks rather than detection. Even though works on detection techniques in VSNs are very limited, the usefulness of these can be significant in number of situations. For instance, in case of any fabrication attack if prevention mechanism fails, reliable and efficient detection of the fabrication can help drivers in taking the correct action in that situation. In this section, we will give two aspects of techniques (prevention and detection) to security [30].

10.4.3.3.1 Prevention Techniques

- *Digital signature-based techniques*: Digital signature is the building block of these security mechanisms, which primarily aims at providing message authenticity. Along with the digital signature, these techniques can exploit cryptography with certificate or without certificate.
  - *Without certificate*: In this approach, cryptographic digital signatures are applied to messages or hashes over messages. Digital message signatures are usually formed by asymmetric cryptography, that is by using public–private key cryptography. Messages are signed with the message originators' private keys. This approach can provide three security improvements to communication (message authenticity, message integrity protection, and nonrepudiation). The advantage of this approach is that it is simple to realize with small requirements. Mechanisms based on this approach are widely deployed. However, attacks like message forging, DoS, and Sybil are still possible. Moreover, this approach does not prevent attackers to create fake warning messages.
  - *With certificate*: In order to enhance earlier approach, the signatures can be combined with digital certificates provided by a trusted certificate authority (CA). The basic notion with certificates is that nodes, which include such certificates in their messages, are trusted by other nodes by verifying the certificates. The signed messages include a certificate, which is cryptographically linked to the public key that belongs to the private key the message issuer uses to sign messages. The advantage of the certificate concept lies in the possibility to exclude external attackers from the system, as well as in the ability to remove malicious or defective nodes.

- *Proprietary system design*: This category of security mechanisms aims to exploit nonpublic (proprietary) protocols or hardware to control the unauthorized access to the networks. Similar to the certificate approach, this concept prevents nonauthorized nodes from participating in the network. Ultimate objective of this concept is to increase the required effort an attacker has to put in order to enter into the system. This scheme does not prevent him from doing so, nor do they prevent any attack from an insider.

- *Temper proof hardware*: In order to complement the aforementioned mechanisms, tamper-resistant device (TRD) or tamper-proof device (TPD) hardware is meant to provide secure input to the communication system, by securing the in-vehicle communication system and by protecting it from manipulation. Along with the storing secret information, this device will be also responsible for signing outgoing messages. To protect itself of being compromised by attackers, the device should have its own battery, which can be recharged from the vehicle, and clock, which can be securely resynchronized, when passing by a trusted roadside base station. The access to this device should be limited to authorized people. For instance, cryptographic keys can be renewed at the periodic technical checkup of the vehicle. Usually, the TPD contains a set of sensors that can detect hardware tampering and erase (self-destructive) all the stored keys to prevent them from being compromised. This sophisticated feature makes the TPD too sensitive for VSNs conditions as well as too expensive for nonbusiness consumers. A trusted platform module (TPM) that can resist to software attacks but not to sophisticated hardware tampering can be an alternative option to a TPD.

10.4.3.3.2 Detection Techniques   By detecting security threats along with the common signature-based and anomaly-based detections, we can exploit the contexts of a VSNs and its application to detect attacks on it.

- *Signature-based detection*: In signature-based detection, attacks can be detected by comparing network traffic to known signatures of attacks. As soon as an attack is detected, appropriate countermeasures can be initiated. The primary concern of this approach is to realize a mechanism that is capable of detecting known attacks on a communication system.

- *Anomaly detection*: This approach is based on a statistical approach that defines normal communication system behavior. Any deviation from that behavior is statistically analyzed and as soon as they reach a defined level, the security system concludes that there is an attack ongoing.
- *Context verification*: Context verification is an approach that specifically considers the properties and applications of VSNs. The notion is to collect as much information from any information source available by each vehicle and to create an independent view of its current status, its current surrounding (physical) environment, and current or previous neighboring vehicles. In order to do the evaluation of the situation, this approach requires to define the rule sets that determine what is to be expected with which probability in which situation. Situation evaluation mechanisms can be either application independent or application dependent. In application-independent case, it can exploit position as well as time-related information. On the other hand, application context-dependent evaluation exploits parameters specific to a certain application.

10.5 Conclusion

This chapter first gives a brief description of vehicle social networks including main physical components, network architectures, and special characteristics. Then, various VSN applications are summarized. Finally, the existing challenges are discussed, including mobility model, routing protocol, and security. Since experimental evaluation of VSNs is expensive, most current VSN schemes are evaluated through simulations. However, these simulators only provide simple mobility model and poor physical radio models. Thus, simulation techniques for VSNs should be improved. Besides, even though many routing protocols for VSNs have been proposed, how to design and implement efficient routing mechanisms in partitioned networks and how to handle dynamic topology in VSNs still need to be deeply investigated.

References

1. Vegni, A.M. and V. Loscri. A survey on vehicular social networks. *IEEE Communications Surveys & Tutorials*, 2015; 17(4): 2397–2419.
2. Hu, X., T.H.S. Chu, and V.C.M. Leung. A survey on mobile social networks: Applications, platforms, system architectures, and future research directions. *IEEE Communications Surveys & Tutorials*, 2015; 17(3): 1557–1581.
3. Fei, R., K. Yang, and X.Q. Cheng. A cooperative social and vehicular network and its dynamic bandwidth allocation algorithms. In: *Proceedings of the 30th IEEE International Conference on Computer Communications (IEEE INFOCOM 2011)*, Shanghai, China, April 10–15, 2011.
4. Li, F. and Y. Wang. Routing in vehicular ad hoc networks: A survey. *IEEE Vehicular Technology Magazine*, 2007; 2(2): 12–22.
5. Etemadi, N. and F. Ashtiani. Throughput analysis of IEEE 802.11-based vehicular ad hoc networks. *IET Communications*, 2011; 5(14): 1954–1963.
6. Hu, X.P., W.H. Wang, and V.C.M. Leng. VSSA: A service-oriented vehicular social-networking platform for transportation efficiency. In: *Proceedings of the 15th ACM International Conference on Modeling, Analysis and Simulation of Wireless and Mobile Systems (MSWiM'12)*, Cyprus Island, October 21–25, 2012.
7. Liu, N., M. Liu, J.N. Cao, G.H. Chen, and L. Wei. When transportation meets communication: V2P over VANETs. In: *Proceedings of the 30th International Conference on Distributed Computing Systems (ICDCS)*, Genova, Italy, June 21–25, 2010.
8. Al-Sultan, S., M.M. Al-Doori, A.H. Al-Bayatti, and H. Zedan. A comprehensive survey on vehicular ad hoc network. *Journal of Network and Computer Applications*, 2014; 37: 380–392.
9. Liang, W.S., Z. Li, H.Y. Zhang, S.L. Wang, and R.F. Bie. Vehicular ad hoc networks: Architectures, research issues, methodologies, challenges, and trends. *International Journal of Distributed Sensor Networks*, 2015; 2015(17).
10. Faezipour, M., M. Nourani, A. Saeed, and S. Addepalli. Progress and challenges in intelligent vehicle area networks. *Communications of the ACM*, 2012; 55(2): 90–100.
11. Wang, Y. and F. Li. Vehicular ad hoc networks. *Guide to Wireless Ad Hoc Networks*. London: Springer, 2009, pp. 503–525.
12. Vegni, A.M., M. Biagi, and R. Cusani. Smart vehicles, technologies and main applications in vehicular ad hoc networks. INTECH Open Access Publisher 2013.
13. Microscopic vehicular mobility trace of Europarc roundabout. Available online: https://hal.inria.fr/medihal-01148989/.
14. Vehicular Mobility Trace of the City of Cologne, Germany. Available from: http://koIntracE.project.citi-lab.fr/#availability.

15. Uppoor, S., O. Trullols-Cruces, M. Fiore, and J.M. Barcelo-Ordinas. Generation and analysis of a large-scale urban vehicular mobility dataset. *IEEE Transactions on Mobile Computing*, 2014; 13(5): 1061–1075.

16. Kumar, R. and M. Dave. Mobility models and their affect on data aggregation and dissemination in vehicular networks. *Wireless Personal Communications*, 2014; 79(3): 2237–2269.

17. Akhtar, N., S.C. Ergen, and O. Ozkasap. Vehicle mobility and communication channel models for realistic and efficient highway VANET simulation. *IEEE Transactions on Vehicular Technology*, 2015; 64(1): 248–262.

18. Oliveira, R., M. LuíSa, A. Furtadoa, L. Bernardo, R. Dinis, and P. Pinto. Improving path duration in high mobility vehicular ad hoc networks. *Ad Hoc Networks*, 2013; 11(1): 89–103.

19. Alam, M., M. Sher, and S.A. Husain. Integrated mobility model (IMM) for VANETs simulation and its impact. In: *Proceedings of International Conference on Emerging Technologies (ICET)*, Islamabad, Pakistan, October 19–20, 2009.

20. Silva, F.A., T.R.M.B. Silva, R. Vicente, L.B. Ruiz, and A.A.F. Loureiro. On the improvement of vehicular macroscopic mobility models. In: *Proceedings of the 78th International on Vehicular Technology Conference (VTC Fall)*, Las Vegas, NV, September 2–5, 2013.

21. Li, Y., D.P. Jin, Z.C. Wang, P. Hui, L.G. Zeng, and S. Cheng. A markov jump process model for urban vehicular mobility: Modeling and applications. *IEEE Transactions on Mobile Computing*, 2014, 13(9): 1911–1926.

22. Altayeb, M. and I. Mahgoub. A survey of vehicular ad hoc networks routing protocols. *International Journal of Innovation and Applied Studies*, 2013; 3(3): 829–846.

23. Farooq, W., M.A. Khan, S. Rehman, and N.A. Saqib. A survey of multicast routing protocols for vehicular ad hoc networks. *International Journal of Distributed Sensor Networks*, 2015; 2015(8).

24. Al-Doori, M. Directional routing techniques in VANET. Ph.D Thesis, De Montfort University, United Kingdom, 2011.

25. Lee, K.C., U. Lee, and M. Gerla. Survey of routing protocols in vehicular ad hoc networks. *Advances in Vehicular Ad-Hoc Networks: Developments and Challenges*, 2010, pp. 149–170.

26. Yokoyama, R.S., B.Y.L. Kimura, L.M.S. Jaimes, and E.D.S. Moreira. A beaconing-based opportunistic service discovery protocol for vehicular networks. In: *Proceedings of the 28th International Conference on Advanced Information Networking and Applications Workshops (WAINA)*, Victoria, Canada, May 13–16, 2014.

27. Maia, G., A. Boukerche, A.L.L. Aquino, A.C. Viana, and A.A.A.F. Loureiro. A data dissemination protocol for urban vehicular ad hoc networks with extreme traffic conditions. In: *Proceedings of the International Conference on Communications (ICC)*, Budapest, Hungary, June 9–13, 2013.

28. Cunha, F.D., G.G. Maia, A.C. Viana, R.A. Mini, L.A. Villas, and A.A. Loureiro. Socially inspired data dissemination for vehicular ad hoc networks. In: *Proceedings of the 17th ACM International Conference on Modeling, Analysis and Simulation of Wireless and Mobile Systems (MSWiM' 14)*, Montreal, Canada, September 21–26, 2014.

29. Luan, T.H, X.S. Shen, and F. Bai. Integrity-oriented content transmission in highway vehicular ad hoc networks. In: *Proceedings of the 32nd IEEE International Conference on Computer Communications*, Turin, Italy, April 14–19, 2013.

30. Razzaque, M.A., A. Salehi, and S.M. Cheraghi. Security and privacy in vehicular ad-hoc networks: Survey and the road ahead. *Wireless Networks and Security*, Berlin, Germany: Springer, 2013, pp. 107–132.

31. Gillani, S., F. Shahzad, A. Qayyum, and R. Mehmood. A survey on security in vehicular ad hoc networks. In: *Proceedings of International Workshop on Communication Technologies for Vehicles*, Villeneuve d'Ascq, France, May 14–15, 2013.

32. Mejri, M.N., J. Ben-Othman, and M. Hamdi. Survey on VANET security challenges and possible cryptographic solutions. *Vehicular Communications*, 2014; 1(2): 53–66.

11

DEVICE-TO-DEVICE-BASED TRAFFIC OFFLOADING

11.1 Introduction

Wireless access techniques have developed in a so fast speed and varieties of bandwidth-hungry applications and services such as web browsing, video streaming, gaming, and social networking are gradually shifted to mobile networks, which leads to an exponential increase in data traffic in mobile networks. Meanwhile, modern mobile systems are upgrading at a tremendous speed. Devices of this date such as smartphones and tablets deploy powerful multicore CPUs, fast memory, and smart sensors, which means that the performance of the devices is getting better. Therefore, mobile users are able to complete demanding tasks and enjoy convenient mobile services. For better user experience, mobile operators are upgrading the infrastructure to long-term evolution (LTE) and LTE-advance (4G) for better access capacity and services. However, currently cellular networks, even like 4G networks, do not have enough capacity to accommodate an exponential growth of data. According to Cisco forecasts [1] and practical experiences of mobile operators, we are now facing the *mobile data apocalypse*. The number of mobile-connected devices has already exceeded the number of people on Earth, while by 2017 there will be nearly 1.4 mobile devices per capita. Thus, there is urgency for the research community to look for new solutions [2].

Mobile data offloading is the use of complementary network technologies for delivering data originally targeted for cellular networks. From the survey [3], there are two main approaches to offload in cellular networks: access point (AP)-based offloading and device-to-device (D2D) offloading (see Figure 11.1). Diverting traffic through fixed Wi-Fi APs represents a conventional solution to reduce traffic

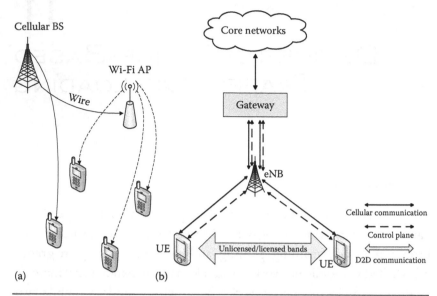

Figure 11.1 Two main approaches to cellular data offloading: (a) AP-based offloading and (b) D2D offloading.

on cellular networks. End users located inside a hotspot coverage area might use it as a worthwhile alternative to the cellular network when they need to exchange data. Hotspots generally provide better connection speed and throughput than cellular networks. However, coverage is limited, and mobility is in general constrained within the cell (Figure 11.1a). D2D communications are now being pursued as an important feature [4] for the next generation cellular networks (LTE-advanced). As an underlay to cellular networks, D2D aims to leverage the physical proximity of communicating devices to improve cellular coverage in sparse deployments, to provide connectivity for public safety services, and to improve resource utilization in conventional deployments [5]. So, it is capable to offload traffic from cellular networks with D2D communications. In addition, there exist in the literature many approaches that exploit D2D content sharing between neighboring nodes. Therefore, this chapter mainly puts emphasis on offloading strategies exploiting D2D technologies in this chapter.

D2D technology aims to, without increasing the cost of central infrastructure, improve transmission speed for cell-edge users by supporting connection to a cellular network for terminals in a shadow region and by increasing system capacity by reducing interference.

Specifically, the D2D technology in a cellular communication system is becoming more important due to its advantages of offloading cellular traffic to the D2D link. Figure 11.1b shows an example of traffic traversing in the LTE-A system [6].

In literature, there exist various D2D technologies, including Bluetooth, Wi-Fi Direct, LTE Direct, and so on. Wi-Fi Direct is built upon the IEEE 802.11 protocol stack and enables efficient D2D connections in unlicensed bands. In comparison with wide cellular technology, it is a far simpler protocol, thus it consumes less energy. In addition, since it operates over shorter links, it achieves a better level of spatial reuse. LTE Direct is a new and innovative D2D technology that enables discovering thousands of devices and their services in the proximity of less than 500 m, in a privacy sensitive and battery efficient way. LTE Direct uses licensed spectrum, allowing mobile operators to employ it as a way to offer a range of differentiated applications and services to users. It relies on the LTE physical layer to provide a scalable and universal framework for discovery and connecting proximate peers. Currently, most mobile devices have multi-radio capabilities, that is, they can flexibly use both LTE and Wi-Fi interfaces [7].

Figure 11.2 illustrates the content organization of this chapter. Section 11.2 expounds the architecture of D2D offloading system. Then, various strategies of D2D offloading are discussed in Sections 11.3 through 11.5. The usage cases of traffic offloading

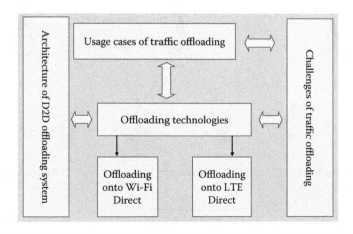

Figure 11.2 Content organization of this chapter.

are summarized in Section 11.5. The challenges of cellular traffic offloading are discussed in Section 11.6. Finally, we briefly conclude this chapter.

11.2 Architectures of D2D Based Traffic Offloading

As shown in Figure 11.3, UE1 (user equipment 1) and UE2 are engaged in a cellular data communication, which is being routed through the eNodeB (eNB) and the core network infrastructure. The packet data network gateway (PDN-GW) in the network keeps a routing table and routes IP packets to the proper eNB, which is connected to the destination user equipment (UE). The gateway is able to detect potential D2D traffic for the fact that it processes the IP headers of the data packets.

The gateway informs the candidate UEs that they need to perform a discovery process after detecting D2D candidates (UE1 and UE2). One party transmits a known reference signal sequence and the other party receives this signal sequence, and then the process of the discovery is done. In that time, the serving eNB coordinates that the two peer UEs meet in space, time, and frequency for the reference signal.

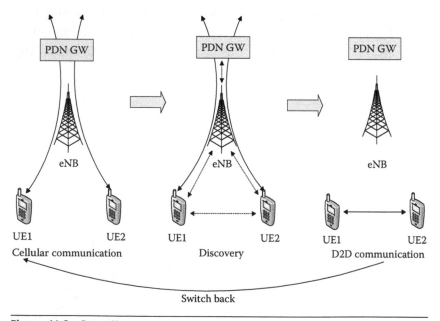

Figure 11.3 Data-offloading process using D2D communications.

As the UE1 and UE2 come into proximity of each other, each detects the peer in its proximity. Then the cellular data session between UE1 and UE2 is switched to a D2D communication path.

The D2D data session will keep going unless one party lost connection. When the D2D communication path is no longer feasible, the D2D session is going to switch back to the cellular path. In the proposed data-offloading process, the user will not perceive the switching of user traffic sessions between the cellular communication and D2D communication paths [6].

Basically, if the network is capable of tracking client locations, it can significantly reduce the amount of time clients spend in discovery by informing them when they are in proximity. Also, with network assistance a client can simply enter a search for the desired content/service, and the network will inform it if/when there is an authenticated device in the proximity offering the desired content service. In terms of safety, the network assistance enables client anonymity during discovery and D2D communication by masking their permanent device IDs. Also, UEs communicate with others or download data through cellular link when no other UEs exist in proximity. However, it can communicate with other UEs on the D2D channel when they are getting closer.

Usually, most P2P applications have some sort of content tracker (i.e., the third-party AS), which is a trusted entity in the Internet to which all registered users have access. The AS in Figure 11.4 plays the role. The content tracker logs all available (i.e., offered) user content, authenticates users, and authorizes content access. In conventional cloud-based services, this content tracker functionality is coupled with a content delivery network that acts as a relay between users for content exchange. Nearly all social networking applications work like this way, examples include Facebook, YouTube, and many more.

Figure 11.4 shows a more detailed communication mechanism. When a user clicks on the content link provided by the AS (the content link references a peer with D2D capability), the user's device contacts its D2D server for assistance with D2D connection establishment. Then, depending on the information provided by the network's location services, the D2D server decides which path will be used for the P2P session (i.e., infrastructure or D2D). Although the P2P session is

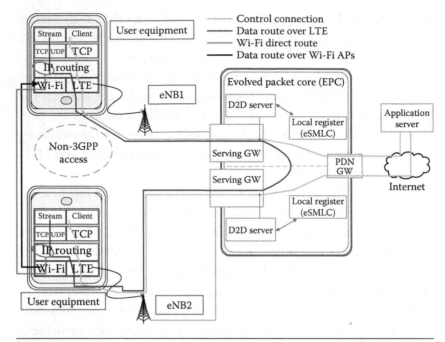

Figure 11.4 Architecture of the envisioned D2D offload system. (From Pyattaev, A. et al., Network-assisted D2D communications: Implementing a technology prototype for cellular traffic offloading, in *Proceedings of 2014 IEEE Wireless Communications and Networking Conference* (*WCNC*), Istanbul, Turkey, pp. 3266–3271, 2014.)

active on the D2D link, the D2D server monitors the users' locations (and potentially their D2D links) to determine when/if the session should be moved back to the infrastructure. Note that D2D server should possess special mechanism to allow for some hysteresis in the decision to offload onto D2D or fall back to the infrastructure, preventing the client devices from powering interfaces on and off due to small channel or position changes. The offloading procedure can be implemented in Figure 11.4 [8].

In Figure 11.4, evolved packed core (EPC), the core of LTE system, mainly includes four elements: D2D server, serving GW, local register, and PDN GW. The D2D server acts as a trusted connection manager for devices engaged in D2D discovery and/or communication. The D2D server performs the following functions:

- Maps client device identifiers to their users' appIDs (the unique application-specific user IDs in the form of username@ domain)

- Tracks client device positions based on available positioning services
- Provides clients with temporary link layer IDs to enable anonymous discovery
- Automates D2D connection establishment (including security key exchange)
- Manages active D2D connections (e.g., initiating fallback to infrastructure to guarantee service continuity, providing guidance for improved radio resource management)

The local register provides positions of UEs to D2D Server.

The serving GW routes and forward user data packets, while also acting as the mobility anchor for the user plane during inter-eNB handovers and as the anchor for mobility between LTE and other 3GPP technologies. For idle-state UEs, the serving GW terminates the downlink data path and triggers paging when downlink data arrives for the UE.

The PDN GW provides connectivity from the UE to external packet data networks by being the point of exit and entry of traffic for the UE. A UE may have simultaneous connectivity with more than one PDN GW for accessing to multiple PDNs. The Package data network gateway (PGW) performs policy enforcement, packet filtering for each user, charging support, lawful interception, and packet screening.

In Figure 11.4, the lines with different colors (totally four) represent different kinds of communications interactions. UEs can communicate with each other through three ways: LTE network, Wi-Fi APs, and Wi-Fi Direct (i.e., the light dark line, the dark black line, and the dark gray line). In addition, the light gray line represents control plane, which means that network will help UEs to discover others, create communication links, and so on.

11.3 Traffic Offloading onto Wi-Fi Direct

Wi-Fi Direct is a mature technology, which has been successfully applied into smartphones. However, it is still not formulate large scales of products, especially in D2D communication to offload traffic. And in this section, some strategies to offload traffic onto Wi-Fi Direct are summarized and compared.

11.3.1 Preliminaries

As this chapter mainly deals with the network-assisted D2D offloading, the functionalities of network assistance should be reflected. Common D2D communication requires two basic operations: one is device discovery and the other is connection establishment. It is known that the existing D2D technologies working on unlicensed bands, that is, Wi-Fi Direct, support these functions. However, it is very difficult to guarantee particular (good) conditions on D2D links, and the quality of such links may vary significantly over time and with movements of peers. Network assistance can provide session continuity by moving D2D traffic sessions back onto the cellular infrastructure network when/if the D2D link fails.

There is a prime issue that a connection that is not yet established cannot be represented or managed in any conventional way. After careful consideration for network assistance possibilities, the conclusion is reached that there cannot be a single entity that would handle the tracking of content and security as well as the link management. The content tracking needs scalability and rich functionality (which are available at the service infrastructure level), whereas link management requires real-time decisions based on position and radio resource availability (such information is only collected by operators for the access network management). Therefore, to manage all of these assistance features, we introduce a new entity into the operator's core network called the D2D server and an AS that manages the features specific to content tracking.

The D2D server stores the subscribers' UE identifiers along with any P2P application IDs the users have provided, and it communicates with other entities in the operator's core network for the purposes of UE location estimation and with UEs to assist with device discovery, D2D connection establishment, and session continuity. It also communicates with third-party P2P ASs for the purpose of updating user profiles with their associated D2D server names. P2P ASs maintain a database of users' P2P application IDs and offer P2P content and/or services. Registered users can access these P2P ASs to search for peers who are offering their desired P2P content/service.

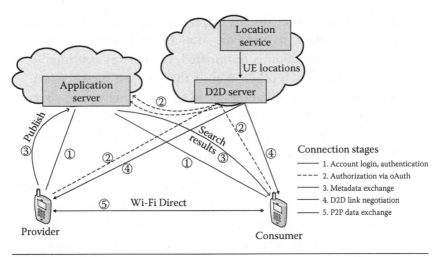

Figure 11.5 The processes of assisted D2D connection establishment via the D2D server.

Figure 11.5 shows the processes of the assisted D2D connection establishment via the D2D server and it works as follows:

1. Each UE, upon user's command, uses the application-layer credentials to authenticate itself with the P2P AS (e.g., Facebook). This allows it to perform operations with content as well as authorize third-party access.
2. The UE also authorizes its D2D server to update the user's profile on the P2P AS with the D2D server name (necessary information for any user that wants to connect via D2D with another peer). This process also verifies that the UE indeed belongs to the owner of the associated P2P application ID.
3. The user may publish or search for the locations of P2P application content and/or services on the P2P AS.
4. The UE asks the D2D server to assist in establishing a D2D connection by resolving the application-layer content location provided by the P2P AS into an actual link-layer connection and IP address to which sockets can be bound.
5. Finally, the P2P data exchange begins. Note that the P2P AS is not involved at this point. The D2D server, however, may monitor and adjust the properties of the D2D link as necessary. Interestingly enough, the devices participating

in the data transfer do not require any prior authorization or direct contact. In addition, after the transfer is complete and D2D connection is terminated, the devices continue to be unaware of each other, even if they are just meters apart. This interesting detail allows us to envision anonymous sharing services, which are not possible with distributed systems, where devices give away their identity on regular basis by broadcasting discovery beacons [9].

This solution requires only one new network entity, the D2D server, which will reside in the EPC of the network. This position allows the D2D server to communicate with the location center to learn the UE positions, while also allowing it to interact with the outside world ASs effectively. From the information illustrated earlier, therefore, architectures, models or frameworks, and so on can take this method as underlying support.

11.3.2 System Strategies

11.3.2.1 Subscribe-and-Send Architecture In this section, an architecture called subscribe-and-send is introduced, which aims to offload the mobile Internet data traffic from cellular networks through Wi-Fi Direct communications.

First, some promises will be stated before introducing the architecture. The smart mobile terminals (SMTs) are equipped with wireless broadband connectivity and are also able to communicate via Wi-Fi. Both the wireless broadband connectivity and Wi-Fi interface are active. Software is installed on the SMT to subscribe to contents on the content service provider (CSP) and to send files to encountering nodes. The software can get a file through either the wireless broadband connection or the Wi-Fi connection. We suppose that the SMTs are Wi-Fi Direct devices. Wi-Fi Direct devices allow the creation of peer-to-peer connections between Wi-Fi client devices without requiring the presence of a traditional Wi-Fi infrastructure network (i.e., AP or router).

There are two main stages in subscribe-and-send: (1) subscribe and (2) send:

1. In the subscribe stage, a user accesses the CSP and subscribes to some interesting contents. The subscription consists of the name of subscribed content, the user's ID, and the deadline of the subscription. Before the deadline of the subscription, the software does not download the subscribed content through the cellular network. It prefers to receive the content from others via its Wi-Fi connection.

2. Some users would like to download the interesting contents through the cellular network. These users can be called source nodes. Source nodes could be the members of a CSP, or the one who would like to pay for downloading contents through 3G. A node accesses the CSP and checks whether someone has subscribed to the content it has. If the node has content that is subscribed to by others, it starts the sending process and delivers the content to the subscriber through opportunistic Wi-Fi connections. When two nodes meet, they exchange their respective subscription tables. If a node has the content subscribed by the encountering node, it sends the content to the node. Otherwise, the opportunistic routing protocol working in the routing layer determines how a node delivers its files. Various opportunistic forwarding protocols can be employed in the routing layer.

Figure 11.6 shows that a user subscribes a movie, and then the source node reads the subscription table and sends the movie to the subscriber one by one. In subscription table, each record consists of the name of subscription resource, subscribed node, and subscribed time. Specifically, node_1 contacts the CSP and downloads movie_1 through 3G, and node_7 subscribes to movie_1 on the CSP as shown in Figure 11.6. Node_1 accesses the CSP and knows that node_7 subscribes to movie_1. Then node_1 begins to send movie_1 to node_7 through opportunistic Wi-Fi connections. Because the Wi-Fi connection cannot always be created, a relay has to carry the content and waits for future Wi-Fi connections. In Figure 11.6, the solid line means the connection created by 3G interface and the dotted line means peer-to-peer Wi-Fi communication. After node_3 gets movie_1 from node_1, as it has no other Wi-Fi connections, node_3 has to carry movie_1 all the time. When node_3 meets node_7 in the future, it sends movie_1 to node_7.

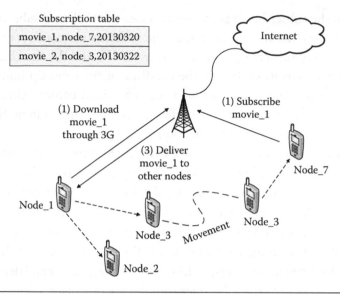

Subscription table

| movie_1, node_7,20130320 |
| movie_2, node_3,20130322 |

Internet

(1) Download
movie_1
through 3G

(1) Subscribe
movie_1

(3) Deliver
movie_1 to
other nodes

Node_7

Node_1

Node_3   Movement   Node_3

Node_2

Figure 11.6   Illustration of subscribe-and-send architecture.

Next, the subscription management is stated as follows:

1. When a subscriber receives the content subscribed on the CSP, it sends a response message to the CSP and removes the subscription of this content from the subscription table.

2. If a subscriber does not receive the subscribed content after the deadline of the subscription, the CSP prompts the user about the failure of the subscription. Then the user can download the content through 3G, or extend the deadline of the subscription. If the user does not extend the deadline, the CSP removes the subscription of the content from the subscription table on CSP. Meanwhile, the subscription is removed from the subscription table on the user's device, too.

3. As the deadline of a subscription is created by the subscriber, the deadlines of different subscriptions are different. And it is not necessary that the clocks on all devices are synchronous because the CSP maintains the total subscription table.

4. Each node accesses the CSP to check the subscription table every 10 minutes or 30 minutes through 3G, so the 3G traffic load of a SMT is very low [10].

In subscribe-and-send architecture, the function of CSP is similar to the AS. However, some parts of the CSP's function cannot be realized by AS, which is introduced before. For example, the CSP provides subscription table for subscribers. If these functions can be realized in AS, and with the coordination of D2D server, the effectiveness of subscribe-and-send architecture could be better embodied.

11.3.2.2 Social Tie-Based Offloading Framework Reference 11 pointed out that a large portion of the traffic load is due to duplicated downloads of the same popular files. Recently there have been many studies to exploit the D2D opportunistic sharing during intermittent meetings of mobile users for traffic offloading in mobile social networks (MSNs), which is a special form of the delay-tolerant networking (DTN) with more consideration of the social relationship of users. However, there are still several important issues, which are not fully elaborated:

- How to know or how to predict the dissemination delay of each user for each content?
- How to design the seeding strategy to minimize the cellular traffic while satisfying the delay requirements of all users?
- Why mobile users share content with others?

Targeting at the aforementioned issues, the framework of traffic offloading assisted by social network services (TOSS) via opportunistic sharing in MSNs is proposed to offload social network services (SNS)-based cellular traffic by user-to-user sharing [12].

The TOSS framework entails both an online SNS and an offline MSNs. Supposing that there are a total of N mobile users, U_i, $i = 1, 2, \ldots N$, who have corresponding SNS identities; for any two users, U_i and U_j, if U_j follows U_i, U_j is one follower of U_i, and U_i is one followee of U_j. At any time, a user may find or create a new interesting article, image, or video, and share it in the SNS as an initiator of the content. A short message posted by a user containing the content (or link to the content) is defined as a microblog (e.g., a tweet in Twitter or a post on Facebook), and a content file is called a content object. All his or her followers will then be able to access the content, and some of them will further reshare on their timelines. Making comments will not induce any information spread; thus, resharing is only considered. Afterward, while the microblog is being spread to other users in the online SNS, the content object will be accessed and spread in the offline MSN.

In the TOSS framework, several factors are defined for each user U_i: two for the online SNS, (1) the outgoing spreading impact, $I_i^{S \rightarrow}$, and (2) the incoming spreading impact, $I_i^{S \leftarrow}$, both of which indicate how important the user is in propagating the microblog (to others or from others); two for the offline MSN, (3) the outgoing mobility impact, $I_i^{M \rightarrow}$, and (4) the incoming mobility impact, $I_i^{M \leftarrow}$, which indicates how important the user is in sharing the content object (to others or from others) via physical encounters.

From the illustrated scenario of TOSS in Figure 11.7, in the online SNS, Cindy shares a video (link) with Eva and Alex, who may in turn share with Bob and David, respectively. Meanwhile, the video content is first downloaded via a cellular link and stored in Cindy's phone. However, in the offline MSN, Cindy is geographically distant from other people but David is in proximity. Although David may not know Cindy, TOSS detects that the $I_i^{S \rightarrow}$ impact of Cindy to David via Alex is also very strong, and thus lets Cindy share the video with David via local Wi-Fi connectivity. Furthermore, TOSS evaluates

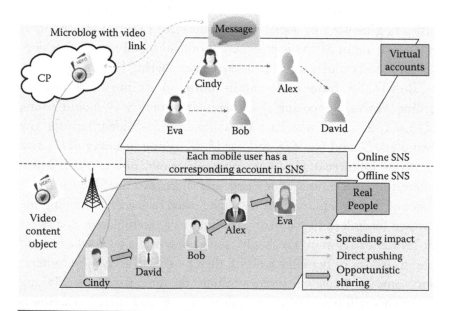

Figure 11.7   Illustration of TOSS framework.

the I^M impact of Alex and pushes another copy to him via a cellular link, because Alex is likely to meet Bob and Eva in the offline MSN frequently, and Bob and Eva often access content with some delays. Then the content object will be propagated by local connectivity from Alex to Bob and to Eva later [13].

Now, some parameters are introduced to help understanding the framework. A vector $\vec{p}$ is defined to indicate whether to push the content object to a user via cellular links or not. The factor $I^{S\to}$ of user U_i to user U_j is denoted by γ_{ij}, which is the probability that U_j will reshare the microposts from U_i. From another perspective, γ_{ij} can indicate how often U_i's microposts will appear in U_j's timeline. Hence, $I_i^{S\to}$ and $I_i^{S\leftarrow}$ of U_i to and from the whole user base can be respectively calculated by $I_i^{S\to} = \sum_{j=1}^{N}\gamma_{ij}^*$, $I_i^{S\leftarrow} = \sum_{j=1}^{N}\gamma_{ji}^*$, where γ_{ji}^* is to denote the impact from user U_i to user U_j via all possible paths with less than or equal to H hops. Here, H is less than or equal to the maximal diameter of the SNS graph. γ_{ij} is an forwarding probability to U_j, whereas γ_{ji}^* depicts how influential from U_i to U_j between H hops. $A_i(t)$ is denoted as the probability to access the content at time t. $A_i(t)$ can be considered as the access utility function, in order to calculate the user satisfaction performance. λ_{ij} is denoted as the opportunistic contact rate of user U_i with user U_j. Note that $I_i^{M\to}$ is actually the same as $I_i^{M\leftarrow}$ since $\lambda_{ij} = \lambda_{ji}$ for any U_i and U_j due to the symmetric nature of contacts. Hereby, the I^M factor for U_i is defined as $I_i^{M\to} = I_i^{M\leftarrow} = \sum_{j=1}^{N}\lambda_{ij}$.

To evaluate the effectiveness of TOSS framework, SNS trace data is needed to quantify the spreading impact factors and access delays, as well as MSN trace data to analyze the mobility impact. First, the most popular online SNS in China, Sina Weibo is selected, and 2,223,294 users are kept track for four weeks during July, 2012. A total of 37,267,512 microposts generated (and partially reshared) by the users are collected and further the list of all the resharing activities for each micropost is obtained. The online SNS spreading impact can be evaluated by γ_{ij} and I^S. The results show that a smaller number of people have significant outgoing impact ($I^{S\to}$) to the whole SNS, whereas many users have very small impact. Also, many users are more likely to be impacted rather than impacting others ($I_i^{S\to} < I_i^{S\leftarrow}$). Measurement results of the access delays on the whole user base show that a large portion of users access the SNSs with sufficiently large delays, which TOSS can utilize to disseminate the content object by offline opportunistic sharing.

Second, four mobility traces: Massachusetts Institute of Technology (MIT), Infocom, Beijing, and SUVnet are chosen to evaluate the performance of the offline MSNs according to λ_{ij} and I^M. These traces record either direct contacts among users carrying mobile devices or GPS-coordinates of each user's mobile route. The four traces differ in their scales, durations, and mobility patterns; the MIT and the Infocom traces are collected by normal people, but the Beijing and the SUVnet traces are collected by vehicles. The results show that the Infocom trace has the highest contact rate because users are at a conference spot, and thus have high contact rates. The Beijing and the SUVnet traces have large intercontact intervals because they have relatively low frequency of GPS records and large user base.

11.3.2.3 Offloading Cellular User Sessions onto D2D Links The current existing literature fails to provide a unified framework for modeling the intricate interactions between a cellular network in the licensed bands and a D2D network in the unlicensed bands. In this section, an integrated system model of cellular and D2D networks is introduced, which is capable of offloading user sessions onto D2D connections in the unlicensed spectrum.

It concentrates on a cellular network in the licensed bands coupled with a D2D network in the unlicensed bands both serving data from wireless users. The considered traffic corresponds to real-time sessions with the target bitrate r. For each session i, we differentiate between the transmitting user T_i, which is data originator and the receiving user R_i, which is the respective destination. Due to nonoverlapping frequency bands, transmissions on the two networks do not interfere with each other and every T_i may send its data to R_i via either the cellular network (infrastructure path) or the D2D network (direct path) as shown in Figure 11.8. Here, an assumption is given as follows:

Assumption: The transmitting users are distributed as a Poisson point process (PPP) in the three-dimensional space (two-dimensional spatial locations and time) with time-independent rate function $\lambda \cdot f(x)$, where λ means the arrival rate. It is further assumed that this user location distribution is uniform within a circle of a particular radius R.

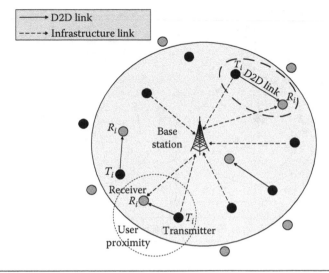

Figure 11.8 Offloading of cellular traffic with celullar operator's assistance.

For a transmitting user T_i, the corresponding receiving user R_i arrives simultaneously with T_i, such that the location of R_i is distributed uniformly within the same circle of radius R. The aforementioned assumption implies that the locations of transmitting users are also distributed uniformly within the same circle R. We additionally assume that the duration of a real-time session by each T_i is exponentially distributed with mean $1/\mu$.

General system operation is illustrated in Figure 11.9. The following consecutive services are assumed when a new data session arrives into the system. First, cellular network attempts to offload the newly arrived session onto the D2D network. In case the session is accepted, it is served by the D2D network without interruption until when it successfully leaves the system. Otherwise, the cellular network attempts to serve this session. Finally, if the session cannot be admitted by the cellular network as well, it is considered blocked and permanently leaves the system.

It is reminded that the arrival rate on the D2D network is λ (see previous Assumption). Due to the Poisson property of thinned flow, the arrivals on the cellular network (those rejected by the D2D network) also follow a Poisson process of density $\lambda \cdot (1 - P_a)$, where P_a is the D2D network accept probability. Abstracting away the point locations for analytical tractability, we assume that the

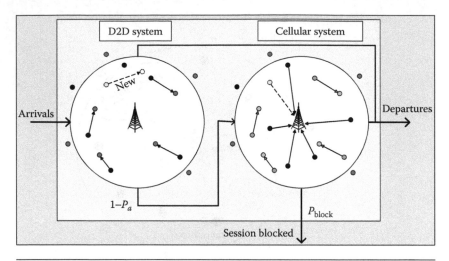

Figure 11.9 General system operation procedure.

arrivals on the cellular network are also uniformly distributed within the circle of radius R.

Consequently, the system-blocking probability P_{block} may be established as follows:

$$P_{block} = 1 - \left(P_a + (1 - P_b) \cdot (1 - P_a) \right)$$

where P_a is the D2D network accept probability, and P_b is the cellular network blocking probability.

Another important metric considered by this work is the energy consumption ε of a typical session given that it satisfies the bitrate requirement. This follows from the Little's law and the definition of the average energy consumption as

$$E[\varepsilon] = \frac{E[P]}{\lambda \cdot P_a}$$

where P_a for the D2D network may be replaced by $1 - P_b$ for the cellular network [14].

Here gives the test scenario that mimics LTE-assisted offloading of user sessions from cellular onto Wi-Fi Direct. This scenario concentrates on an area of interest, in which colocated cellular and D2D networks cover a limited region with many users requiring service (e.g., shopping mall, business center). In particular, we consider an isolated circle cell of radius $R = 100$ m and disregard interference coming from the neighboring cells as assumed before. In this area, the users need to exchange small multimedia fragments with the required bitrate of $r = 4.8$ Mbps. Given that the session duration is distributed exponentially with the mean of 3 seconds, an average transmission carries about 2 MB of information.

As assumed earlier, the session interarrival times are exponential with λ new sessions arriving every second and requesting service. All sessions have specific destinations within the considered area of interest. However, a particular transmitting user may either be successfully accepted by the D2D network, or rejected and need to attempt the LTE base station (BS) instead. If cellular resource is insufficient to admit this user, it is blocked permanently.

Simulation results and discussions are given in the following. In close connection with the capacity goes the blocking probability or the proportion of service requests that cannot be served by the network. It is demonstrated how system-blocking probability P_{block} and D2D-rejection probability evolve with an increase in the load on the network.

When the cellular system is empty, it can afford accepting all links, no matter the quality. Under such conditions, the link quality for arrivals and accepted links is similar, and there are almost no discards. When the cellular system becomes loaded, however, it takes only shorter links in—as those have significantly better chances to fit into the schedule. An empty D2D system cannot afford such luxury—the links are overall much worse, and it has to be very selective to ensure connectivity. From the results, it can be seen that irrespective of the arrival rate, the D2D system consistently remains highly selective to the links based on their length, with almost identical distributions for both empty and overloaded conditions. The reason for this is that the survival of a D2D link is primarily determined by its interference at higher loads. Indeed, shorter links have somewhat better chances of not getting blocked, but combined with other effects it does not reflect in the final statistics.

In addition, it can be clearly seen that at low arrival rates the D2D connections have very high impact on the energy efficiency of the system, improving it by up to 14%. However, as the system gets loaded, the D2D can no longer take over any significant portion of the links, and the energy savings become less significant.

Finally, it can be generally concluded that network-assisted offloading of LTE data onto Wi-Fi Direct D2D connections may significantly improve session blocking probabilities, as well as boost energy efficiency of wireless transmitters.

11.3.3 Analysis of the Aforementioned System Strategies

Subscribe-and-send architecture proposes a basic way to make users share contents through Wi-Fi Direct connection in which UEs can use CSP to download contents with Wi-Fi, or contents can be delivered by relays using Wi-Fi Direct. The CSP, which plays a role of storing information used by UEs, provides the chance that a particular UE can check whether other UEs have the information, which is subscribed by him or her. The problem is that it lacks an effective opportunistic routing protocol when the source node starts the sending process and delivers the content to the subscriber through opportunistic Wi-Fi connections, though in the article [13], there is a forwarding protocol termed High PRobability Opportunistic forwarding (HPRO) playing a part in the sending process. The HPRO algorithm is inefficient to find the destination node.

The most attractive part of the TOSS framework is that it utilizes social relationship to offload traffic. User relationships and interests in online SNSs have significant homophily and locality properties, which is similar to those of offline MSNs. Homophily here means that the online and offline users are both clustered by regions and interests. People with similar interests like to share and transfer the interesting information with each other. The locality here means that people who are geographically close may have similar trends of accessing the content and sharing with each other. Therefore, this social relationship could be used by subscribe-and-send architecture in the forwarding process to help to find the destination nodes. In addition, using the $I_i^{M\rightarrow}$ of a particular UE, the TOSS determines whether push to a copy of content via a cellular link, which maximizes the function of an active UE to offload traffic.

The integrated system model of cellular and D2D networks, capable of offloading user sessions onto D2D connections in the unlicensed spectrum, mainly analyzes from mathematics how likely that D2D networks could be accepted. Also, it only utilizes a system-level simulator to support diverse deployment strategies, traffic models, channel characteristics, and wireless protocols for evaluating D2D connectivity. However, it lacks a concrete implementation. Therefore, this proposed model could be the basis for the aforementioned subscribed-and-send architecture and the TOSS framework.

Actually, there could be some improvements in the subscribe-and-send architecture and the TOSS framework. In the forwarding process of the subscribe-and-send architecture, the HPRO algorithm is inefficient to find the destination node, whereas the TOSS framework can greatly facilitate to find destination with the help of social relationship. Therefore, the subscribe-and-send architecture and the TOSS framework should be combined together in delivering contents to the destination nodes in some particular environments.

11.4 Traffic Offloading onto LTE Direct

LTE Direct is a synchronous system in licensed spectrum under the control of the operator. However, given many technical challenges and disjoint opinions of 3GPP member companies, *product* is not expected for several years. Therefore, at current status, there exists no commercially available applications/implementation offloading cellular traffic based on LTE Direct. This section first introduces a system design for D2D users communicating on licensed cellular spectrum, which uses uplink channel. Then, a scheme to balance load between different eNBs is described in the next section.

11.4.1 Use of Uplink Channel

There are some reasons to use uplink resources for D2D communication:

- The use of uplink resources is less impacting on the performance of the network as a whole.
- The use of downlink resources can create exclusion zones within which network-connected UEs are unable to receive eNB transmissions [15].
- The use of the uplink allows the network to exercise the needed control to meet the performance requirements of both network-controlled and directly communicating UEs.

- The use of uplink resource rather than downlink resource is advantageous in the viewpoint of implementation complexity of mobile terminals [16].

A system design is proposed to enable D2D communications on the LTE uplink spectrum, wherein the uplink frequencies alone are used for the bearer link between the devices while control of the link is provided by a control path through the eNB(s) to which the UEs are associated [17].

Figure 11.10 depicts a general operation of D2D service and it requires minimal changes to the existing signaling mechanisms in the LTE specifications. However, in a frequency division duplexing (FDD) system, such an approach would require the UEs to add a receive capability on the uplink, whereas in time division duplexing (TDD) systems, such a hardware requirement does not exist. The eNB uses downlink to transmit control signals, whereas uplink is used to transmit data before UEs get to communicate in D2D way.

The system design contains four parts: the selection of waveform and associated channel structure, device discovery, resource allocation and management, and simulation models and results.

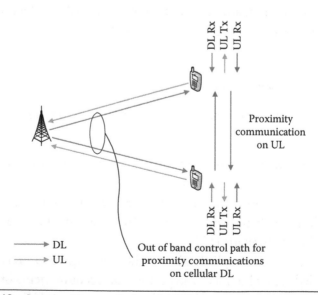

Figure 11.10 Control and bearer paths for uplink D2D.

- *Selection of waveform and associated channel structure*: In keeping with the tenet of maximizing reuse of the existing LTE air interface, it is evaluated that the options of reusing either the eNB Downlink waveform/channel structure or the UE uplink waveform/channel structure, for the UE to UE transmissions on the uplink. As the uplink channel single-carrier frequency-division multiple access (SC-FDMA) waveform has a lower peak to average power ratio (PAPR) and therefore will support greater D2D communication range, it is selected along with the current LTE uplink channel structure.

- *Device discovery*: The device discovery process is illustrated through the scenario as shown in Figure 11.11 in which UE1 and UE2 wish to communicate over the D2D link. Each D2D UE is then configured to report back the measured signal strength of its peer's sounding reference symbols (SRS) along with its own power headroom (from which its SRS transmit power can be derived). The eNB is able to combine the reports of the two UEs (link quality derived from

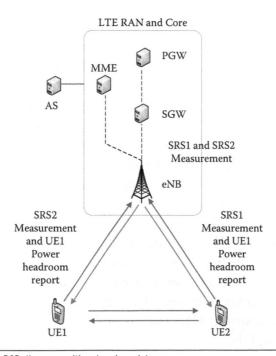

Figure 11.11 D2D discovery with network assistance.

SRS measurements) to make a determination of the path loss between the UEs (or the broadband link quality, if signal-to-interference-plus-noise ratios [SINRs] are reported). It is also able to further derive the uplink rates achievable by the UEs based on the power headroom information and compare the direct communication data rate with the rates achievable if the communications were to be routed through the network. This allows the eNB to make a decision on both the viability and relative benefit from direct communication. In Figure 11.11, the application server (AS) uses a control plane interface to the network control element, the mobility management entity (MME) that then uses its own knowledge of UE point of attachment to the network to determine coarse proximity, and uses eNB assistance, described later, to determine fine-grained proximity.

- *Resource allocation and management*: Once the network has made a determination to support direct communications between two UEs, it remains to make a resource allocation. An augmented eNB scheduler is best suited to accommodate this progression of increasingly spectrally efficient solutions. However, the key challenge of eNB resource allocation is that the scheduler is now separated from both the data source and the data sink (i.e., the transmitting and receiving UE, respectively). The simple resolution envisaged to this problem is a transmission of the acknowledgment feedback by the receiving UE at a power level sufficient for accurate decoding at both the scheduling eNB and the transmitting UE. Therefore, UE reception of the direct transmission requires only an additional format for the scheduling grant from the (scheduling) eNB, which would function as an uplink listen grant and include the D2D transmission parameters determined by the eNB scheduler, for the packet in question. The D2D transmission and reception are then enabled by the simultaneous communication of an uplink transmit grant (to the transmitting UE) and an uplink listen grant (to the receiving UE).

- *Simulation models and results*: Based on the system design, a standard 57 cell FDD system is considered with uplink bandwidth of 20 MHz, operating at 700 MHz, and with an intersite distance of 1.732 km. Sixty UEs uniformly dropped within the center cell of the cluster are allowed to communicate directly or via the eNB, enabling a comparison between the rates of direct and eNB-routed communication for these UEs. Different numbers of UEs are randomly selected as candidates for D2D pairing. The pairing for each of these selected UEs with their respective peers is completed by random selection of peering UEs within a nominal 100 m radius of the previously selected UEs. The UE's transmit powers are selected according to the fractional power control rule so as to not disturb the performance of network-routed UEs in neighboring cells. UEs operate in one of two modes: orthogonal or group reuse. In the orthogonal mode, a directly communicating UE pair uses physical resource blocks (PRBs) different from those used by other directly communicating UEs or UEs transmitting to the eNB. In group reuse, the D2D pairs fully reuse the frequency resources assigned to them while maintaining orthogonality with the resources assigned to the UEs transmitting to the eNB.

The results of simulation indicate that LTE D2D operation in the UL radio frequency resources demonstrates significant throughput gains. As expected, the group reuse scheme is more spectrally efficient than the orthogonal D2D scheme, and the gains of both schemes increase with the number of D2D pairs.

11.4.2 Load Balancing

Usually the typical application of direct D2D offloading is the mobile P2P-style content sharing, where each mobile UE acts as a mobile P2P server, installs a big memory block and stores inside lots of popular contents, and registers its available contents to the operators. This application, although able to offload data from the serving eNB, is subject to the limitation that the receiver of the outgoing data (or the holder of the requested data) is just in close proximity to the transmitter. Furthermore, it cannot detour traffic from congested macro eNBs or pico eNBs to adjacent lightly loaded (uncongested) eNBs. So, in the following parts, a D2D communication-based load-balancing algorithm is proposed to achieve efficient load balancing among different tier cells in LTE-A networks. In this algorithm, the D2D communication adopts the licensed band. When the technique of LTE Direct comes into service, the D2D UE pairs can take LTE Direct to communicate.

In this algorithm [18], three-tier heterogeneous network was considered: macrocells, picocells, and femtocells. The algorithm has four steps in total, and it will proceed to the next step only after it fails in the current step. Assumption: suppose in the congested macrocell, an associated mobile UE is requesting for access to Internet. As the serving macro eNB is already fully loaded, there is no available PRB for the requesting UE within the macrocell. The macro eNB operates as follows:

Step 1: The macro eNB tries to offload the requesting UE to an uncongested cell adjacent to the UE via a D2D relay. Figure 11.12 shows an example of Step 1. The UE1, UE3, and UE5

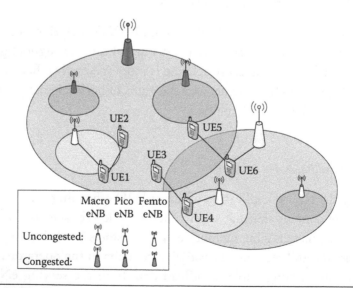

Figure 11.12 Illustration of Step 1 in the D2D communication-based load balancing algorithm.

request the congested macro eNB to provide Internet access, and the macro eNB manages to offload their traffic to adjacent lightly loaded picocells or macrocells via the D2D relays UE2, UE4, and UE6, respectively. But, if there is no other eNBs around the requesting UE, or all neighboring eNBs are fully loaded, or the uncongested neighboring eNBs fail to find an eligible D2D relay, the macro eNodeB proceeds to Step 2.

Step 2: The macro eNB tries to release some occupied PRBs for the requesting UE, by offloading a currently being served macro-tier UE and its ongoing traffic to an adjacent uncongested cell via combined D2D link and cellular link. Figure 11.13 shows an example of Step 2. The congested macro eNB first offloads a currently being served UE, that is, UE2, and its ongoing traffic to the adjacent uncongested pico eNB1 via D2D relay UE3, then allocates the newly released PRBs (occupied previously by UE2) to the requesting UE1. The dashed line denotes a newly released link. But, if the requirements cannot be satisfied and the macro eNB fails to offload any being served macro-tier UE, that is, no PRBs can be released from the macrocell for the requesting UE, the macro eNB proceeds to Step 3.

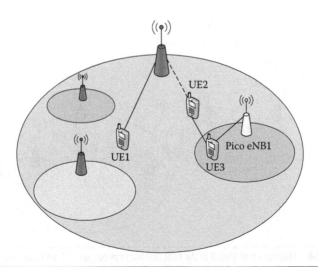

Figure 11.13   Illustration of Step 2 in the D2D communication-based load balancing algorithm.

Step 3: The macro eNB tries to offload the requesting UE to a congested eNB, which is close to the UE and able to release some PRBs by offloading a currently being served UE to its nearby uncongested cell. Figure 11.14 shows an example of Step 3. After UE3 is offloaded from the congested pico eNB1 to the uncongested pico eNB2 via D2D relay UE4, the requesting UE1 is offloaded from the congested macro eNB to the pico eNB1 via UE2. The dashed line denotes a newly released link. If the macro eNB fails to offload the requesting UE to an adjacent congested eNB, due to lack of good channel quality, D2D relay(s), or sufficient PRBs, and so on, it proceeds to the last step.

Step 4: The macro eNB tries to allocate to the requesting UE PRBs newly released by offloading a currently being served macro-tier UE to a nearby congested eNB, which is able to offload a being served UE to an adjacent uncongested cell. As shown in Figure 11.15, the macro-tier UE2 can be offloaded via D2D relay UE3 to the congested pico eNB1, after UE4 is offloaded to the uncongested pico eNB2 via UE5 [18].

In the aforementioned algorithm, there could be more steps in theory, but if transmission is through too many UEs or eNBs, it

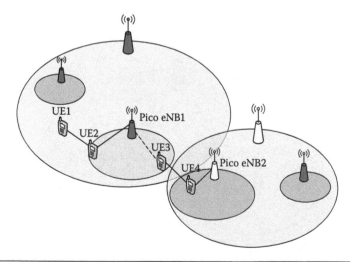

Figure 11.14 Illustration of Step 3 in the D2D communication-based load balancing algorithm.

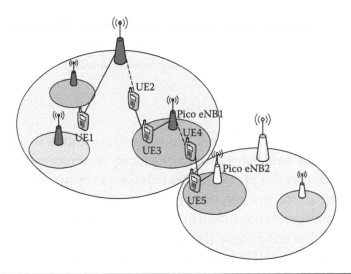

Figure 11.15 Illustration of Step 4 in the D2D communication-based load balancing algorithm.

will be less efficient. In addition, LTE Direct has longer transmission distance than Wi-Fi Direct, which is fit for the aforementioned algorithm. It is known to all that, in LTE-A networks, some areas may have severe interference among nodes depending on the node density there. Due to the limited channel resources at the eNB and the overly crowded nodes in the area, the wireless link (uplink or downlink) between a mobile UE and the eNB usually has very poor SINR. For such scenarios, the proposed D2D-based traffic offloading algorithm can also be utilized to alleviate the air interface congestion, to increase system throughput, and to improve user experiences. However, one thing should be paid attention to, that is, besides the basic requirements for setting up combined D2D link and cellular link in detouring traffic, the offloaded served UEs should be seamlessly connected to an adjacent uncongested cell to ensure users' satisfaction.

11.5 Usage Cases of Traffic Offloading

In this section, some D2D usage cases are summarized, which can be classified into two broad categories. The first category is referred to as the peer-to-peer case in which the D2D devices are the source and destination of the exchanged data. The second category is the relay case, which means that one of the communicating D2D devices has to relay the exchanged information to the BS or another D2D device, which further forward the data to the destination device.

11.5.1 Peer-to-Peer

Local voice service: D2D communications can be used to offload local voice traffic when two geographically proximate users want to talk over the phone, for example, people in the same large meeting room want to discuss privately, or companions get lost in a supermarket, as shown in Figure 11.16. However, this usage case is rare according to the operators' current market statistics.

Local data service: D2D communications can also be used to provide local data service when two geographically proximate users or devices want to exchange data, as shown in Figure 11.17. Some applications of D2D communications are given as follows:

- *Content sharing*: Friends exchange photos or videos through their smartphones, or people attending a conference download materials from a local server.
- *Multiplayer gaming*: The famous Japanese game *Dragon Quest IX* has a cooperation mode consisting of up to four players using local wireless connections to play together. The three guests join the host system's world and can go anywhere that the host has explored.

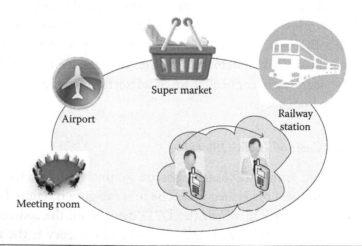

Figure 11.16 Local voice service.

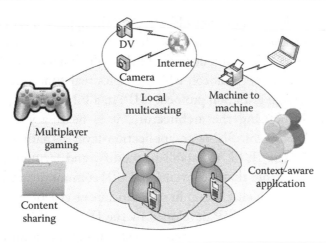

Figure 11.17 Local data service.

- *Local multicasting*: The shops advertise the sale promotion information to the customers.
- *Machine-to-machine* (*M2M*): A laptop connects to a printer, or a smartphone connects to a television for the photo or video display.
- *Context-aware application*: It is a driving factor for the D2D technologies and is based on the people's desire to discover their surroundings and communicate with nearby devices (machines or people). An example is location-aware social networking, such as Foursquare, where users *check in* at venues using a mobile website, text messaging or running a device-specific application, and selecting from a list of venues that the application locates nearby. Therefore, context-aware applications may be based on any of the aforementioned four types of D2D communications [16].

11.5.1.1 P2P Content Sharing One of the typical local data service applications based on D2D offloading is mobile P2P-style content sharing, where each mobile UE acts as a mobile P2P server, installs a big memory block and stores inside lots of popular contents, and registers its available contents to the operators.

In this part, an architecture is presented that allows users in a group to experience the ubiquitous and real-time sharing service of resource like image, video, and website address. And the approach is motivated by taking advantage of the novel industrial standard Wi-Fi Direct, which allows Wi-Fi terminals to establish a Wi-Fi network without a real AP and P2PSIP protocol, which enables real-time communication using session initiation protocol (SIP) in a P2P fashion [19].

Before introducing the architecture, it is necessary to explain the P2PSIP protocol. SIP is an application-layer signaling protocol developed by the IETF to establish, modify, and terminate multimedia session. Note that a conventional SIP communication system is not easily and quickly set up in spontaneous or emergency scenarios. Nevertheless, P2PSIP developed by the IETF P2PSIP working group addressed those weaknesses by distributing the binding information over P2P overlay. The P2PSIP network consists of a number of nodes, which play different roles like a bootstrap peer, an adapter peer, or an admitting peer. By using P2PSIP over Wi-Fi Direct layer, we can deal with the dynamic scenarios in which any peer easily has a desire to join or leave the group, even in the ongoing sharing process.

The following picture is a sharing-service P2P overlay architecture that allows for establishing SIP sessions between members without centralized servers in Wi-Fi Direct environment.

In this architecture, there are two types of entities (excluding the normal peer and the joining peer):

Bootstrap/presence peer (B/P/P) provides initial configuration information to newly joining nodes and is able to receive the SUBSCRIBE and PUBLISH message from peers and notifies others the change of presence status by sending NOTIFY message. The reason for combining the bootstrap and presence function within one node is the small group created by Wi-Fi Direct.

Push-to-talk peer (PTT/P) is the first peer initializing the conference. Its IP address is registered on overlay and can be retrieved by others. The function of this peer is to control the media burst control protocol (MBCP) message from participant in conference.

The client-server-based model is applied for signaling plane and the P2P one is for data. In this architecture, both B/P/P and PTT/P are the main entities maintaining two kinds of signaling message: one presented by the continuous light-dark line for presence information-related message, and the remaining presented by the dotted line for media handling-related message. As shown in Figure 11.18, all signaling messages

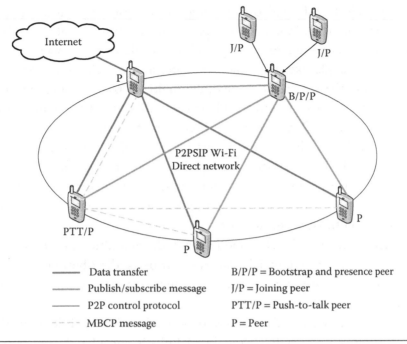

Figure 11.18 A sharing-service P2P overlay in Wi-Fi Direct.

must go through a central point and then be broadcast to others in the overlay whereas the data, presented by the continuous thick-dark line, is exchanged among participants in direct style. In addition, a Wi-Fi Direct device can connect to two different networks at the same time with the hardware support as mentioned earlier. With the support of dual mode networking, Internet user might communicate and send data with other participants in Wi-Fi Direct network, or he or she is able to download resource from outside before deciding to share.

Since every peer can quit the group at any time, P2PSIP is intentionally utilized to address this issue. If a peer who is the first cannot retrieve the address of PTT peer from the overlay, it can change itself to play a role as a server for other clients by registering the corresponding uniform resource identifier (URI) on overlay. By doing so, the robustness and flexibility of sharing service can be improved, especially in a high-churn environment.

11.5.1.2 Multicast D2D Transmission   In D2D-enabled cellular networks, direct multicast transmissions, where the same packets from a UE are sent to multiple receivers, are important for scenarios such as local file transfer/video streaming, device discovery, cluster head selection/coordination, and group/broadcast communications. The aforementioned scenarios are typical, especially in some places with abundant people like shopping mall, entertainment chambers, and so on. And in these scenarios, one direct multicast transmission reduces overhead and saves resources. However, unlike the more commonly studied unicast D2D, multicast D2D has its own challenges. For example, due to the heterogeneous locations of receivers and complicated radio environment, link quality may vary significantly over receivers in each multicast cluster; thus, retransmissions are often required to cover more or all the receivers, which degrades the whole point of multicast versus unicast.

Here, a tractable baseline model for studying multicast D2D transmissions is proposed, which contains four parts: distributions of network nodes, multicast transmission, channel model, and performance metrics.

First, in the part of distributions of network nodes, a hybrid network consisting of both cellular and D2D links is considered. The positions of BSs form an independent PPP Φ_b with intensity λ_b. The PPP model for BS locations has been recently shown to be about as accurate in terms of both SINR distribution and handover rate as the hexagonal grid for a representative urban cellular network. Similarly, the positions of multicast D2D transmitters form an independent PPP Φ_m with intensity λ_m. It is going to further assume that for each D2D transmitter, the positions of its intended receivers form a point process Φ_{m,x_i}. Note that it does not assume any specific distribution for the receiver point process except the first-order intensity measure.

Then, there are some introductions about multicast transmission. For each D2D transmitter x_i, they have a common message for all the intended receivers in Φ_{m,x_i} and the message can be sent for τ_m times, which is a preconfigured system parameter. Compared to one shot transmission, sending the message $\tau_m > 1$ times enables more intended receivers to successfully decode the message. Further, there is an assumption that multicast transmitters are static during τ_m transmissions. When D2D UEs are in coverage, the ground cellular network can assist D2D communications. Specifically, each in-coverage multicast D2D transmitter has a serving BS; normally, the serving BS is the BS providing the strongest reference signal receiving power (RSRP). And in this model, it chooses the nearest BS as the serving BS. It assumes that D2D is overlaid with cellular networks, that is, D2D transmitters and BSs use orthogonal transmission resources, and thus there is no mutual interference between cellular and D2D transmissions. In addition, the multicast message of each D2D transmitter is known by its serving BS. Note that when cellular network coverage is available, D2D transmissions are under relatively tight network control. So the coordination between cellular and D2D transmissions can be easily achieved by communication through the BS control channels [20].

In the channel model, the SINR of the link from the typical D2D transmitter to D2D receiver and the SINR of the link from the nearest BS of the typical D2D transmitter to D2D receiver are given by mathematical expression.

The last part of the model is performance metrics. Focusing on the typical cluster, it is interested in the probability that an arbitrary receiver can decode the multicast message of the typical D2D transmitter and this is termed as coverage probability. Although coverage probability characterizes the performance of an individual receiver in the typical cluster, it is also desirable to have a metric to measure the performance of the typical cluster as a whole. Thus, another metric studied is the mean number of covered receivers in the typical cluster.

In the following parts, analysis of multicast performance will be given in two aspects:

- *With network assistance*: With network assistance, the probability that the receiver is covered is increased as long as either the BS or the multicast transmitter covers it. Further these two events are independent. Network assistance can significantly reduce the number τ_m of transmissions to achieve the same mean number of covered receivers in the absence of network assistance. Thus, network assistance is very useful.
- *Without network assistance*: For coverage probability, the farther the potential receiver away from the multicast transmitter, the smaller the coverage probability is. Further, repetitive transmissions are instrumental in improving the coverage probability, especially for far away receivers. But the gain diminishes as τ_m increases.

For mean number of covered receivers, repetitive transmissions are instrumental but the gain quickly diminishes as τ_m increases. This implies that if a D2D transmitter would like to cover far away receivers, other approaches rather than simple repetitive transmissions are expected; such approaches may include increasing transmit power and interference cancellation.

11.5.2 Relay

In this part, one of the communication D2D devices has to relay the exchanged information to the BS or another D2D device, which further forward the data to the destination device (Figures 11.19 and 11.20).

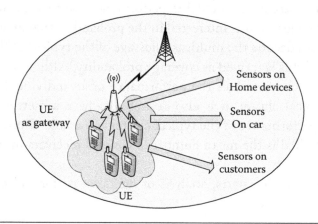

Figure 11.19 UE as gateway to sensor networks.

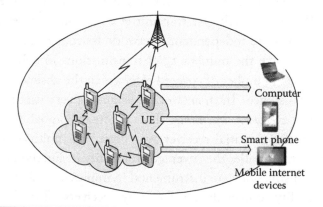

Figure 11.20 UE cooperative relay.

UE as gateway to sensor networks: Most M2M devices are not *directly cellular*. In other words, M2M devices usually first connect to an M2M gateway using wireless personal area network (WPAN), for example, ZigBee, and the M2M gateway connects to a cellular network. For many consumer M2M devices, for example, sensors on home devices, cars, or even the on-body health care devices, cell phones on the consumers are the most suitable M2M gateways. The communications between these sensors and UEs can use the D2D technologies.

UE cooperative relay: In the wireless telecommunications systems that have a large number of subscribers, it is well known that one efficient communication method is to break a long path into a number of smaller hops so that the information is relayed among a number of terminals.

However, the UE relay faces a number of business model difficulties apart from the technical challenges. The biggest obstacle is the users' concern on the information security, wireless radiation, and excessive consumption of their battery power, all of which are due to opening up their mobile devices to other users.

These problems still exist today, and a proper business model with enough incentives for users needs to be designed if the D2D technologies are to be applied to this scenario.

The benefits, marketing challenges, and potential business models for different operator-controlled D2D usage cases are summarized in Table 11.1.

Actually, D2D communications can be applied in many aspects: traffic offloading, efficient content distribution and sharing, 3D video and photography, portable surround sound system, and so on [21]. Many works are with traffic offloading in nature. For example, Reference 22 investigates the problem of cooperative energy-efficient content dissemination among a number of cellular UEs, with the assumption that these UEs are seeking to receive the same content from a common wireless AP and/or eNB. This work integrates the traffic offloading and content distribution together with cooperative D2D communication in which UEs play both roles of relay and source or destination node.

Table 11.1 Usage Cases for Operator-Controlled D2D Communications

	PEER-TO-PEER			RELAY
USAGE CASES	LOCAL VOICE SERVICE	LOCAL DATA SERVICE	UE AS GATEWAY TO SENSOR NETWORKS	UE COOPERATIVE RELAY
Benefits	Enhance capacity	Provide new services		Enhance capacity
Marketing challenges	Rare occasion	Competition from traditional free D2D techniques (like Bluetooth)		Users' concern on information security, and so on.
Potential business models		1. Attract users to pay for identity, QoS and security, context information, management, and so on. 2. Charge the users based on how many minutes or how much bandwidth they use in fully controlled D2D communications and charge a certain amount of fee per month irrespective of the actual D2D data flow in loosely controlled D2D communications		

Source: Zhang, Y. et al., *Canadian Journal of Electrical and Computer Engineering*, 39(1): 2–10, 2016.

11.6 Challenges of Cellular Traffic Offloading

There are paramount challenges and active research activities regarding D2D communications underlaying cellular networks. First, interference management is critical since cellular networks need to manage new interference scenarios by supporting D2D communications, especially for LTE Direct. In cellular networks, traditional cellular UEs (CUEs) can be considered as primary UEs, and additional D2D UEs (DUEs) should not degrade the performance of CUEs. On the other hand, the interference from current cellular networks may also hurt the quality of service (QoS) requirements of DUEs [23]. Second, multihop D2D communication, which allows a UE to be a relay to help other UEs, has not been fully investigated yet. Network coding can be attempted in such scenarios and help improve the throughput of multihop D2D communications underlaying cellular networks. Third, it is critical to inspire people to participate in D2D communications. Therefore, proper incentive mechanisms are imperative for D2D-based traffic offloading. Fourth, the problem of energy consumed in devices that take part in D2D communications is worthy to be considering.

In D2D networks, interference mainly exists in the following aspects: If D2D communications use unlicensed spectrum, the interference mainly exists among the DUEs, which are in proximity. If D2D communications use licensed spectrum, and when DUEs shared downlink cellular resources, the interference sources consist of interference from the eNB in the same cell, interference from other cochannel DUEs in the same cell, and interference from eNBs and cochannel DUEs from other cells. If D2D communications use licensed spectrum, and when DUEs shared uplink cellular resources, the interference sources consist of interference from all cochannel CUEs at the same cell and other cells, and interference from all cochannel DUEs at the same cell and other cells. Solutions to solving interference have been discussed in other chapters of this book.

11.6.1 Congestion Control

In a sense, multihop D2D communication networks can be regarded as a special type of DTNs. Leveraging some smart devices as delay to transmit data will create corresponding problems, such as congestion in the network. As numerous routing algorithms for DTNs employ the *store-carry-forward* scheme, and when intermediate devices buffer messages for others, for their store space is limited, their caches can be easily overwhelmed. Therefore, congestion will therefore be incurred.

There are some articles that proposed several solutions to solve the congestion problem. In the article [24], it designs an efficient context-aware congestion control approach (CACC), which could cope with the congestion without causing any side-effect to the network. It first analyzes the congestion phenomenon and specifies several major context-aware congestion factors, mainly including social strength, message size, and free buffer size. But it only considers contact number and average duration to evaluate social strength. Then it models the congestion control problem as multiple criteria decision-making (MCDM) to identify the messages with the least potential congestion effect on the recipient. In MCDM, the weight of congestion factors is measured by the criteria importance through intercriteria correlation

(CRITIC) way, and a metric (called gain) is defined to characterize the congestion effect. After that, it presents a CACC mechanism that decides the forwarding set.

Among the context factors, message size and free buffer are size social-oblivious factors, whereas contact number and average duration are social-aware factors. As the name implies, contact number $C_{i,j}$(num) is the total encounter number of a node pair in the elapsed time. Average duration $C_{i,j}$(dur) is the mean contact time of each encounter in the past time.

Before the introduction of the context-aware congestion control approach, multiple criteria decision-making is stated as follows: It can help to make a good decision in a complex decision-making environment and mainly contains two parts: deciding the weight of criteria and making the best decision. The first part is a critical stage and plays a key role in the process of decision-making. However, different criteria always have various meanings. It is difficult to measure the weight of criteria. Therefore, CRITIC is adopted to measure the weight of various context-aware congestion factors.

The process is given in Figure 11.21.

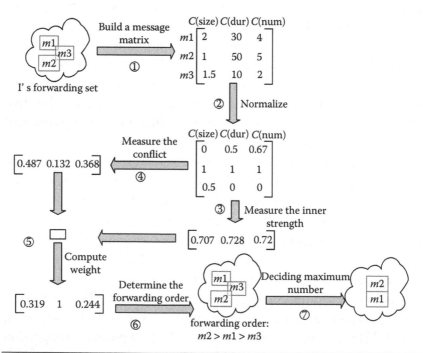

Figure 11.21 Procedure of evaluating the weight of context-aware congestion factor and determine forwarding set.

Step 1: Build a matrix for the messages in the forward set to measure their weight, as shown in step ① of Figure 11.21. In the matrix, row i represents the context-aware congestion factors of message m_i: C(size), C(dur), and C(num).

Step 2: Normalize the message matrix to eliminate the anomalies of data in different measures, as shown in step ② of Figure 11.21. This process transforms different scales and units among various congestion factors into common measurable units such that comparisons of different congestion factors are allowed. Each entry is denoted by $r_{i,j}$.

Step 3: Measure the inner strength of each congestion factor, which is represented by the standard deviation δ of the scores in each congestion factor. This is achieved by step ③ of Figure 11.21.

Step 4: Measure the conflict between congestion factors, which is obtained by summing up the correlation coefficients of a congestion factor with others, as illustrated by step ④ of Figure 11.21.

Step 5: Compute the weight W_j of each congestion factor $C(j)$.

Step 6: Determine the forwarding order according to weights calculated through Step 5.

Step 7: Deciding maximum number based on peer's free buffer.

After acquiring the weight of each congestion factor, the congestion effect of a message to the candidate custody node could evaluated, which is characterized by a new metric gain (G_{mi}). The larger the value of the gain, the smaller the congestion effect of the message. In order to decide the final forwarding set, it first sorts the order of messages in the initialization forwarding set in descending order and then considers the free buffer size of the receiver to decide the maximum number of messages that could be transmitted.

Based on the aforementioned work, the CACC approach is presented, and the flowchart is summarized in Figure 11.22.

In the CACC approach, each node maintains a social table, including the social strength of the current node to other encountered nodes. When the CACC approach is triggered, they first exchange their buffer states and social tables. If the buffer of a peer node is insufficient, the current node gives up forwarding messages to that

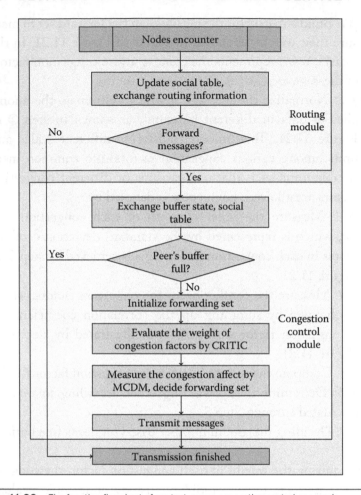

Figure 11.22   The function flowchart of context-aware congestion control approach.

peer; instead, it initiates the forwarding set by including these messages the destinations of which have stronger social strength (e.g., contact duration and contact number) with the peer than current custody, After that, the weight of each context-aware congestion factor is measured by CRITIC and the congestion effect caused by the message in the forwarding set to the peer node is evaluated. Next, the forwarding set and its order are determined by the following two factors: Peer node's free buffer size, and the metric gain, as shown in steps ⑥ and ⑦ Figure 11.21.

Finally, nodes start to transmit messages until the connection is interrupted or there is no message left in the forwarding set.

The experiment scenario and results will be given from this paragraph. From previous discussion, three representative routing schemes, Epidemic, Spray&Wait, and Prophet are considered; their congestion-aware versions are denoted as Epidemic-c, Spary&Wait-c, and Prophet-c, respectively. Without loss of generality, it assumes that each node has the same buffer size. Messages are generated periodically at the source, and thus the network lifetime determines the total traffic generated. Messages' destinations are selected uniformly at random from the whole network, and their sizes are also uniformly distributed. The results will be depicted from three aspects including delivery ratio, delivery overhead, and delivery delay. In addition, two different ways are adopted to mimic various congestion levels of the network: changing the buffer size and changing the message generated interval.

For delivery ratio, the results show that the larger the buffer capacity, the higher the delivery ratio. The reason behind is that the buffer size of a node determines the maximum number of messages stored. Larger buffer space means more messages will be stored in intermediate nodes, so these messages have a greater chance of being delivered to their destinations. By comparison, all congestion-aware algorithms are of superior performance to the original ones under different scales of buffer size.

Due to the large span of the delivery overhead of different algorithms, it is difficult to show their absolute values on one figure clearly. Instead, the results are showed by delivery overhead percentage, that is, a ratio of overhead generated by the original algorithm (or its congestion-aware version) over their sum, to better illustrate the delivery overhead of all algorithms. It is obvious that all congestion-aware algorithms perform better than their original ones. On the one hand, CACC adopts the context knowledge to determine forwarding set and forward messages to the nodes who are closest to destinations, which can shorten the forwarding hops, and thus reduces network overhead. On the other hand, CACC can significantly decrease message loss by forwarding messages to the congestion free nodes, and eventually reduce network load.

For delivery delay, it is clearly seen that all congestion-aware algorithms have a shorter delivery delay than the original ones. This phenomenon comes from two factors. The first is that CACC optimizes the forwarding messages and only delivers messages to a node that has greater social strength with the destination, which could greatly shorten the transmission time. The second is that CACC only delivers messages to a node that has sufficient storage space for newly arrived messages. This could avoid message loss caused by overflow, which will induce the source node to retransmit the original data.

11.6.2 Incentive Mechanism

In the process of D2D communications, the network cannot provide continuous connectivity because of users' mobility, thus leading to the decrease in the user experience. Therefore, it is necessary to propose an incentive mechanism to inspire user to participate in the D2D communication. In this part, there is an investigation on the trade-off between the amount of traffic being offloaded and the users' satisfaction, and a novel incentive framework is proposed to motivate users to leverage their delay tolerance for traffic offloading.

Figure 11.23 illustrates the main idea of the win-coupon framework. Here, the coupon means discount. In detail, in the proposed framework, the user act as sellers to send bids, which include the delay that he or she is willing to experience and the discount that he or she

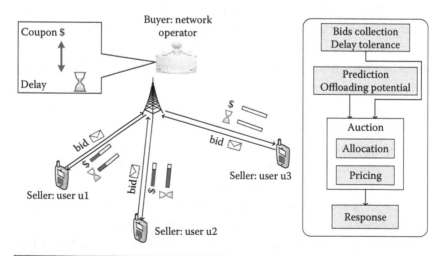

Figure 11.23 The main idea of win-coupon. (From Zhuo, X.J. et al., *IEEE Transactions on Mobile Computing*, 13(3): 541–555, 2014.)

wants to obtain for this delay, and such discount requested by users is called coupon. The network operator acts as the buyer, who offers coupons to users in exchange for them to wait for some time and opportunistically offload the traffic. When users request data, they are motivated to send bids along with their request messages to the network operator. Each bid includes the information of how long the user is willing to wait and how much coupon he or she wants to obtain as a return for the extra delay. Then, the network operator infers users' delay tolerance. In addition, users' offloading potential should also be considered when deciding the auction outcome. Based on the historical system parameters collected, such as users' data access and mobility patterns, their future value can be predicted by conducting network modeling, and then based on the information, users' offloading potential can be predicted [25].

The optimal auction outcome is to minimize the network operator's incentive cost subject to a given offloading target according to the bidders' delay tolerance and offloading potential. The auction contains two main steps: allocation and pricing. In the allocation step, the network operator decides which bidders are the winners and how long they need to wait. In the pricing step, the network operator decides how much to pay for each winner. Finally, the network operator returns the bidders with the auction outcome that includes the assigned delay and the value of coupon for each bidder. The winning bidders (e.g., users u1 and u2 shown in Figure 11.23) obtain the coupon and are assured to receive the data via cellular network when their promised delay is reached.

However, with the increase of downloading delay, the user's satisfaction decreases accordingly, the rate of which reflects the user's delay tolerance. To flexibly model users' delay tolerance, a satisfaction function $S(t)$ is introduced, which is a monotonically decreasing function of delay t, and represents the price that the user is willing to pay for the data service with the delay.

Figure 11.24 shows an example of the satisfaction function $S(t)$ of a specific user for a specific data, where t_{bound} is the upper bound of the user's delay tolerance, p is the original charge for the data service, and the satisfaction curve represents the user's expected price for the data as the delay increases. With this satisfaction function, dynamic characteristics of users' delay tolerance can be captured.

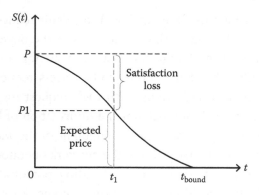

Figure 11.24 Satisfaction function.

11.6.3 Business and Commercial Perspectives

Data offloading in general provides business prospects to everyone in the value chain including device manufacturers, service providers, and hardware and chip vendors. In order to overcome the problem of data overload on cellular networks, vendors, solution providers, and equipment manufacturers are adopting different strategies [26].

Although the product with LTE Direct is not expected in several years, this technology still gets a lot of public attention. Here, we tell about the business and commercial perspectives of LTE Direct. LTE Direct application channels include communication channel and discovery channel. Communication channel enables mobile devices to communicate directly with each other and also to secure connections by controlling the radio resource allocation. But, 3GPP is complied with a full range of requirements in supporting communication channel for national security and public safety. Thus, it is not proper for business propose. However, discovery channel can be used for commercial business. For example, contents provider can send necessary information to customer nearby, send commercial interests, or match information with advertisement. It is also expected to be used in many other services, such as information delivering service geo fencing, network games, social matching, and targeting advertisements in the future.

In mobile operators' perspective, LTE Direct will decrease the maintenance costs of wireless network. In addition, by expanding coverage, data provision in coastal areas and mountainous regions becomes possible in stable speed. With LTE Direct, mobile operators can utilize the acceleration of data speed for marketing and can have competitive advantage in the LTE quality competition. However, mobile operators may face two major problems: first, data usage of the users may decrease in the short term, and average revenue per user (APRU) may be reduced in the long term. Second, mobile operators may lose their control and domination of the network, because information is not stored in the communications company as the communications between devices do not pass through BSs.

The hardware and chip vendors play an important role in LTE Direct. They can develop devices with embedded technology or just insert a chip into existing devices. In addition, mobile operators and hardware and chip vendors are able to cooperate. When LTE Direct is commercialized, hardware and chip vendors will have the benefit of expanded market. Thus, the development of chips must fit their scale of profit and loss. If the government allows LTE Direct only to be used for public disasters, hardware and chip vendors will have difficulty in making profit with disaster network. Therefore, hardware and chip vendors will expect to conduct new commercial service such as advertisement by using discovery channel of LTE Direct.

Besides the aforementioned two aspects, there are opportunities for service and content providers. But, service and contents providers are including not only content owners/creator and mobile app developers but also advertisers. If service and contents providers use discovery channel, all device users in 1 km could become targets for promotion, and they will become advertisers. Location based service (LBS) providers and mobile app developers can participate in this kind of business. Various businesses are engaged in marketing activities such as coupon mailing by using location information of the smartphone users. In the past, this kind of target marketing required passing through specific service platforms. In the case of LTE Direct, information transmission is possible through built-in chips without passing through service platforms. Thus, small businesses or advertisers can deliver their information in cheap cost. Various businesses can participate in the business utilizing LTE Direct by using the network of the mobile operators [27].

11.7 Conclusion

In this chapter, we first give a general introduction to the basic concept of D2D communication. Then, an architecture of D2D offloading system with network assistance is illustrated in Section 11.2. Sections 11.3 and 11.4 mainly talks about different schemes from the technical angles including Wi-Fi Direct and LTE Direct. Section 11.5 depicts the usage cases of traffic offloading, which is divided into two parts: peer-to-peer and relay. At last, Section 11.6 gives the challenge of cellular traffic offloading. It mainly consists of three aspects: congestion control, incentive mechanism, and business and commercial perspectives.

References

1. Forecast C V N I. Cisco visual networking index: Global mobile data traffic forecast update 2009–2014[J]. Cisco Public Information, February 9, 2010.
2. Dimatteo, S., P. Hui, B. Han, and O.K. Li. Cellular traffic offloading through WiFi networks. *In: Proceedings of IEEE Eighth International Conference on Mobile Ad-Hoc and Sensor Systems*, Valencia, October 17–22, 2011, pp. 192–201.
3. Rebecchi, F., M.D. De Amorim, V. Conan, A. Passarella, R. Bruno, and M. Conti. Data offloading techniques in cellular networks: A survey. *IEEE Communications Surveys & Tutorials*, 2015; 17(2): 580–603.
4. LTE Release 12. 3rd Generation Partnership Project (3GPP), 2013–2014, available online: http://www.3gpp.org/Release-12.
5. Bansal, T., K. Sundaresan, S. Rangarajan, and P. Sinha. R2D2: Embracing device-to-device communication in next generation cellular networks. In: *Proceedings of IEEE INFOCOM 2014-IEEE Conference on Computer Communications*, Toronto, April 27, 2014–May 2, 2014, pp. 1563–1571.
6. Yang, M.J., S.Y Lim, H.J. Park, and N.H. Park. Solving the data overload: Device-to-device bearer control architecture for cellular data offloading. *IEEE Vehicular Technology Magazine*, 2013; 8(1): 31–39.
7. Pyattaev, A., K. Johnsson, S. Andreev, and Y. Koucheryavy. 3GPP LTE traffic offloading onto WiFi Direct. In: *Proceedings of Wireless Communications and Networking Conference Workshops (WCNCW)*, Shanghai, April 7–10, 2013, pp. 135–140.
8. Andreev, S., A. Pyattaev, K. Johnsson, O. Galinina, and Y. Koucheryavy. Cellular traffic offloading onto network-assisted device-to-device connections. *IEEE Communications Magazine*, 2014; 52(4): 20–31.

9. Pyattaev, A., K. Johnsson, A. Surak, R. Florea, S. Andreev, and Y. Koucheryavy. Network-Assisted D2D communications: Implementing a technology prototype for cellular traffic offloading. In: *Proceedings of 2014 IEEE Wireless Communications and Networking Conference (WCNC)*, Istanbul, Turkey, April 6–9, 2014, pp. 3266–3271.

10. Lu, X.F., P. Hui, and L. Pietro. Offloading mobile data from cellular networks through peer-to-peer WiFi communication: A subscribe-and-send architecture. *China Communications* 2013; 10(6): 35–46.

11. Scellato, S., C. Mascolo, M. Musolesi, and J. Crowcroft. Track globally, deliver locally: Improving content delivery networks by tracking geographic social cascades. In: *Proceedings of the 20th International Conference on World Wide Web*, New York, 2011, 457–466.

12. Wang, X.F., M. Chen, Z. Han, D.O. Wu, and T.T. Kwon. TOSS: Traffic offloading by social network service-based opportunistic sharing in mobile social networks. In: *Proceedings of IEEE INFOCOM 2014-IEEE Conference on Computer Communications*, Toronto, ON, April 27, 2014–May 2, 2014, pp. 2346–2354.

13. Wang, X.F., M. Chen, T. Kwon, L.H. Jin, and C.M. Leung. Mobile traffic offloading by exploiting social network services and leveraging opportunistic device-to-device sharing. *IEEE Wireless Communications*, 2014, 21(3): 28–36.

14. Andreev, S., O. Galinina, A. Pyattaev, K. Johnsson, and Y. Koucheryavy. Analyzing assisted offloading of cellular user sessions onto D2D links in unlicensed bands. *IEEE Journal on Selected Areas in Communications*, 2015, 33(1): 67–80.

15. Janis, P., C.H. Yu, K. Doppler, C. Ribeiro, C. Wijting, K. Hugl. et al. Device-to-device communication underlaying cellular communications systems. *International Journal of Communications, Network and System Sciences*, 2009; 2(3): 169–178.

16. Lei, L., Z.D. Zhong, C. Lin, and X.M. Shen. Operator controlled device-to-device communications in LTE-advanced networks. *IEEE Wireless Communications*, 2012; 19(3): 96–104.

17. Vasudevan, S., K. Sivanesan, S. Kanugovi, and J.L. Zou. Enabling data offload and proximity services using device to device communication over licensed cellular spectrum with infrastructure control. In: *Proceedings of Vehicular Technology Conference (VTC Fall)*, Las Vegas, NV, September 2–5, 2013, pp. 1–7.

18. Liu, J.J., Y.C. Kawamoto, H. Nishiyam, N. Kato, and N. Kadowaki. Device-to-device communications achieve efficient load balancing in LTE-advanced networks. *IEEE Wireless Communications*, 2014, 21(2): 57–65.

19. Duong, T.N., N.T. Dinh, and Y.H. Kim. Content sharing using P2PSIP protocol in Wi-Fi Direct networks, In: *Proceedings of Communications and Electronics (ICCE), 2012 Fourth International Conference*, Hue, August 1–3, 2012, pp. 114–118.

20. Lin, X.Q., R. Ratasuk, A. Ghosh, and J.G. Andrews. Modeling, analysis, and optimization of multicast device-to-device transmissions. *IEEE Transactions on Wireless Communications*, 2014, 13(8): 4346–4359.
21. Moiz, A. Design, implementation and testing of a mobile cloud. University of Oulu, Department of Communications Engineering, Degree Program in Wireless Communications Engineering. Master's thesis, 57 p. available online: http://jultika.oulu.fi/Record/nbnfioulu-201512082287.
22. Zhang, Y., F. Li, X.L. Ma, K. Wang, and X.H. Liu. Cooperative energy-efficient content dissemination using coalition formation game over device-to-device communications. *Canadian Journal of Electrical and Computer Engineering*, 2016; 39(1): 2–10.
23. Wei, L.L, R.Q. Hu, Y. Qian, and G. Wu. Enable device-to-device communications underlaying cellular networks: Challenges and research aspects. *IEEE Communications Magazine*, 2014; 52(6): 90–96.
24. Wei, K.M., S. Guo, and K. Xu. CACC: A context-aware congestion control approach in smartphone networks. *IEEE Communications Magazine*, 2014; 52(6): 42–48.
25. Zhuo, X.J., W. Gao, G.H. Cao, and S. Hua. An incentive framework for cellular traffic offloading. *IEEE Transactions on Mobile Computing*, 2014; 13(3): 541–555.
26. Aijaz, A., H. Aghvami, and M. Amani. A survey on mobile data offloading: Technical and business perspectives. *IEEE Wireless Communications*, 2013, 20(2): 104–112.
27. Kim, T., S.K. Park, and B.G. Lee. LTE D2D—Challenges and perspective of mobile operators. In: *Proceedings of the Second International Conference on Electrical, Electronics, Computer Engineering and their Applications (EECEA 2015)*, Manila, Philippines, 2015.

Index

Milton Keynes UK
Ingram Content Group UK Ltd.
UKHW021911071024
449327UK00022B/1650

9 780367 573348